D0894272

Strike from the Sky

Smithsonian History of Aviation Series

On December 17, 1903, on a windy beach in North Carolina, aviation became a reality. The development of aviation over the course of little more than three-quarters of a century stands as an awe-inspiring accomplishment in both a civilian and a military context. The airplane has brought whole continents closer together; at the same time it has been a lethal instrument of war.

This series of books is intended to contribute to the overall understanding of the history of aviation—its science and technology as well as the social, cultural, and political environment in which it developed and matured. Some publications help fill the many gaps that still exist in the literature of flight; others add new information and interpretation to current knowledge. While the series appeals to a broad audience of general readers and specialists in the field, its hallmark is strong scholarly content.

The series is international in scope and will include works in three major categories.

Smithsonian Studies in Aviation History: works that provide new and original knowledge.

Classics of Aviation History: carefully selected out-of-print works that are considered essential scholarship.

Contributions to Aviation History: previously unpublished documents, reports, symposia, and other materials.

Strike from the Sky

THE HISTORY OF
BATTLEFIELD AIR ATTACK
1911–1945

Richard P. Hallion

SMITHSONIAN INSTITUTION PRESS
Washington and London

This book was edited by Nancy Dutro.

Library of Congress Cataloging-in-Publication Data

Strike from the sky: the history of battlefield air attack, 1911–1945 / by Richard P. Hallion.
p. cm.—(Smithsonian history of aviation series)
Bibliography: p.
Includes index.
ISBN 0-87474-452-0
1. Close air support—History. 2. Air interdiction—History. I. Title. II. Series.
UG700.H35 1989

British Library Cataloguing-in-Publication Data is available.

Cover illustration: Martin B-26 Marauders attack German positions during World War II. Photo courtesy U.S. Air Force Museum.

Acronyms in the photo credits are listed in the Glossary.

The paper used in this publication meets
the minimum requirements of
the American National Standard
for Permanence of Paper for
printed Library Materials Z39.48-1984.

Printed in the United States of America
10 9 8 7 6 5 4 3 2 1
98 97 96 95 94 93 92 91 90 89

This study is dedicated to the memory of Harvey Victor. He served as a young P-47 pilot in the ETO in 1944–45, attained the rank of Captain, USAAF, and earned the Distinguished Flying Cross; he died in 1986 in an aircraft accident at Van Nuys Airport, California. Harvey Victor was typical of thousands of Allied fighter-bomber pilots: largely anonymous men who flew with courage and resolve day after day against Hitler's Wehrmacht. He, at least, lived to see his actions contribute to final victory; he retained for the rest of his life a love for the Air Force, an admiration for his comrades who flew Republic's massive "Jug," and a detestation for all forms of totalitarianism.

Contents

Foreword

An account of the first thirty-five years of the contribution of air power to land warfare has an obvious appeal to military historians. It also has an intrinsic interest to those who are fascinated by the evolution of military aircraft and their emergence as a major influence on twentieth-century warfare. It is easy to forget that air operations before 1945 considerably exceeded in numbers, locations, and sheer geographical scale those that have occurred in the second half of the century. In an age of rapidly advancing technology, and especially in an environment in which the contribution of technology enjoys the highest visibility, it is easy to overlook underlying continuities in warfare.

Of course there have been abundant innovations in the relationship of air power to land warfare since 1945. Not necessarily in order of significance, they have included the widespread exploitation of the electromagnetic spectrum, which in turn has fostered electronic warfare; the combat helicopter; surface-to-air missiles; precision guided munitions; standoff, air-launched missiles; surface-to-surface missiles with accuracies measured in yards; night and all-weather target identification and acquisition systems; aircraft that by vertical takeoff and landing can be independent of traditional runways; unmanned aircraft; and an array of technological advances that collectively have been labeled "stealth." In this same period much of the new technology has actually been employed in battlefield-related combat, in Korea, Southeast Asia, the Middle East, the South Atlantic, Afghanistan, and the Persian Gulf, while in several low-intensity conflicts aircraft have provided firepower in support of ground forces. The details of these conflicts have not only sustained the interest of the historian and the enthusiast, they have been analyzed exhaustively by military staffs in an attempt to draw "lessons" from them that might point the way to the most effective future applications of air power to the battlefield.

There is no shortage of evidence, and there is no shortage of technology. There is even a large measure of international agreement on the authenticity of most of the evidence and on a great deal of the current or potential impact of specific innovations. There, however, the consensus stops, and the debates are not academic. They affect the provision of defense for superpowers, medium powers, and emerging nations alike. They influence the disposal of millions of dollars, rubles, marks, yen, and riyals in procurement programs. They dictate force structures, training programs, deployments, and ultimately the disposition of governments to select, or reject, military force as an instrument of policy.

In any examination of the past to seek guidance for the future, and especially in the history of warfare, it is essential to distinguish between those circumstances that are transient or unique to one occasion and those that persist or recur in slightly different forms but maintain a consistent substance. When the details of the air-land interaction in the first half of the century are distilled and added to those of the second, the likelihood of identifying reoccurrence is increased and confidence in decision making for future air-land warfare can be strengthened. Even the most cursory examination of that future suggests that the decision makers are going to require all the confidence they can muster.

For example, in the central confrontation in Europe between the superpowers and their allies, the contribution of air power will be reevaluated. If the conventional arms negotiations at Vienna produce an agreement to reduce ground and air forces in Europe to the levels proposed by President Bush in May 1989, the operational environment will be transformed by far more than any technological innovation. Ground forces will be smaller, with a consequent reduction in ability to mount a large-scale offensive or to concentrate defensive forces in depth. Conversely, war of maneuver and encirclement may become less problematic. Numbers of potential targets will decrease, but the proportional impact of weapons with high kill probabilities will increase. "Battlefields" will become diffuse, possibly fragmented and fluid. The scale of logistic support may be reduced, but ensuring its availability when and where required by highly mobile forces may become more difficult. Time, and the battlefield diminutive of tempo, may increase in relative significance. In the foreseeable future, surface-to-air defenses will not only be excluded from negotiations, but may well be enhanced to make good the deficiencies left by the withdrawal of manned fighters and fighter bombers from the European theater. Even if no formal agreements are reached in Vienna on air power, the residual impact of air power on both sides after ground-force reductions would need unilateral reassessment as a result of economic pressures in the Soviet Union and both political and economic pressures in the western democracies. At the center of those reevaluations and reassessments would be the question of how much air support would be required by ground forces, and how best, and most cost-effectively, to provide it. On the other hand, comprehensive arms reductions may be achieved in Vienna and accompanied by such political circumstances that east-west military conflict in Europe becomes so unlikely that major procurement programs and operational doctrines need no longer be driven by such a possibility.

Such a judgment would, however, be likely to be accompanied by an assumption that while the location of, and possible opponent in, conflict might have changed, the possibility of conflict itself had not so unequivocally diminished. Indeed, the only certainty would be the disposition of nation states to resort from time to time to military force as a political instrument. Battlefields would reoccur. Most states possess combat aircraft. The interaction of the two would continue. Aircraft, weapons, doctrine, and tactics would all continue to be required.

Dictatorships of whatever hue could place any priority they wished on arms purchases and military organizations. Democracies would continue to seek a level of defense expenditure commensurate with a government's ability to meet other obligations within acceptable levels of taxation. After forty-five years of regarding the Soviet Union as the primary source of instability and the major threat to the West, it may not be too easy to persuade the western democracies that conflicts in the third world would justify similar levels of defense commitment. That in turn would force hard decisions on projected weapon procurement programs, many of which are directly related to air attack in and around the battlefield. Those countries that were disposed to look to force to solve local regional difficulties, and that in many cases might not have the economic resources nor the indigenous technological capability to procure widely across the aircraft/weapons spectrum, would be faced with similar decisions, if not exactly the same choices.

In reflecting on the Spanish Civil War, Richard Hallion observes in this book that "intelligent, well-informed critics could reach often diametrically opposite conclusions or, interpreting data correctly, nevertheless reach a flawed conclusion." Perhaps, he suggests, "critics took away from the Spanish war what they wished to believe, and they searched its lessons carefully to selectively acquire data for their own particular viewpoint." Those comments should be displayed prominently in every military college and every procurement office worldwide.

With the obvious reservation that writers of forewords can be equally susceptible to predispositions, several circumstances have been repeated so often during eight decades of battlefield-associated air power that the failure to identify and apply certain "lessons" to modern battlefield air operations verges on the incomprehensible.

First and paramount is that without achieving at least local air superiority, the attrition suffered by friendly aircraft will neutralize their contribution to the land battle and leave friendly ground forces vulnerable not only to the opposing armies but to their air power as well. In 1945 the U.S. Army Air Force manual listed three priorities for tactical air forces. The second may be summarized in modern terms as interdiction; the third as close air support. The first was "to gain the necessary degree of air superiority." To all soldiers who had been subjected to enemy air attack in Europe during the previous four years it needed no explanation—least of all to those troops who had been demoralized in the Kasserine Pass in North Africa.

Since 1940 only a handful of western troops have endured air attack. For too long the pattern of unimpeded close air support introduced by the Royal Air Force in the western desert and perfected in Europe in 1944 and 1945 has been regarded as a model. Hallion reminds us that Rommel's air force in North Africa was outnumbered, short of fuel, and ultimately technologically inferior. In the six months before the Normandy invasion, the Luftwaffe lost 2,262 pilots and, in March alone, 56 percent of its available fighters. The massacre of Falaise in August 1944 was uninterrupted by any counter air activity. In the previous year at Kursk, Soviet fighters outnumbered their opposition by two to one while Luftwaffe sortie rates were further restricted by fuel and weapon shortages.

In the subsequent forty-five years air superiority has not been contested in conjunction with a land campaign, either because an opponent lacked the capability or, as in the Israeli examples of 1967 and 1982, opposing air forces were annihilated at the outset of a conflict. This may be just as well, because a recurring theme for eight decades has been the severe psychological impact on raw troops of sustained air attack, quite apart from the operational impact of destruction and dislocation.

When army commanders urge that the overriding priority of aircraft must be to redress inferiority on the ground, they should be reminded that unless air superiority is first achieved, no assistance of any sort is likely to be forthcoming. Only if an air force has the residual strength or, as with the Israeli Air Force in 1973, recourse to rapid reinforcement can heavy attrition be accepted temporarily and locally as a price for close air support to retrieve a critical position on land. Otherwise, attrition rates will again be exponential, not linear; once lost, air superiority is a difficult advantage to regain, especially in a short conflict.

In this context there is sufficient evidence to debate the allocation of resources to specialist ground-attack aircraft whose survivability in air-to-air combat is questionable. The JU 87 is the most notorious example, while the success of the IL 2 Sturmovik must be placed in the eastern front context of massive Russian air superiority and vast theaters of operations; it is an exception that reinforces the general deduction. Similar principles apply to airborne Forward Air Control aircraft, tactical transports and, in a modern scenario, airborne early warning and control systems. Conversely, the most successful contributors to the air-land battle began life as fighters, with the characteristics of speed, maneuverability, and self-defense common to the type.

Undoubtedly, achievement of air superiority has become complicated by the deployment of surface-to-air missile systems, but the basic principles of physics, in the relationship between omnidirectional targets at high speed at variable heights and static or relatively slow-moving reactive forces on the ground, have not changed.

Quite rightly, however, attrition exchange rates between aircraft and ground forces have continued to be questioned, introducing a further perennial question of the balance of air effort to be applied, after air superiority has been gained, between application of firepower against forces directly engaged with friendly troops and the "interdiction" of forces with an imminent impact on that engagement. Here, the extraction of general lessons demands even greater caution, but clear guidelines are nevertheless discernible. Interdiction of any sort is unlikely to be productive if the opponent's ground forces are not compelled either to consume resources or to maneuver. It is "either or" because disruption, delay, and dislocation as well as destruction have habitually upset the timing and coordination of offensive action. The greater the dependence on momentum, the greater the impact of delay. Conversely, the greater the dependence of defensive forces on reinforcement, the greater the impact of both delay and destruction.

The greater the depth of the interdiction, the less the immediate impact on the battlefield engagement, the greater the opportunities for the opponent to redeploy, the fewer the aircraft sortie rates, the higher the risk of attrition to the interdictors, and the greater the demand for air-to-air escort and suppression of ground-to-air defenses. In addition, the further away the targets are located, the more difficult to find and the greater the need for updated, real-time target information. Whatever the nature of air superiority, or the extent of "smart" weapons availability, a moving tank presents a difficult target to hit, let alone destroy. When all these factors are taken into account, they constitute a very complex equation of cost-effective exchange rates for aircraft and individual ground targets.

The answer has been sought in increasingly close coordination between army and air force staffs, now happily accepted as a sine qua non to achieve the synergism of combined arms operations. The natural desire of a battlefield commander to control his own firepower led to too many fruitless debates clouded by vested interests of resource allocation and force structure. It was not just the dissipation of air power into uncoordinated penny pockets that was the problem, but the failure to appreciate that any weapons system should be deployed not according to its firepower, but commensurate with its combat radius and flexibility. Thereby its impact could be directed in accordance with the requirements of the army group, corps, division, battalion, or company. When the firepower of aircraft could be directed against so many different kinds of targets—armor, artillery, logistics, airfields, and command and control centers—failure to concentrate effort with the greatest effectiveness to a theater of operations as a whole was profligate.

The search for effective command and control has been continual—not just to provide accurate and timely information, but to ensure that the enemy and not friendly forces should be destroyed. Current emphasis on computerized real-time data flow together with sophisticated IFF systems would lead to greater confidence but for the neutralizing effects of electronic warfare, largely absent from the pre-1945 battlefield.

However one interprets events between 1911 and 1945, there are two certainties about future air support and land warfare, certainties shared with all other aspects of combat. The first is that this area will remain the province of uncertainty, and sooner or later commanders will be called upon to deal with the unexpected. The second is that the pendulum in the advantageous application of technology will continue to swing between offense and defense, even against expectations at the outset of a conflict. When, before 1945, technology was harnessed to doctrine, tactics, strategy, and political objective, the unexpected was more readily absorbed. When technology was allowed to dominate theory, or when theory disregarded technology and its limitations, the ensuing doctrine hardened rapidly into a fatally dogmatic thrombosis.

Strike from the Sky not only presents a fascinating historical record, it offers a considerable amount of sobering relevance to all those concerned with, and responsible for, the application of modern air power in support of ground forces.

Air Vice Marshal R. A. Mason
CB CBE MA

Acknowledgments

A study of this sort cannot be completed without the support and assistance of a number of organizations and individuals. First, I thank the staff of the Military History Institute, Carlisle Barracks, Pennsylvania, for their unfailing courtesy and help during my year as the Harold Keith Johnson Visiting Professor of Military History. In particular, I acknowledge the assistance of Col. Rod Paschall, USA; Lt. Col. Martin Andresen, USA; Judith Meck, Linda Brenneman, Dave Keough, and June Rhodes. The staff of the U.S. Army War College offered a congenial body for discussion and analysis. I thank Maj. Gen. Howard D. Graves, USA, for his support of my research and study efforts. In particular, I appreciate the comments and thoughts of my students in Advanced Course 22, *Battlefield Applications of Air Power: An Historical Survey*, especially Lt. Col. Mohammed Aberchane (Morocco); Capt. Dale Brown, USA; Lt. Col. John Dallager, USAF; Lt. Col. Richard Fousek, USA; Lt. Col. James Gallivan, USA; Col. Arthur Patterson, USAF; Col. Juan Peltzer (Argentina); Col. Wilhelm Romatzeck (Federal Republic of Germany); and Lt. Col. Alfonso Zawadski, USA (ret.). In addition, I acknowledge the comments of five other faculty members and students: Col. John Lewis, USAF; Col. Ed White, USAF (ret.); Col. Dale Streeter, USAF; Col. Bill Kosco, USA; and Mary von Briesen.

I owe a special debt of thanks to my colleagues in the United States Air Force, particularly those of Aeronautical Systems Division, Wright-Patterson AFB, Ohio. Lt. Gen. Thomas McMullen, USAF (ret.), and his successor Lt. Gen. William Thurman, USAF, supported my appointment to Carlisle. Additionally, I thank Lt. Gen. Mike Loh, USAF; Maj. Gen. Peter "Peet" Odgers, USAF (ret.); Brig. Gen. Philippe Bouchard, USAF (ret.); Col. William Diesing, USAF (ret.); Col. Doug McLarty, USAF (ret.); Col. Dave Milam, USAF; Albert Misenko, and Diana Cornelisse for their help and various kindnesses.

The Office of Air Force History deserves special mention for support and assistance, both at Bolling AFB, Washington, D.C., and at the Historical Research Center, Maxwell AFB, Alabama. I thank Richard H. Kohn, the Chief of Air Force History, and his staff, particularly Pat Harahan, Eduard Mark, B. Frank Cooling, and Lloyd Cornett. Bill Stacey of the Foreign Technology Division History Office, Dave Mets of the Armament Division History Office, and Jim Young and Cheryl Hortel of the Air Force Flight Test Center History Office furnished useful materials for the study.

The Air Force Museum offered every consideration and assistance to my research. I thank Col. Richard Uppstrom, USAF (ret.), Dr. Jack Hilliard, and the staff of the research archives for their help.

Present and former colleagues at the Air Force Flight Test Center provided many useful insights, particularly Maj. Gen. Philip Conley, USAF (ret.); Maj. Gen. Mike Hall, USAF; Col. Bob Ettinger, USAF (ret.); Col. John Hoffman, USAF; Col. Cal Jewett, USAF (ret.); Col. Doug Joyce, USAF; Col. William "Pete" Knight, USAF (ret.); Col. Jim McFeeters, USAF (ret.); Col. Jim Manly, USAF (ret.); Col. Mike Marks, USAF; Col. John Taylor, USAF (ret.); Lt. Col. Jim Doolittle, USAF; Lt. Col. Bill "Flaps" Flanagan, USAF (ret.); Lt. Col. Dave Spencer, USAF; Maj. Jack Hudson, USAF; and Maj. Harry Walker, USAF.

Col. Ken Alnwick, USAF (ret.); Col. Tom Fabyanic, USAF (ret.); Col. Jack Schlight, USAF (ret.); Lt. Col. Price Bingham, USAF, Lt. Col. Bob Ehrhart, USAF (ret.), and Lt. Col. George Reed, USAF, offered provocative insights, materials, and thoughtful comments, as did Bob Perry and Caroline Ziemke of the RAND Corporation; Lt. Col. Joe Shipes, USA, of the Army's Training and Doctrine Command (TRADOC); and Capt. Kevin Stubbs, USAFR, of the Center for Strategic Technology, Texas A & M University.

Danny Crawford, Head, Reference Section, Marine Corps Historical Center, Washington, D.C., furnished a very useful package of material relating to Marine close air support (CAS) development. I benefited from discussions with Lt. Gen. Willard W. Scott, USA (ret.), Col. John Benson, USA, and Richard "Hap" Miller of the Institute for Defense Analysis.

Von Hardesty, Dom Pisano, and Brian Nicklas of the National Air and Space Museum, Smithsonian Institution, were, as usual, most helpful.

Brig. Gen. Jay Hubbard, USMC (ret.) offered personal insights into the application of Marine CAS in World War II, Korea, and Vietnam. I also benefited from discussions with Bill Bettis; Brig. Gen. Bob Strauss, USAF (ret.); and Col. Walter Flint, USAF (ret.). The late Harvey Victor furnished much useful material and recollections of flying the Republic P-47 Thunderbolt in combat against the Nazis; he is missed.

I express my appreciation to the U.S. Army Aviation Engineering Flight Activity and the U.S. Air Force's F-16 Combined Test Force, both at Edwards AFB, California, for the opportunity to obtain somewhat of an "operator's" perspective on air-to-ground operations via familiarization sorties in the Bell AH-1S Cobra gunship and the General Dynamics F-16B Fighting Falcon. In particular, I wish to thank Col. Alan Todd, USA; Maj. Austin Omlie, USA; CW4 Mark E. Metzger, USA; and Col. Bob Ettinger, USAF (ret.).

Finally, this study should not be construed as representing an official position or viewpoint of the Department of Defense or any of its personnel.

Richard P. Hallion
Military History Institute
Carlisle Barracks, Pennsylvania

Introduction

Throughout the history of warfare, new developments in technology have had a profound effect upon the conduct of military affairs. The twentieth century has witnessed many of these: the tank, the airplane, the submarine, the atomic bomb, radar and electronic communications, and the long-range guided missile, to name just a few.

Traditionally, the air forces of the world have set three major missions for themselves and have rank-ordered these in a specific priority. First are missions to gain and retain *air superiority*—control of the air—by destroying any opposing air force and thus inflicting a policy of air denial upon opposing aviation. Second are *strike missions*. These may take the form of long-range strategic missions directed against strategic targets, such as industrial facilities located deep within an opponent's heartland, or they may be so-called "interdiction" missions, conducted over shorter distances in the hope of isolating the battle area and denying an opponent the ability to move, provision, and support his forces in the field. Third are missions involving actual *battlefield support*: the application of air strikes to support one's own ground forces engaged—or soon to be engaged—in close combat with an enemy. Variously termed through the years—beginning with "trench strafing" and "ground strafing" during the Great War—these are generally broken down into two subcategories of air support: *close air support* (CAS), and *battlefield air interdiction* (BAI).

The difference between close air support and air interdiction has existed primarily in requirements for integrating air action with the action of ground forces. For example, the Department of Defense's *Dictionary of Military and Associated Terms* (JCS Pub. 1, January 1, 1986) defines close air support as air action against hostile targets which are in close proximity to friendly forces and which require detailed integration of each air mission with the fire and movement of those forces.

Air interdiction, on the other hand, is defined as

air operations conducted to destroy, neutralize, or delay the enemy's military potential before it can be brought to bear effectively against friendly forces at such distance from friendly forces that detailed integration of each air mission with the fire and movement of friendly forces is not required.

However, this latter definition, while useful for characterizing traditional interdiction operations (such as those that preceded the invasion at Normandy) does not completely satisfy the needs for a definition for battlefield air interdiction, which differs from traditional air interdiction in that it is conducted within the battle area, and differs from traditional close air support in that detailed integration of these missions with the fire and movement is not generally required. Battlefield air interdiction is a subcategory of air interdiction and is generally considered that portion of air interdiction used to support close-in battle. (Contemporary views on BAI are discussed in the epilogue to this work.) It is in the sense of these definitions and understandings that the terms "close air support" and "battlefield air interdiction" are used in this study.

This study is an attempt to trace the emergence of battlefield air support and to show how it evolved from trial and error in the midst of the First World War to a military necessity by the time of the Second World War. Too often two equally fallacious viewpoints concerning battlefield air support have reigned: that air support has not been of significance to the land battle and that air support has been decisive in land warfare. The actual answer, of course, is in the middle. As might be expected, the advent of air attack did not immediately radically reshape the battlefield or, for that matter, radically influence it. But over time air attack evolved from an unpleasant exception to an even more unpleasant common occurrence. By the end of 1918, the airplane had already demonstrated its natural partnership with the tank in mechanized warfare and the havoc that it could wreak against unprotected troops. It had likewise demonstrated its vulnerability to light antiaircraft fire, particularly when operated across the front at low altitudes. These traits of "attack" aviation would be accentuated in years to come, and new missions would be added to the airplane's roster of duties, notably its use against tanks and road transport during the Second World War and its use in so-called counterinsurgency warfare. It is this role, perhaps, that is the most familiar, courtesy of the long conflict in Southeast Asia, as well as other conflicts, both past and present, around the world.

There are four significant concerns that must be considered when examining the history of battlefield air support: *doctrine, command and control procedures, operational circumstances,* and *aircraft technology.* All of these have been controversial, and all have the potential for either making a battlefield air support system that works or one that is seriously flawed. *Doctrine* traditionally has been an area in which the air forces of the world have been most weak. Too often air forces allow the state of technological research and development to push them down acquisition paths that may or may not be appropriate. In the history of ground attack, for example, the existing air doctrines of the world's air arms often rejected any real need for it: the major missions would be strategic, operating deep within an enemy's territory, in classic Douhet or Mitchell fashion. The realities of war, specifically the wars of the 1930s, quickly revealed the fallaciousness of such thought, and the Second World War demonstrated the absolute necessity of appropriate doctrine to address ground-attack needs.

Hand-in-hand with doctrine is the problem of *command and control*: who should control ground-attack aviation, and what kind of mechanism should exist so that ground forces have an expectation that when they need air support it will be forthcoming? In the era of "Army" air forces, around the time of the First World War, this question was of less significance than in the era of separate air forces wedded increasingly to a doctrine of long-range strategic air war. But the problems here have not been exclusively those of air forces reluctant to support their comrades on the ground. In virtually every case where a ground commander has exerted direct command authority over subordinate air units, the quality of the air support has proven deficient: the commander is typically so concerned about what is happening on his section of a front that he concentrates the available air assets at that portion of the front to the detriment of what is happening along other portions of the front. The French made this mistake in 1940, as did the United States early in the Second World War, during fighting in Tunisia. (We, fortunately, had the opportunity to profit from our mistakes and refine what would become a masterfully smooth air-ground operation by the time of the breakout across France in 1944.) Though it was Nazi Germany that graphically demonstrated the impact that systematic battlefield air operations could have on the progress of a ground war (notably during the period of the blitzkrieg in 1939–40), it was the British who, in the hot crucible of

Western Desert fighting, refined and honed a system of air support that served as the basis of the subsequent Anglo-American system that devastated the Third Reich.

The *operational circumstances* under which air forces employ battlefield air support are crucial to its success. Traditionally, air support has worked best when the side employing it also has air superiority. Under these circumstances the attack aviation forces can go about their job with minimal concern about being subject to attack from enemy aircraft. They can concentrate on the "air to mud" mission in evading and suppressing enemy air defenses, locating the target, and destroying it. Such conditions existed for the Luftwaffe in 1939 and 1940 when the Stuka dive bomber was able to get away with attacks that would have been suicidal in other circumstances (as the Battle of Britain would reveal). Such conditions also existed for the Royal Air Force and the U.S. Army Air Forces during the 1944-45 campaign in Western Europe when, arguably, the Allies had not merely air superiority, but air supremacy as well. In fluid air war situations, such as existed on the Russian front during the Second World War, opposing forces were able to get away with air operations that simply would not have worked, or which would have worked only with exorbitant cost, on other fronts. For example, both Nazi Germany and the Soviet Union operated relatively slow aircraft on ground-attack missions, often without fighter escort, and managed to do so without experiencing devastating losses. In part, this was because the Russian front was so vast in relation to the forces employed across it that both German and Soviet air commanders could generally achieve local air superiority when they chose by massing air units over some selected portion of the front. At the battle of Kursk, for example, both Soviet and German antiarmor aircraft operated with great efficiency, not experiencing the kinds of losses that one would have expected had the air war been less fluid and more dominated by one or the other side over a smaller section of front. Such conditions did not exist in Western and Southern Europe in 1944; Germany had to withdraw its close-support aircraft from both those areas in 1944 and transfer them to the more conducive climate of the Russian front, because they simply could not survive in the air, owing to the Allies' overwhelming air superiority.

This latter problem—the ability of the close-support aircraft to survive the "threat environment"—is one in which *aircraft technology* is critical. During the First World War, when the disparity between the performance of single-seat fighters and larger two-seat ground-attack airplanes was slight, the technological and performance differences between fighter and attack aircraft were largely insignificant. But by the time of the late 1930s, this had changed dramatically. The "dedicated" attack airplane—the airplane intended from the outset for attack of enemy forces on the battlefield—had grown appreciably in size and weight, and had evolved in two directions: as a twin-engine design (such as the Douglas A-20 or the French Breguet 693) that was largely indistinguishable from the twin-engine medium bomber, and as the two-seat overspecialized dive bomber (such as the vaunted Stuka). Both kinds of aircraft were very vulnerable to enemy defenses, namely light antiaircraft fire and fighters. By the end of the Second World War, the best kind of ground-attack aircraft for battlefield air support had proved to be the modified bomb-loaded fighter (which had first appeared in 1917 over the Western Front)—the fighter-bomber of Great Britain and the United States, the *Jagdbomber* ("fighter bomber") of Nazi Germany. In the postwar years, this trend predominated—and predominates to the present day, with certain exceptions—with the development of jet fighter-bombers such as the Republic F-84 family, Hawker Hunter, Dassault Mystère, Sukhoi Su-7, McDonnell F-4, Mikoyan MiG-23, General Dynamics F-16, and numerous others. Why the fighter? Simply put, such aircraft offer a degree of flexibility, performance, and safety unknown to the special-purpose attack aircraft. The fighter is a swing-role aircraft that can turn from an air-to-ground mission to an air-to-air mission if need be; it has performance, agility, and handling qualities that give it a far better chance of evading both ground-to-air and air-to-air threats than the single-purpose attack aircraft. Even the Soviet Union, which had profound success in operating special purpose attack aircraft over the Eastern Front (the famous Ilyushin Il-2 Shturmovik), recognized that the era of such aircraft had passed with the introduction of genuine jet fighter-bombers in the early 1950s, typified by the Su-7.

The final section of this study consists of a brief epilogue on how battlefield air support adjusted to the changes of the postwar world and how it is constrained and enhanced by the available technology and doctrine of the present day. This section, of course, offers only an indication of the kind of detailed, incisive examinations of postwar tactical air power that are needed so that decision makers can adequately plan for the defense needs of the late 1990s and beyond.

PART ONE

The First World War

1

The Military and the Airplane

For many years, military aviation in the First World War has been regarded as a mere sideshow to the war on land and at sea. This attitude has generated a belief that the roles and missions of modern military airpower grew out of the experiences in the Second World War, and over time has resulted in a more generalized feeling that any military airpower experience prior to 1939–45 has little to offer the student of present-day military affairs. In point of fact, of course, the roles and missions of modern military airpower were first promulgated, explored, verified, and undertaken during the First World War. This is particularly true of the missions of air superiority, interdiction, and close air support. The air leaders of the Second World War were, by and large, conditioned by their experiences in the First World War; they elaborated upon trends and doctrines first undertaken in that earlier conflict, and what differences existed between the nature of military airpower in 1914–18 (particularly during the last two years of that conflict) and 1939–45 were more in the nature of technological change than in the nature of doctrine and strategy.

The Birth of Military Aviation, 1903–1914

The advent of the man-carrying powered airplane did not immediately revolutionize military affairs. Rather, for five years after the Wrights' success at Kitty Hawk, the secretive tendencies of the brothers, coupled with popular skepticism and bureaucratic inertia, tended to repress any effort to apply the new technology to the military. (In the United States, this may have stemmed in part from the embarrassment the Federal government suffered by the well-publicized failure of Smithsonian Institution Secretary Samuel Langley's man-carrying "Aerodrome" in two unsuccessful flight attempts in October and December 1903; the Aerodrome had been built with a $50,000 grant from the War Department, the first governmentally financed heavier-than-air flying machine in the world.) This situation, incidentally, was not unique to the United States; following Santos-Dumont's hop off the ground in France in 1906—one hesitates to call it a genuine flight—a similar pattern was followed in Europe. In 1905, the Wrights had approached both the British and American governments with a view to selling them aircraft, but met with official rebuffs. In December 1907, following intensive behind-the-scenes efforts by a small number of enthusiasts, the Army Signal Corps issued a development specification for a flying machine capable of carrying two crewmen at a speed of 40 mph. In 1908, Orville Wright demonstrated the value of the Wright Flyer for military reconnaissance in a series of test and demonstration flights at Ft. Myer, Virginia. Though propeller failure led to a fatal crash that claimed the life of Army Lt. Thomas Selfridge, the undoubted value of the Flyer resulted in an order for a slightly improved model that appeared the following year, the world's first "operational" military airplane. In 1911, Army aviators tested a bombsight during experimental bombing flights at College Park, Maryland, and in 1912, they experimented with firing a Lewis gun from an airplane against ground targets. In both cases, the results were promising, but official views—one hesitates to call it doctrine at this primitive stage of aeronautical development—that the airplane had value only as an instrument of reconnaissance resulted in the work not being followed up. The bombsight inventor subsequently went to France where he competed in an influential bombing competition sponsored by France's air-minded Michelin brothers, winning a $5,000 prize. Further development of the airplane as an instrument of war shifted from the United States to Europe in the remaining years of peace before the outbreak of the First World War. European aviation owed its rapid growth to the example of the Wright brothers. At the same time that Orville was busily pirouetting about the Virginia countryside, brother Wilbur was in Europe revolutionizing European aviation, which had largely lain dormant since the death of glider pioneer Otto Lilienthal in 1896. As a result, European aviators drastically reshaped their approach to aviation, adopting a Wright approach that emphasized controllability and maneuverability, yet adding innovations—such as ailerons for roll control in place of the Wrights' penchant for wing-warping, and replacing the Wrights' canard pusher configurations with tractor biplanes and monoplanes. Very quickly, European aviation matched that of the United States, and by 1912 it had actually shot ahead; in that year, for example, the French Deperdussin company manufactured a high-speed racing aircraft—what would today be termed a technology demonstrator and technology integrator—that combined a so-called "monocoque" wooden shell fuselage structure with a monoplane wing configuration and a then-powerful rotary piston engine. This design, the Deperdussin Monocoque Racer, exceeded 100 mph in level flight and anticipated the highly streamlined aircraft of the later war years and the 1920s.[1]

By 1910, then, two years after the Wrights had first demonstrated their aircraft before military audiences in the United States and Europe, there was a growing recognition that the airplane possibly could revolutionize reconnaissance; such sentiments were appearing in professional military journals. But when the airplane first entered combat the following year, it quickly added bomb-dropping to its duties; although these first bombing sorties were more in the nature of operational experiments, they nevertheless pointed to an increasing belief that the airplane could take part in actual air-to-ground combat. During fighting in Tripoli against Turkey, Italian aviators flying French-designed Bleriot and Austrian-designed Taube ("Dove") monoplanes performed both reconnaissance and strike missions against Turkish positions. On October 23, 1911, Capt. Carlos Piazza flew a reconnaissance mission from Tripoli to Aziza in a Bleriot, and, just over a week later, 2d Lt. Giulio Gavotti of the *Squadriglia di Tripoli* dropped four small bombs from a Taube on the towns of Taguira and Ain Zara. Subsequent bombing sorties became commonplace, and although they had a negligible effect upon the war, the nascent potential of the airplane impressed military correspondents who witnessed its employment.[2] The experiences of aviators and soldiers in the Libyan war were repeated in the various Balkan wars that preceded the First World War, and in other conflicts as well. France in particular showed a noteworthy grasp of the airplane's potential as a weapons platform. During the Balkan war of 1912, for example, France sent aircraft and a contingent of aviation officers to the assistance of Serbia; during colonial fighting in Morocco, French airmen undertook limited bombing missions in support of the capture of Taza in 1914. Earlier, the French Inspectorate of Aeronautics had ordered antipersonnel darts *(fléchettes)* that could be dropped on troops, duplicated the earlier American experience of gun-firing tests from aircraft (including 37mm cannon), and actually manufactured aircraft armed with Hotchkiss machine guns, exporting six to Rumania in 1913. Closer to the United States, the Mexican revolution offered a limited but nevertheless revealing glimpse at the military potential of attack aviation. In Mexico, rebel and governmental aviators alike flew reconnaissance and bombing missions against enemy forces, using aircraft purchased from France and the United States; at the end of 1913, for example, the government forces under Gen. Victoriano Huerta included no less than fifty aviators, and Huerta had a force of twenty-four Bleriot and Deperdussin monoplanes in service or on order. His opponent, Venustiano Carranza (leader of Mexico's Constitutionalist movement), relied on American and Mexican pilots (including some of his relatives) piloting American-built Moisant biplanes for reconnaissance and bombing purposes.[3]

The Nature of Conflict in the Great War

The First World War offers a classic example of a conflict that few wanted, but which quickly grew beyond the control of governments to contain. What is most interesting about the First World War is the degree to which it was technologically innovative. The American Civil War introduced the military community to the new face of war: heavy battlefield casualties caused by increasingly accurate and rapidly firing rifles and cannon. The experience of the Franco-Prussian War and the various imperialist/colonial wars that followed, from the Sudan to the veldt of South Africa, had confirmed that of the Civil War. Yet, the revolution in technology and the increasing lethality of the battlefield made little impact upon the general staffs of the European nations. Instead, on the eve of the war, blind faith in outdated doctrine and notions of national supremacy (such as French reliance on the "doctrine" of the *offensive à outrance*—the "unlimited offensive," with its extraordinary notion that "what the enemy intends to do is of no consequence") held sway over careful and thoughtful analysis of contemporary and future trends.[4] The First World War introduced the systematic use of mechanized warfare, typified by the airplane, the tank and truck, and the ocean-going submarine, to modern high-intensity conflict. Each of these had been experimented with before, in rudimentary form, but the First World War offered the opportunity for military technologists and strategists to employ these weapons for the first time on a mass scale, supported by a fully mobilized war economy of major industrial states. All proved spectacularly successful, and, to a greater or lesser degree, all succeeded in confounding their critics and surpassing the expectations of even their most enthusiastic supporters. Commanders (with few exceptions) did not envision the impact that integration of these new systems could force upon combat operations—the synergy, for example, of the airplane and the tank. Instead, they tried and repeatedly failed to restore the war of mobility that had characterized the opening phases of the conflict on both the Western and Eastern fronts. By and large, they generated conditions leading to horrendous casualties and the subsequent emergence of a disturbed and reform-minded group of highly experienced junior officers (of all nationalities) determined not to fall victim to the generally unimaginative mind-set of their seniors. This group would be responsible for many of the doctrinal and tactical innovations of the interwar years that would find their fruition upon the battlefields and in the skies of the Second World War.

One can go further and state that it was the junior officers and (rarer) more senior "mavericks" who were responsible for what innovative changes *did* take place on the Western Front during the Great War. These were the individuals who experimented—often in direct violation of higher headquarters—with (for example) aerial photography, air fighting, bombardment, offensive use of tanks, and air-ground cooperation.

The Air War and Aviation Technology

To understand the impact of aerial operations in World War I, it is necessary to examine the nature of the air war and aviation technology from 1914 through the Armistice in 1918. The intensity and pace of the air war in theaters other than the Western Front were so different from the level of warfare over France and Belgium that it is best, for the purposes of this study, to concentrate primarily upon the Western Front, pointing to important examples from other theaters as they occur and have relevance to the text. Basically, there was a major strategic difference between the Allies and Germany in the way that each combatant nation viewed the air war in the West. By and large, through the First World War, the Allies fought an offensive air war emphasizing operations against German forces over the German side of the line. For this reason, Allied fighter forces, for example, undertook air superiority sorties deep within German territory. On the other hand, Germany tended to follow a defensively minded air war, attempting to control the air over its side of the line by dealing with Allied intruders as they entered German airspace. German fighter pilots, for example, referred to the importance of "letting the customer come into the shop," and this notion resulted in an emphasis on fast-climbing short-range interceptors scrambling in response to calls from a front-line air warning network. For its part, the Allied air forces tended to emphasize much more capable multimission fighter aircraft, and by 1917–18, when Germany was confronted not merely with larger fighter forces than its own, but with swing-role forces capable of undertaking both air-to-air and air-to-ground missions with aircraft qualitatively superior to those of the German air service, the signs of ultimate German collapse were all too clear. While the fighter forces of Germany envisioned their role as defensive, Germany undertook ambitious "strategic" bombing campaigns directed against targets in France and England. In 1917, goaded by the British example of air-ground operations over the Somme, Germany increasingly emphasized development of ground-attack aviation; by mid-1918, at the time of the last major offensives of the war, German attack aviation was extremely well developed, qualitatively equal to that of the Allies (and in some technical respects more advanced, particularly with regard to aircraft types), though not, on the whole, as effective as the combined British-French-American tactical air forces operating on the Western Front. Implicit in the development of new military capabilities is the notion of appropriate technology: the demand to field equipment suitable to the desired task. To fully appreciate the "air" contribution to the First World War, one must consider the technological development of aviation during the war.[5]

The airplane underwent a profound transformation during the First World War, and, as it transformed itself from a prewar sporting craft to a full-fledged instrument of war, its offensive capabilities increased as well. Prewar technology demonstrators such as the Deperdussin Monocoque Racer pointed toward flight speeds on the order of 100 mph. As is typical both then and now, however, the advances shown by technology demonstrators often take a decade or more to be incorporated in mainstream aeronautical practice, and thus the airplanes that first appeared in wartime skies tended to be much less sophisticated craft. An advanced "high-performance" operational airplane of 1914 had a maximum speed of approximately 70 mph, an endurance of one hour, a crew of one, and was powered by a single 80-hp engine. By war's end, the typical high-performance operational single seater had an endurance of over two hours, had a maximum speed of 120 mph, relied upon a 180–200-hp engine, and could carry a mix of weaponry including twin forward-firing synchronized machine guns and light bombs. Expressed another way, the nature of aircraft design up to the First World War had emphasized development of single-seat aircraft of limited military utility. By the time of the Armistice in 1918, the technology was in place for constructing not only multimission single-seat,single-engine aircraft, but a wide range of multiplace and multiengine aircraft capable of extremely long flights. Bombing operations from the Continent against targets in southern England had become commonplace, and the Allies were gearing up for an offensive campaign against targets in Germany, including Berlin. In 1919, no fewer than three transatlantic flights would take place: one by a large American flying boat, the Curtiss NC-4, the second by a modified British Vickers Vimy bomber, and the third by a British dirigible airship. Less obvious changes occurred in the nature of aviation technology. For example, powerplants grew more reliable and increased in horsepower, thus leading to larger and heavier aircraft and to aircraft that had much more favorable power-to-weight ratios. Handling qualities improved as the aileron replaced the prewar Wright notion of wing warping for lateral (roll) control. Designers paid greater attention to flight loads criteria and aircraft stress analysis, abandoning the previous "by guess and by God" methods of construction in favor of engineering rationalism. Prewar all-wooden structures covered with doped fabric were replaced by mixed wood-and-metal fabric-covered structures, and even by early production aircraft having all-metal structures, including the skin covering. Scientific flight testing emerged, leading to an improved understanding of why aircraft behaved in flight as they did, and thus resulting in not merely safer and more mission-capable new designs, but in greater pilot knowledge and confidence. Improved training methods for both pilots and mechanics appeared. Put in modern terms, advanced technological approaches generated aircraft having greatly improved operational capability. Coupled with training, this resulted in increasing standards of reliability and maintainability, and an overall stronger force structure.

As discussed earlier, when prewar military officials and planners thought of the airplane's uses in war, they invariably thought in terms of the airplane influencing the land battle via reconnaissance. As has been seen, however, no sooner did the airplane appear in combat in a reconnaissance role than airmen themselves modified it for more offensive tasks, specifically bomb dropping. When the First World War broke out, the evident success of reconnaissance aircraft and the threat of hostile reconnaissance sorties resulted in the evolution of the fighter to prevent the acquisition of intelligence information by intercepting and shooting down reconnaissance machines. As time went by, the fighter itself evolved from a simple interceptor to a complex aircraft intended for primary air superiority missions and secondary light attack missions as well. This evolutionary trend, then–interceptor to multimission fighter to "dedicated" ground-attack fighter—was one important influence on the emergence of the true ground-attack aircraft, and was the development path followed by the British, whose late-war ground-attack fighters (such as the Sopwith Salamander) sprang from an earlier lineage of air superiority designs. Imperial Germany followed a very different development path. German ground-attack aircraft traced their lineage to earlier two-seat cooperation and liaison aircraft, with a strong inspirational input from Allied offensive fighter operations in 1917. German fighter development per se played little role in determining the design of ground-attack machines, and, largely because Germany continued to think of the fighter as a defensive aircraft right up to the moment of defeat in 1918, German fighters to the war's end tended to be strictly single-purpose interceptor and air superiority machines. Thus, the second important trend influencing development of attack aircraft in the First World War was the development of the two-place "cooperation" aircraft (typified by the German "C" class of aircraft—armed two-seaters intended for reconnaissance and liaison duties). Inspired by the success of Allied ground-attack fighter sorties in 1917, German designers combined the technology of the late-war fighter (good power-to-weight ratio, efficient armament system, good maneuverability and agility) with that of the "C" class of aircraft (two-seat layout with close placement of crew for good tactical coordination and communication, rear defensive gun, good endurance). They generated a new class of design, the "Cl" airplanes, which became acceptable fighters in their own right, optimized for air-to-ground sorties and with a secondary air-to-air offensive mission. Using small bombs, stick grenades, and machine gun fire, these aircraft—typified by the Hannover and Halberstadt ground-attack biplanes of 1917–18—flew extensive close air support and battlefield interdiction sorties during the war's final months.

Thus, for the Allies, the resulting ground-attack aircraft was largely a product of the *single-seat fighter experience.* For the Germans, the ground-attack aircraft was largely a product of the *two-seat battlefield cooperation airplane*. Which was more ideally suited to the battlefield of 1918 and the anticipated battlefield of 1919? To partially answer this, a comparison of two operational aircraft representing the differing approach taken by Great Britain and Germany offers important clues: the Sopwith Camel and the Junkers J I (Table 1).

It may be said that, having been "second" in the race to develop a genuine ground-attack capability, Germany was a "fast second." The Junkers J I can clearly be seen as the antecedent of later heavily armored ground-attack aircraft. Many subsequent heavily armored attack aircraft have been failures, not so much because they could not survive the intense ground fire thrown at them, but because they could not survive air-to-air threats. What made the J I a success was that it had a performance closely approximating that of its fighter opponents, and other attributes (rearward defensive guns, armor) that more than offset any deficiency in performance. Later overspecialized attack aircraft had much greater performance disparities between themselves and contemporary fighters. By mid-1918, Germany had a total of 189 J I's in service at the front, together with hundreds of the Hannover and Halberstadt Cl ground-attack aircraft. It must be remembered, however, that the total air superiority that the Allies possessed over the Western Front at this time meant that many of these aircraft were unable to pursue their ground-attack missions without being harassed and shot down by marauding Allied fighters. Thus, even though Germany had technologically superior ground-attack aircraft in service by war's end, it was unable to take fullest advantage of their capabilities. Since the Allies were able to operate more freely than the Germans, there was little apparent difference in the efficacy of Allied vs. German air support efforts.[6]

In another important catagory of aircraft—light bombers—the Allies reigned supreme; interestingly enough, the German air service never seriously competed, despite developing some aircraft which could have served in the role. The two best examples of such aircraft were the British De Havilland D.H. 4 and the French Breguet Bre. 14B2. Both were hefty two-seat single-engine aircraft, and both went on to years of military and civilian service after the First World War. Both could carry relatively large bombloads; the D.H. 4 typically carried two 230-lb bombs or four 112-lb bombs, while the Breguet could carry over 500 pounds of bombs, typically thirty-two 17.5-lb (8-kg) antipersonnel bombs. While the Allies utilized aircraft such as the D.H. 4 and the Bre. 14B2 for what would now be termed "deep interdiction" and strategic

Table 1. Comparison of the Sopwith Camel with the Junkers J I

	Camel	J I
Engine	130-hp rotary	200-hp inline
Crew	pilot	pilot + gunner
Guns	2 forward-firing	2 forward-firing; 1 hand-held rear-firing, occasionally 2 angled to fire downward
Bombs	4 × 25-lb bombs	approx. 150 kg of 4.5-kg to 50-kg bombs; stick-grenades
Speed	104.5 mph	96 mph
Structure	wood	all metal (steel and aluminum)
Armor	none	5mm chrome-nickel steel "bathtub" shell encasing crew, engine, and fuel
Radio	none	air-to-ground radio for liaison with friendly ground forces (downward-firing guns deleted when radio carried)

bombing sorties, they also made extensive use of these aircraft for tactical air operations, particularly what would now be considered battlefield air interdiction. When Allied light bombers were added into the force structure available to Allied commanders, the Allied air superiority over the Germans was sealed.[7]

We turn now to two cases in the use of battlefield airpower in the First World War: The Western Front in 1917–18, and the final stage of the war in Palestine.

2

Ground Attack on the Western Front, 1917–1918

From the Somme to Cambrai

From the earliest days of combat on the Western Front, aviators felt compelled to attack opposing forces on the ground. Until 1916, such attacks were sporadic, reflecting more on the initiative of the individual than on any policy, plan, or doctrine. Such attacks proved demoralizing to troops, but also revealed the vulnerability of aircraft to ground fire. For example, Roland Garros, the first "fighter pilot," was shot down and subsequently taken prisoner on one ground-attack sortie. In the battle of the Somme, July 1916, the British employed eighteen aircraft on what would be termed "armed reconnaissance" in later wars; the aircraft were to fly over the trenches for observation purposes, and then undertake (as opportunity arose) attacks on enemy trenches. There was no doctrinal or policy directive underlying these attacks, rather, they stemmed from an ad hoc approach to war. The results were startling; German troops were unnerved by the attacks, which occurred from altitudes as low as fifty meters. As a subsequent German commentary noted:

> Infantry lacked schooling in defence against low-flying aircraft and did not have the confidence that they could shoot them down. The result was a condition akin to panic, heightened by the presence of the enemy's airplanes.[8]

As a result of the Somme, German interest in ground-attack aircraft was awakened, and further British efforts were encouraged. Despite the evident success of these initial attacks—a success due as much to the unfamiliarity of troops to aircraft attack as it was to the offensive ability of the machine-gun armed aircraft themselves—formulation of policy for ground attacks still remained over a year in the future. Interestingly, during the Somme attacks, German infantry dubbed the British air strikes as "punishment," using the verb *strafen*. "Strafe," of course, is now an accepted part of the fighter and attack pilots' lexicon.

The ground-attack airplane—at first a modified air superiority fighter and then, subsequently, a specialized aircraft type of its own—appeared in strength in mid-1917, in the Somme fighting following the disastrous French Aisne offensive. In early May 1917, the Royal Flying Corps experimented with using aircraft attacks to break up German troop concentrations along the front. On May 11, F.E. 2b and Nieuport fighters, carrying small 20-lb bombs in addition to machine guns, attacked German troops in support of advancing British infantry. As the planes exhausted their ammunition, they returned to base, replenished, and returned to the front. This effort to use aircraft in direct support of advancing troops marked the first steps toward a British ground-attack doctrine.[9]

Eventually, the British recognized two forms of ground attack: *trench strafing*, which corresponded to today's concept of close air support, and *ground strafing*, which was roughly equivalent to today's notion of battlefield air interdiction. During the battle of Messines, British fighter pilots had instructions to "cross the line at Armentières very low and then shoot at everything."[10] Loss rates were high, but would be much higher on subsequent operations as troops overcame their fear of aircraft and concentrated on shooting them down; of the fourteen RFC aircraft dispatched across the front on ground-attack duties on June 7, two were shot down and two others badly damaged. For their part, German two-seater crews roamed over the front, shooting at British troops. Already then, the basic dichotomy between British and German close air support aircraft operations had been established: the British relied primarily on single-seat fighters, and the Germans on heavier two-seaters having, in addition to the pilot's forward firing guns, an observer manning a machine gun on a rotating mount from an aft cockpit.

To further enhance the destructive potential of ground-attack operations, the RFC added bomb racks under the wings of fighters such as the Sopwith Camel, the S.E. 5, and the Bristol Fighter. A single-seater like the Camel or S.E. 5 would typically carry four 25-lb Cooper bombs, but a more powerful two-seater such as the Bristol Fighter would carry up to twelve.

By the battle of Cambrai, in the fall and winter of 1917, both Allied and German ground attack from the air was commonplace. With the mystique somewhat gone, casualties among ground-attack aircraft soared, particularly for those on trench-strafing (CAS) missions. At no time during the battle did casualties among ground-attack aircraft drop to less than 30 percent, and during the opening of the battle, 35 percent of the attack aircraft sent across the front failed to return. Fighter pilots preferred risking themselves in combat against the enemy's airplanes; one bitterly wrote that "with few exceptions," trench-strafing missions constituted "a wasteful employment of highly trained pilots and expensive aeroplanes," adding that "rather than face a single trench-strafing foray, I would much prefer to go through half a dozen dogfights with Albatroses."[11]

Despite the losses inflicted on attacking aircraft, aerial attack of frontline troops appeared, on the whole, to be quite effective. On November 23, 1917, for example, RFC D.H. 5 fighters (a type used almost exclusively for ground-attack duties) cooperated with advancing British tanks, attacking artillery positions at Bourlon Woods as the tanks advanced. Subsequent analysis concluded that "the aeroplane pilots often made advance possible when the attacking troops would otherwise have been pinned to the ground."[12] The critical problem affecting the quality of air support in the First World War was, interestingly, one that has appeared continuously since that time as well: communication between the air forces and the land forces. During these early operations, communication was virtually one-way. Infantry would fire flares or smoke signals indicating their position, or lay out panel messages to liaison aircraft requesting artillery support or reporting advances or delays. For their part, pilots and observers would scribble messages and send them overboard (on larger aircraft, crews carried messenger pigeons for the same purpose). Though by 1918, radio communication was beginning to make an appearance in frontline air operations—as evidenced by its employment on German ground-attack aircraft such as the Junkers J I and on Col. William Mitchell's Spad XVI command airplane—it was still of such an uncertain nature that, by and large, once an airplane had taken off, it was out of communication with the ground until it had landed. Thus, attack flights—both Allied and German—tended to operate on what would now be termed a "prebriefed" basis: striking targets along the front on the basis of intelligence information available to the pilots before the commencement of the mission. The "on-call" and "divert" CAS operations associated with the Second World War and subsequent conflicts were not a feature of First World War air command and control, though attack flights often loitered over the front watching for suitable targets of opportunity, as would their successors in the Second World War.

In 1917, the first of Germany's powerful two-seat Cl class attack fighters entered service. By the end of the year, Germany had two such designs, the Hannover Cl II and Halberstadt Cl II, in service, as well as Junkers' formidable all-metal J I. While both the Halberstadt and Hannover had originally been intended to operate with "protection" flights *(Schutzstaffeln)*, escorting more vulnerable two-seat reconnaissance and liaison aircraft, they now were assigned to a new aerial formation, attack squadrons *(Schlachtstaffeln*, abb. *Schlasta)* which, in turn, were organized into attack wings *(Schlachtgeschwader)* in a manner similar to the organization of fighter squadrons *(Jagdstaffeln)* into fighter wings (*Jagdgeschwader*). By March 1918, in preparation for the final offensives of the war and in response to an anticipated increase in Allied forces at the front due to the contributions of the United States, the *Luftstreitkräfte* (the Imperial German Air Service) had added thirty Schlasta (a total of 180 attack aircraft) to its forces on the Western Front. Already, at the end of November 1917, the Schlasta had received their baptism, when Halberstadt attack biplanes had cooperated with German infantry in savage attacks on British positions at the end of the battle of Cambrai. British troops wasted so much machine gun ammunition trying to hit the low-flying attack fighters that they had little left to confront German infantry forces.[13]

The 1918 Offensive

In March 1918, Germany began its long-anticipated offensive to end the war. As with the Ardennes thrust of 1944, slightly over a quarter century later, it at first achieved surprising success; as with the Korean offensive of 1950, the German forces advanced so rapidly that airfields close to the front were overrun, thus forcing the evacuation of all flightworthy aircraft and limiting the amount of Allied air power that could be mustered to confront the German assault. Memoirs of individuals exposed to German Schlasta attack abound with references to air attack; it is clear that such attacks had a profound demoralizing effect upon newly arrived and inexperienced troops.[14] In his memoirs, the architect of Germany's last-ditch offensive, Gen. Erich von Ludendorff referred to the intense CAS and BAI attacks that accompanied the German ground assault, perceptively noting the combined arms nature of the air-land struggle in 1918. He wrote that the German airmen

> hurtled from the heights and flew at low level, machine-gunning and dropping light bombs on the enemy's infantry lines, artillery, and, eventually, the enemy's reserve formations and supply and troop columns. . . . As with the infantry, artillery, and other combatant branches, they participated in the ground battle.[15]

Ludendorff's stroke fell upon the British Fifth Army which fought as best it could, but which fell back in disarray. Aloft, the RFC attempted to come to the aid of its comrades on the ground, as the French rushed infantry replacements to the British section of the front. But close-flying British airmen had problems mistaking the powder blue uniforms of the French with the *feldgrau* of the German army, and false reports abounded that the Germans were operating deep behind the British lines.[16] Meanwhile, German two-seater attack fighters—and even some single-seaters not really suited for ground-attack work—roamed over the front harassing infantry and artillery formations.

By March 1918, both German and British air doctrine called for the mass employment of close air support aircraft. German doctrine stipulated that strikes should consist of no less than four machines for concentrated attacks with machine-gun fire and small bombs and grenades on the forward line of enemy troops and artillery. RFC ground-attack doctrine stipulated staggering air assets over the battlefield, with S.E. 5's and Sopwith Dolphins used for seizing and holding air superiority high over the battlefield, Camels operating at medium altitudes for protection of reconnaissance, liaison, artillery spotting, and ground-attack flights, and, finally, ground-attack fighters—Camels, Bristol Fighters, and, less frequently, S.E. 5's and Dolphins—operating "on the deck" against ground targets. When attacking ground targets, RFC practice called for one group of aircraft to be attacking, one group to be rolling in preparatory to the attack, and one group to be pulling off target after having completed their attack. During the struggles of early 1918, extremely large formations of attack aircraft appeared over the front. One South African artillery brigade was harassed by no less than thirty German attack fighters, while twenty-seven Camels and S.E. 5's attacked German infantry continuously in the Festubert region. Another brigade commander reporting heavy fighting in Beauvois punctuated by twenty German attack fighters "hovering over the village and diving with extraordinary daring almost into the streets," two of which were shot down "at very close rifle range." A German regiment reported an attack by twenty British fighters flying within "2–3 metres" of the ground; one literally ran over a German company commander with its landing gear wheels. Another German regiment reported heavy losses from a combined machine-gun and bomb attack delivered by British fighters.[17] A British account written three years after the war alleged that the British press and government information officers had downplayed "what German aircraft did against our infantry and artillery."[18] For his part, Ludendorff, in his memoirs, acknowledged that during the March 1918 offensive, "All troops, especially those that were on horseback, suffered greatly from the bombing attacks of enemy airmen."[19] Finally, a captured German document translated in July 1918 stated frankly that "During our offensives, losses through the action of enemy aviators have proved to be extaordinarily high." It concluded with a series of instructions to avoid air attack, including avoiding main roads, limiting the size of troop and supply columns, keeping a sharp lookout for British fighters, maintaining an active antiaircraft defense with machine guns, and constructing dugouts and fortifications.[20] All this might have gratified both British and German airmen, who were taking heavy losses during their sorties. (Casualties for the RFC alone went from ninety-one aircrew KIA/MIA in February 1918 to 245 KIA/MIA in March, many lost on ground-attack sorties.) One S.E. 5a pilot wrote after a particularly grim day, "Our low work today has been hellish. I never did like this ground strafing."[21]

For its part, the French air service in 1918 was at the apex of its wartime strength. Never as badly handled and misused as the French Army had been, the air service entered 1918 with a strong balanced force of fighters, bombers, and observation aircraft, consisting primarily of Salmson and Breguet observation and bombardment aircraft, and Spad VII and XIII fighters. Though it confronted some potentially damaging technical problems—chiefly due to teething difficulties with the geared Hispano Suiza engine powering the Spad XIII—it was in every respect as advanced and offensively minded a force as Britain's newly created Royal Air Force. (The Royal Flying Corps and the Royal Naval Air Service merged to form the Royal Air Force—the first true independent air force—on April 1, 1918.) During the battles of 1918, the French air service, as with the RFC/RAF, pursued three missions: maintaining air superiority, strategic and tactical reconnaissance, and undertaking bombing and machine-gun attacks against enemy forces. When Germany went on the offensive in 1918, the increased pace of war necessarily forced increased German activity in the open with supply and troop columns and the like presenting vulnerable and inviting targets. French bomber and fighter squadrons took advantage of the new conditions to press home low-altitude attacks on such targets, though the major function of French fighter aviation remained keeping Germany's fighter and attack forces on the defensive, preventing them from achieving Ludendorff's desired degree of air cooperation with the rampaging German ground forces.[22]

Several examples give an indication of the heavy involvement of French aviation in the closing months of the war, and, in particular, the emphasis that France put upon what now would be considered battlefield air interdiction missions. For example, on June 4, 1918, 120 Breguet XIV bombers dropped 7000 bombs on German troop concentrations in a ravine near the Villers-Cotterets forest that was sheltered from Allied artillery; the raid devastated the German forces and prevented them from massing for an assault on Allied positions.[23] On June 11, a French counterattack by General Charles Mangin against advancing German forces between Courcelle and Wargemoulin was assisted by no less than 600 aircraft; as H-hour approached, fighters destroyed German observation aircraft, remaining aloft to prevent interference by German fighters with the main attack forces. At H-hour, bombers attacked German troop and artillery positions; the combined air-land counteroffensive broke the German offensive and helped set the stage for the ultimate Allied offensives to follow. On October 27, during the French offensive in Champagne, a Breguet bomber group, covered by two fighter groups, raided Seraincourt. Bombing from 1500 meters, the Breguets struck a German division in bivouac just beyond Allied artillery range, effectively destroying it before it entered fighting at the front. Right up to the Armistice, such raids continued, complemented by low-level attacks against troops at the front and, eventually, against troops retreating toward the German frontier.[24]

St. Mihiel

When the St. Mihiel offensive opened in September 1918, it represented the largest concentration of air power in support of land forces assembled to that time. Air commander Colonel William "Billy" Mitchell had a force of 1481 French, British, Italian, and American aircraft; 701 were fighters operating in the air superiority, close air support, and battlefield air interdiction role. Mitchell's philosophy was simple: the air component would be responsible for "destroying the enemy's air service, attacking his troops on the ground, and protecting our own air and ground troops."[25] Subsequently, U.S. Army Air Service chief Maj. Gen. Mason M. Patrick reported that during the St. Mihiel offensive,

> pursuit pilots also attacked ground objectives or engaged in "ground-strafing," as this work came to be called. On September 12th, American and French pursuit airplanes found the Vigneulles–St. Benoit Road filled with the enemy's retreating troops, guns and transport. This road was a forced point of passage for such of the enemy as were endeavoring to escape from the point of the salient. All day long our pursuit airplanes harassed these troops with their machine gun fire, throwing the enemy columns into confusion. The airplanes of the 3rd Pursuit Group, which were equipped to carry small bombs, did particularly effective work in destroying a number of motor trucks on this important road.
>
> This ground strafing was effectively continued on September 13th and 14th, when good targets presented themselves on the St. Benoit–Chambley and Chambley–Mars–La Tour Roads.[26]

Despite this impressive assembly of air assets under a single air commander, the German air service was still a force to be reckoned with. Unescorted or poorly escorted bomber raids were massacred by Germany's new Fokker D VII fighter, the finest single-seat fighter introduced during the war, pointing to the vulnerability of the unescorted bomber, a vulnerability that would reveal itself in subsequent conflicts through the Korean War as well. German ground-attack forces were not completely deterred by Allied fighters (though the Allies maintained standing fighter patrols to intercept them as they crossed the front, inflicting heavy casualties), occasionally scoring surprising successes. Observers with the A.E.F.'s Fourth Division during the Meuse-Argonne offensive reported that "German aviators were exceedingly active. Not only did they bomb and machine-gun the American infantry and artillery but they daringly attacked the balloon line."[27] In his autobiography, Mitchell himself wrote:

> The Germans had painted their airplanes so that they looked just like the ground when seen from above. The German airmen would get down to within one hundred feet or so of the earth, or even lower at times, fly up ravines or behind forests and pounce on infantry columns or wagon trains and surprise them before they could conceal themselves. We had a couple of engineer companies lined up for mess that were attacked in this way by a flight of German

attack aviation. They killed eighty-seven men and wounded a couple of hundred more before the engineers could get to cover.[28]

So concerned did Mitchell become with the danger of German attack aviation that he was aghast to discover, late on the day of the St. Mihiel attack, that a huge traffic jam had developed on the roads around Avocourt. Immediately sensing the danger to the allies if German attack aviation struck at this bottleneck, he concentrated his bomber forces to strike deep into German territory, forcing the German fighters to remain behind the front defending against the bombers' attack. He directed the 1st Pursuit Group to maintain Spad patrols as low as 300 feet over the jam to intercept any low-flying Schlachtflieger that might sneak across the front. After the war he wrote:

> Never have I seen such a congestion on a European battlefield. . . . If the German aviation had been able to attack this column, it certainly would have destroyed it. This would have put the whole central army corps, the 5th, out of ammunition and supplies. A counterattack then by a small organization of Germans would have broken a hole straight through the center of our line and nothing could have stopped it, because we could not have moved troops up along the roads to help the 5th Corps.
>
> We had received many reports that the German aviation had received armored airplanes that could resist machine-gun fire. Here, in this great congestion of motor trucks, was the target for which they had been waiting. . . . Had we not [undertaken preventive air defense measures] instantly, I believe that this whole mass of transportation would have been destroyed.[29]

Ironically, then, the Schlachtgeschwadern's potentially finest moment passed into history unfulfilled.

Can it be said that the ground-attack air operations of the Western Front played a decisive role in the conduct of the war? The answer, of course, is no. The airplane—and the tank, as well—were still rudimentary weapons, and thus not able to be fully exploited. It would not be until the wars of the 1930s—specifically the Abyssinian War, the Spanish Civil War, and the Sino-Japanese War—that the airplane would exert a powerful influence over the land battle, and it would be the Second World War in which it would demonstrate genuine decisiveness. But as a new (if immature) element of war, the airplane certainly did profoundly impress those it attacked during the war, as both Allied and Central Powers' memoirs and histories indicate. What it did was act as yet another threat, in a war in which infantry faced multiple dangers. It was one more "cross to bear," one more concern to keep in mind. As such it complicated the movement of ground forces and, as the previous text has indicated, occasionally induced panic and confusion. It increased the "fog and friction" of war. Was it, then, worth the effort? The answer here is yes; in an era when aircraft were plentiful and available for such missions, it made perfect sense to use them as ground-attack weapons. Loss rates were made up by rapid replenishment. In a war in which commanders were quite willing to expend thousands of troops for the most meager of gains, the achievements of ground attackers in instilling occasional confusion, panic, and disorder (as well as the actual destruction of enemy forces and material) offset the losses of men and planes.

The war's most destructive example of air-ground attack came not from the Western Front, however, but from the Middle East. In part, it stemmed from the existing conditions—relatively weak defenses and defenders, morale problems among the enemy, a weak enemy air force, excellent flying conditions, and highly motivated attackers—but this should not obscure the significance of this example, which, in its own way, anticipated the kind of devastating air attack that would be seen subsequently. Indeed, looking at the conditions prevailing on the Western Front in the fall of 1918, it may be concluded that only the increasingly bad weather prevented the Allied air arms from inflicting similar casualties upon the German Army as it reeled back toward its frontiers following the collapse of the last German offensives. Because of the clarity with which the Palestine campaign can be examined—itself a noteworthy exception to the usual murk and mire of World War I historiography—it affords a profitable opportunity to analyze what is arguably the first mobile air-land campaign in military history.

3

The Palestine Campaign of 1918

Setting the Stage

Since entering the war, the Ottoman Empire had found itself increasingly on the defensive; in the Middle East, this took the form of fighting on three fronts: against Russian forces in Armenia, against British and Commonwealth forces in Mesopotamia, and against British and Commonwealth forces driving eastward from Egypt. By the end of 1917, the latter forces, under the command of Gen. Sir Edmund H. H. Allenby, had seized control of Jerusalem. It was 1918, then, when Allenby planned a knockout blow aimed at the conquest of all Palestine. During fighting that year in what is now Israel and Jordan, aircraft of the Royal Air Force and the Australian Flying Corps were increasingly involved in ground-attack actions against Turkish and German troops. Allenby planned his offensive to begin on September 19, 1918, and in preparation for this offensive, even before the final date had been established, the airmen attached to his command stepped up the pace of their operations to assert control of the air (thus denying it to German reconnaissance aircraft that might have spotted preparations for the attack) and to bomb and harass German and Turkish forces. Ultimately this set the stage for the truly disastrous collapse of three Turkish armies, the Eighth (located west of the Jordan River, near Et Tire and Tul Keram, within ten miles of the Mediterranean coast), the Seventh (located around Nablus, west of the Jordan and midway between the Jordan and Tul Keram), and the Fourth (located east of the Jordan around Es Salt, midway between the Jordan and Amman).

In July and August 1918, the Bristol Fighters of 1 Squadron, Australian Flying Corps, decimated German air strength over Palestine with a

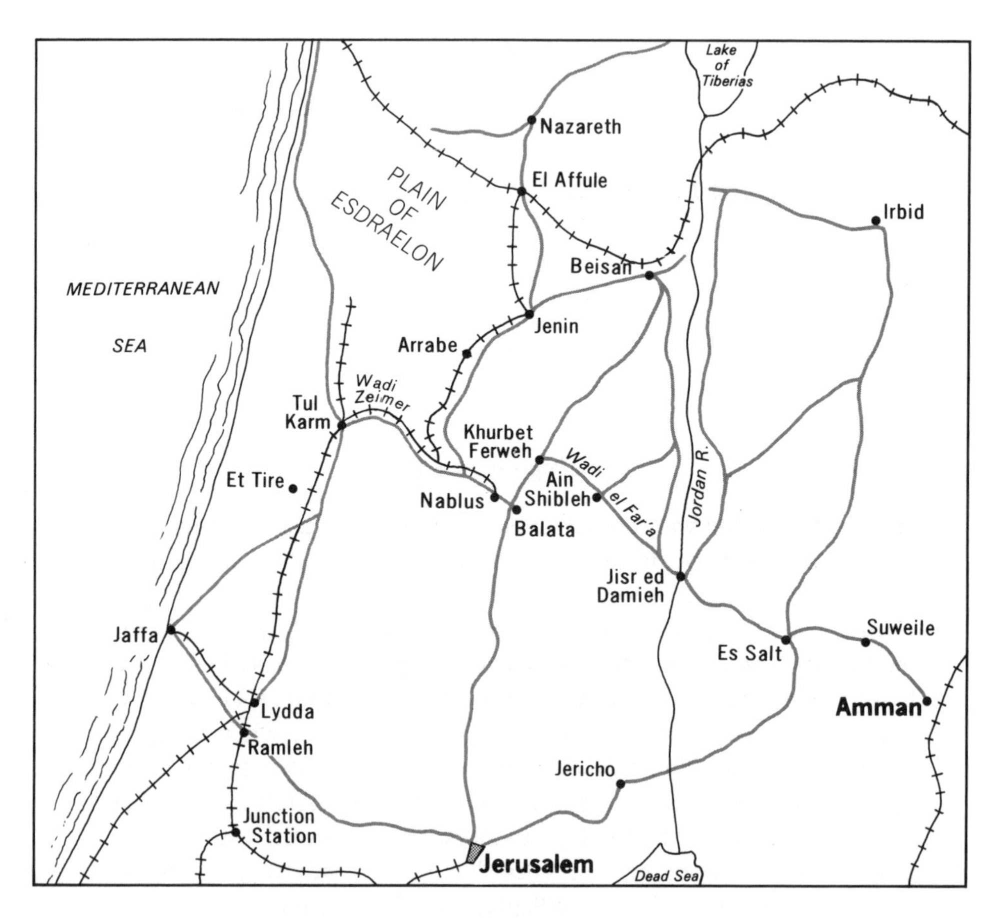

Palestine, 1918

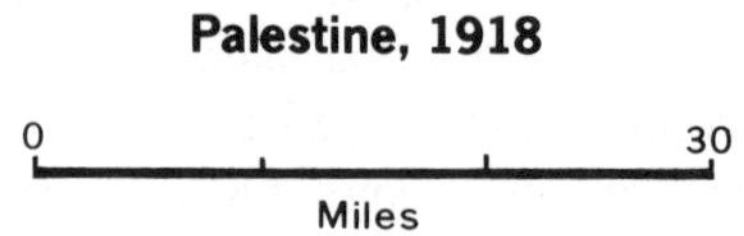

series of air combats and attacks on airfields. Next, in company with squadrons of S.E. 5a's and D.H. 9 bombers (the latter a derivative of the workhorse D.H. 4) from the Royal Air Force, they turned their attention to wide-ranging attacks on Turkish and German positions ranging on both sides of the Jordan River, from Tul Keram in the west to Amman in the east, and from El Affule in the north to Allied lines in the south. Already during these missions, one geographical feature stood out as particularly important: the Wadi el Far'a, running southeast from Khurbet Ferweh above Balata down almost to Jisr ed Damieh on the Jordan River. It formed a natural bottleneck cutting off the Turkish Fourth Army east of the Jordan from the Seventh and the Eighth on the Jordan's west bank, and thus if the Fourth Army moved to assist the Seventh and Eighth once the Allenby offensive began, the Wadi el Far'a could prove a critical chokepoint. More likely, planners thought, the Allenby offensive would result in a massive retreat of Turkish forces east and north—toward Amman and Damascus. Accordingly, the Wadi el Far'a could prove equally as valuable a chokepoint for the entrapment of Turkish forces west of the Jordan. Two other routes were likewise worthy of being watched by air: the twisting rail and road line north from Nablus to El Affule via Arrabe and Jenin (which also offered the option of allowing Turkish troops to turn northeast at Jenin and make their way to Beisan, below Lake Tiberias), and the route from Nablus via Balata, and Khurbet Ferweh to Beisan. Allenby's cavalry planned to swing around the Turkish flank across the Plain of Esdraelon, seizing Jenin, El Affule, and Beisan; at the same time, east of the Jordan, what would now be termed a "special operations force" under the already legendary T. E. Lawrence would disrupt the enemy's rear north of Amman.[30]

Allenby's attack got underway in the early morning hours of September 19, and from the outset his air component played a critical role in the subsequent rout of Turkish forces. Initially, Australian aircraft bombed telephone exchanges at El Affule and Nablus to destroy communication among Turkish forces. (This had results far beyond expectations of the Allies, for as the struggle—generally termed the battle of Nablus, though some more grandiosely dubbed it the "battle of Armageddon"—developed, the Fourth Army remained in bivouac around Es Salt until the late afternoon of the fourth day of the operation! It had been in ignorance of the disaster unfolding on the west bank the entire time.) By noon, the Eighth Army was already in retreat, with elements moving from the vicinity of Et Tire up to Tul Keram, and from Tul Keram east to

Anebta, trying to reach Nablus. Fighters and bombers pounced on them; the size of the force retreating between Et Tire and Tul Keram was estimated at approximately 6000 troops and 500 vehicles. The greatest slaughter, however, took place between Tul Keram and Anebta, where the Bristol Fighters, D.H. 9's, and S.E. 5a's bombed and strafed the retreating columns, trapped in a defile created by the small Wadi Zeimer. German commander General Liman von Sanders subsequently wrote:

> The low-flying British bombing formations, relieved every half-hour, littered the road with dead troops, horses, and shattered vehicles. Officers repeatedly attempted to rally the troops, but in vain as they were concerned only with their own safety. This was the true picture of the Eighth Army's right wing![31]

But worse was to come, two days later. Other air strikes on this first day hit at the Fourth Army's positions east of the Jordan River. While the strikes on the road from Et Tire via Tul Keram to Anebta could be properly considered battlefield air interdiction, since they came within five to ten miles of the rapidly advancing Allied forces driving north toward Tul Keram and on to the Plain of Esdraelon and east toward Nablus, the strikes across the Jordan to Es Salt and near Amman were, generally speaking, more of a classic interdiction mission less likely to have influence on the immediate and short-term battlefield situation.[32]

During the second day of the offensive, September 20, reconnaissance revealed the Eighth Army in full retreat, most elements attempting to make their way north from Anebta to Jenin. To prevent these forces from linking up with forces at El Affule before Allied troops seized that town, the RAF and AFC concentrated its efforts along the Anebta–Jenin–El Affule road. Meanwhile, the Seventh Army was itself retreating from Nablus north to Khurbet Ferweh, preparatory to transiting the Wadi el Far'a, and thus the Bristol Fighters of the AFC launched an attack against this retreating body, which numbered 200 vehicles stretched over approximately five miles. The small bombs and machine-gun fire destroyed a small number of vehicles, but engendered panic among the horses, which bolted; some hurtled over a precipice beside the road. Even this, however, was but a glimpse of the ghastly events to come. By the end of the second day, Allied forces were in possession of Tul Keram, Jenin, El Affule, Nazareth, *and* Beisan; for all practical purposes the Eighth Army had ceased to exist, and was wandering the desert, awaiting (and welcoming) capture. The Seventh Army was making its painful way from Nablus to the Wadi el Far'a, already having experienced what lay in greater store on the morrow. East, across the Jordan, the Fourth Army remained in bivouac.[33]

Wadi el Far'a

On the third day, September 21, an early morning reconnaissance revealed that the Seventh Army was strung out in a long and vulnerable column on the old Roman road from Balata northeast to Khurbet Furweh, then east-southeast along the Wadi el Far'a to Ain Shibleh, and thence northeast again toward the Jordan River and the road that ran north to Beisan. (Little did the Turks know, of course, that their retreat was already cut off and that Beisan was in British hands.) At least 800 horse-drawn vehicles, interspersed with troops and motor transport, was streaming toward the Jordan River. Immediately, the reconnaissance plane, outfitted with radio, called for a bombing strike, and shortly afterward, the first bombers arrived over the target. The account of the Australian squadron—which, it must be remembered, was assisted by six other squadrons, consisting of D.H. 9 bombers and S.E. 5a fighters—gives a clear picture of what happened that day over the Wadi el Far'a:

By 6.30 a.m. the first three Australian bombing machines, sent out in response to Brown's wireless message, arrived, bombed the column, and raked it from end to end. A British formation followed and repeated the attack. All day long air raids were maintained along this S-road by available machines from all squadrons. Towards noon the columns under punishment at that place included thousands of infantry and cavalry as well as transport. No. 1 Squadron made six heavy raids during the day, dropped three tons of bombs, and fired nearly 24,000 machine-gun rounds into the struggling parties in those terrible valleys. Yet this was only half the total ammunition expended, for the British squadrons attacked this same road with another three tons of bombs and 20,000 machine-gun rounds. The panic and the slaughter beggared all description. The long, winding, hopeless column of traffic was so broken and wrecked, so utterly unable to escape from the barriers of hill and precipice, that the bombing machines gave up all attempt to estimate the losses under the attack, and were sickened at the slaughter. In all the history of war there can be few more striking records of wholesale destruction. The passes were completely blocked ahead and behind by overturned motor-lorries and horse-waggons; men deserted their vehicles in a wild scramble to seek cover; many were dragged by the maddened animals over the side of the precipice. Those who were able cut horses from the waggons and rode in panic down the road to Damieh. When British cavalry two days later passed along this road they found abandoned or wrecked 87 guns, 55 motor-lorries, 4 motor-cars, 75 carts, 837 four-wheeled waggons, and scores of water-carts and field kitchens.

The Turkish Seventh Army as a fighting force was destroyed. In the morning of September 22nd broken parties of that army were again attacked from the air along the Shibleh-Beisan road, near the Jordan, and under this further attack a column of several thousand troops first scattered in panic, and then later were seen marching back towards the Wady Fara [sic] under a large white flag.[34]

During the attacks, the aircraft had dropped primarily 20-lb bombs, with the Bristols carrying eight and the S.E. 5a's carrying four; in addition, the D.H. 9's involved carried eight 20-lb bombs and a single 112-lb bomb apiece. Some of the Australian aircraft dropped self-developed experimental weapons; one unloaded a total of 120 Mills bombs (hand grenades) from two four-tube launchers, an early example of a primitive bomblet dispenser. The strike aircraft maintained a shuttle bombing operation from their bases at Ramleh, Jaffa, and Junction Station, each located approximately forty miles from the Wadi el Far'a, and thus roughly twenty-five to thirty minutes flying time away. Flights were staggered so that two aircraft arrived over the Turkish forces every three minutes, with an addiitional six aircraft arriving every half-hour. For the most part, attacks were prosecuted at ranges less than 100 feet; despite what would appear to be ample chance to undertake defensive antiaircraft fire, only two attacking aircraft (and their crews) were lost. Liman von Sanders, busily packing for withdrawal, had many other problems to worry about besides the destruction of the Seventh Army. In his memoirs, written immediately after the war, he dismisses it in a telling passage:

When the troops no longer could retreat over a broad front, they had to pass between the mountains towards Bett Hassan and Beisan, being subjected to unceasing attack from English attack formations, which exacted the most frightful sacrifice from them, severely damaging their morale. The feeling of helplessness in the face of the enemy fliers instilled a paralysis in both officers and men. The columns of salvaged artillery pieces, automobiles, and motor transport, together with shattered wagons, horses, and men, blocked the road in many places.[35]

By the end of the third day of the offensive, all Turkish hopes for salvaging the situation were gone. The collapse of the Turkish forces, materially aided by the air assault, had been so sudden that it taxed the ability of Allied ground forces to advance fast enough to catch and capture the remnants of the Seventh and Eighth armies. Early in the offensive, Turkish and German troops had put up stiff rear-guard resistance, but the harassment of air attack had broken their will to resist. This early fighting had caused some heavy Allied casualties. One exhausted division, in its drive to the Wadi el Far'a, had taken 2000 prisoners, but had suffered 690 casualties killed, wounded, or missing. In this environment, the air attack had eased the pressure on Allied ground forces to remain in contact, as well as sharply reducing the enemy's morale. As Cyril Falls subsequently wrote, referring to the division that had taken 690 casualties on its drive to Wadi el Far'a, "It was the British air arm that had saved these men, who by this time could hardly put one foot before the other, from further exertions."[36]

The fourth day of the offensive, September 22, found the scattered remnants of the Seventh and Eighth armies wandering the west bank of the Jordan, cut off from all retreat, save for those few who were able to cross at Jisr ed Damieh and attempt to reach Es Salt, or for the desperate few who crossed the Jordan farther north and attempted to strike out across the open desert in the direction of Deraa (which, unknown to them, had already been sacked by Lawrence's men); most of these latter few died. Still the Fourth Army remained in bivouac—indeed it remained in its camps until late in the afternoon of the 22d when, alert at last and panicked by the news, it ponderously began an easterly retreat, toward Amman, with hopes of making its way north of Amman toward Mafrak, Deraa, and thence to Damascus. Even this late in the day—by now toward evening—aircraft struck at the Fourth Army, bombing and strafing it as it moved toward Amman. The next day, September 23, the fifth day of the offensive, these bombing attacks continued on the Es Salt–Amman road, converting the retreat "into a rout like that around Nablus."[37]

These strikes—by Bristol Fighters, D.H. 9 bombers, and even R.E. 8 reconnaissance planes modified to carry bombs—occurred within five to ten miles of friendly advancing forces, and, indeed, in the afternoon of the 23d, Es Salt capitulated to a brigade of New Zealand cavalry. While these battlefield interdiction strikes continued, the British undertook longer-range interdiction sorties by Handley Page O/400 twin-engine "strategic" bombers against railway and airfield facilities at Deraa, already suffering from a raid by Lawrence's Bedouins. Thus, the Turkish Fourth Army, like the Seventh and Eighth before it, found it had no place to retreat to. On September 24, the sixth day of the offensive, Allied forces continued to harass the retreating Turks; the next day, September 25, the seventh day of the offensive, Amman surrendered and Allied aircraft fell upon the remnants of the Fourth Army making its way north from Amman to Mafrak. The open country northward from Amman toward Damascus prevented the kind of textbook situations that enabled air power west of the Jordan to devastate trapped columns in ravines and defiles, sparing the retreating Turks from the worst of the attacks that had characterized attacks on the west bank. Eventually, several thousand demoralized survivors straggled into Damascus; the campaign in Palestine had been decisive, and though subsequent fighting occurred, for all practical purposes it was a denouement. Damascus fell on September 30. A month later, on October 31, the Ottoman Empire signed an armistice with the Allies.[38]

4

Great War Air Support in Retrospect

Aircraft and Operations

As with the development of the air superiority fighter and the long-range bomber, the experience of close air support and battlefield air interdiction operations in the First World War influenced not only the conduct of actual combat operations during the war itself, but thinking and planning for the future. Taken as a whole, the record of military air combat operations in the war contributed to what French Marshal Henri Petain referred to as "the capital importance of aviation" in military affairs.[39]

The characteristics of close air support and battlefield air interdiction aircraft and operations had been defined in this war, and would be elaborated upon in the future; basically, the desirable design characteristics for CAS and BAI aircraft pointed to armored aircraft equipped with multiple machine guns and racks for bombs, capable of attaining high speed and operating with great maneuverability and agility (the latter being the ability to transition from one flight condition as rapidly as possible to another). Already a growing split was occurring between proponents of multiseat—and eventually multiengine as well—aircraft "dedicated" to close air support and battlefield air interdiction, and those who favored modifying air superiority fighters for the role. Some of the weaknesses of slavish adherence to a particular military design philosophy were appearing, weaknesses that would become graver and more unacceptable in subsequent conflicts. For example, the slow army cooperation airplane had already demonstrated that it could not survive the hostile air threat and ground-fire environment likely to exist over a "modern" battlefield. Attempts to operate heavily armed but slow and unmaneuverable aircraft over the front—as with efforts to employ strategic bombers over the actual battle—tended to result in aircraft shot down with no appreciable impact on the land battle. For its part, the air superiority fighter had the ability to evade and overcome enemy air threats, but proved vulnerable to the enemy's ground defenses. Additionally, it often could not carry the amount of weaponry that infantry needed for support. What was needed was a compromise: an aircraft having fighterlike agility together with reasonable payload and self-protection features such as armor plating—all attributes found in the more successful fighter-bombers of the Second World War.

The war demonstrated the interrelationship of air superiority with air-to-ground operations. Germany's excellence in support aircraft forces in 1918 was largely offset by the Allies' dominance of the skies; when support aircraft appeared, they had to be constantly vigilant to prevent being shot down not from the ground, but from the air. Then as now, *without air superiority, no other air missions can be performed; pilots are too busy just trying to stay alive.* The war had also demonstrated what air-to-ground air strikes could accomplish over the battlefront. During the 1918 offensives, for example, Allied air strikes against advancing German units had in some critical cases substituted for the lack of meaningful Allied resistance on the ground. In part this was because of the airplane's psychological impact; troops under fire from the air for the first time or two tended to react defensively by seeking cover and fleeing the battle area. Subsequent exposure resulted in a hardening, a "getting used to" process, and the seeking of active defensive measures such as installing machine guns with high-angle mounts on vehicles and around bivouac areas. Accordingly, what pilots were able to get away with in 1917 was no longer possible in 1918.

Identification and Communications

Critical problems revealed themselves in the areas of identification and communications. Infantry that experienced an aircraft attack tended to view all aircraft from that point on as hostile, and to fire upon them. Commanders quickly recognized the importance of aircraft recognition training, but it is easy to understand the tendency of infantry to fire at any plane pointed in their direction. (This, of course, carried over to later wars.) All aircraft of a design generation tend to look alike to nonexperts; for example, most WW-I biplanes tended to look like other WW-I biplanes, as with most WW-II fighters, and indeed, as many fighters look alike today. Identification cuts both ways, however, and the airmen had great difficulty sorting out where the front began and where friendly forces were located during the war of movement in 1918. To help address this, for example, Mitchell imposed stringent guidelines on when aircraft could attack in support of ground forces, stating that a "protective mission" could be undertaken

> if the infantry signaling is efficient, *and in this case only*, an attack may be made by machine guns on the enemy's reserves which are in formation for counterattack. To prevent enemy infantry [i.e., support] planes from entering the battle zone. To help the advance of the tanks.[40]

Communication is another long-standing battlefield air support problem area, and it, too, was first encountered during the Great War. Consider again a quote from Mitchell on his experiences in 1918:

> We were experiencing a great deal of trouble with the ground troops in making them answer the signals from the air and properly maintain their radio stations for communicating with our airplanes. Our pilots had to fly right down and almost shake hands with the infantry on the ground to find out where they were. Many of the infantry and artillery officers left the handling of the wireless stations to the first sergeants of their company. The sergeants said that they got tired of seeing wireless-operators sitting in a hole in the ground, doing nothing and safe from the enemy's fire, so they assigned them to kitchen police or some other duty of that kind. As a result there were no communications at a critical time.[41]

Not all communications problems stemmed from human failing. Rather, early radio gear was primitive in the extreme, and prone to breakdown in the demanding environment of the Western Front. Visual signals worked some of the time; panels and flares worked well, as a rule, as indications of where forces were. But when aircraft flew low over the front to assess signals from the ground, they often fell victim to antiaircraft fire, particularly since many of these planes were slow cooperation aircraft used for artillery spotting duties, such as the British B.E. 2c or R.E. 8. Battle management procedures sometimes got in the way. As mentioned earlier, by the end of the First World War, Germany showed the most sophisticated grasp of air-to-ground aircraft and operational requirements for battlefield air support. For example, Germany actually assigned air liaison officers to work with German ground units and communicate back to air bases to keep air units informed of the ground situation, location of the front, targets, and the like. Despite this, however, the desire to centralize control of ground-attack sorties by having requests transmitted to a central headquarters and then dispatched to an airfield for execution sometimes led to confusion and delays—yet another problem that would be encountered by other nations in other wars.[42]

Aircraft and Tanks

One of the most interesting and prescient uses of air support aircraft in the First World War was in the field of aircraft operations in support of tanks. As mentioned earlier, the airplane and the tank had first worked together in the battle of Cambrai in 1917. Such team efforts became relatively commonplace in 1918. Ground commanders recognized at the outset that the tank was vulnerable to direct-fire artillery, and that the airplane could assist the tank during an assault by attacking antitank positions. Ironically, then, the airplane—subsequently to become the natural enemy of the tank in a manner analogous to the relationship of destroyer and submarine—began its relationship with the tank as a colleague. As historian Brereton Greenhous has pointed out, the relationship of the airplane and the tank was far from a smooth one; many of the problems mentioned earlier, such as communications, proved near-insurmountable. Nevertheless, there were some encouraging signs of progress, including efforts by the British to acquaint air and armor officers with each other's specialty—assigning "tankers" to air duty, and air officers to armored units. When things "clicked," for example during the attack on Flesquieres Ridge in late September 1918, the partnership of the airplane and the tank worked very well indeed; as Greenhous concluded,

> There can be little doubt that, even in 1918, the close ground support aircraft, operating in a counter-anti-tank role was a valuable, if not vital, adjunct of armored operations. But it was a lesson which the British, at least, would have to learn again in 1940, while the Germans, the recipients of the initial experience, remembered only too well.[43]

One final comment on battlefield air support in the First World War must be made: the combatant nations that went to war from 1914 through 1918 did not have independent and coequal air arms. There were no true "Air Forces" in World War I with the exception of the Royal

Air Force, created in mid-1918. As a result—and even including the British experience throughout the war—the air arms of the First World War were dominated by tactical aviation, and by the concerns of the ground commander. By 1918, doctrinal issues such as unified control of air assets, mass employment of air power, colocation of the air and ground commanders' headquarters, and the need for liaison officers between air and ground units, had already appeared, though they did not assume the significance and occasional controversy known in later years.

PART TWO

Small Conflicts of the Interwar Years

5

Emergent Ground-Attack Doctrine and Technology

The ending of the First World War did not, as many hoped, bring to a conclusion the era of human conflict. Rather, the war created great change—the rise of totalitarian regimes, the creation of new central European states, the collapse of an established European order, and a rising anticolonial consciousness—that ignited and fueled a series of small conflicts and wars. All of this change eventually contributed to the eruption of the Second World War in 1939. The trend begun with the First World War—the trend of increasing reliance upon mechanization, and, indeed, what could be hailed as the "birth" of mechanized warfare—continued in the interwar years, characterized by advances in the design and development of new mechanized weapons such as the tank, the airplane, the submarine, and more powerful and effective surface vessels. The small wars in which these embryonic but increasingly mature weapons appeared often did not offer the kind of experiences from which military observers could draw broad lessons for the future; nevertheless there were limited lessons that could be drawn and which, in some cases, were missed. In other cases, the "lessons learned" were misleading or mislearned, and had to be rethought—often at cost—in the Second World War which followed.

Basically, the conflicts and wars of the 1919–39 time period ranged across the spectrum of conflict from low-intensity national and subnational conflict (including terrorism and border incidents), through low-intensity overt limited conflict (such as guerrilla and counter-guerrilla warfare, rebellion, and revolution), and on to minor and major conventional wars involving invasion, seizing of territory, declarations of intent to conquer, attacks on civilian targets, and the full use of available military resources. Some of the more important struggles during this time period include: (1) Russo-Polish war, (2) Third Afghan war, (3) the Rif war, (4) Sandino war, (5) Gran Chaco war, (6) Ethiopian war, (7) Spanish Civil War, (8) Sino-Japanese war, (9) Changkufeng/Khasan incident, and (10) Nomonhan/Khalkhin Gol incident.

In virtually all of these incidents, conflicts, and wars, military aviation—specifically attack aviation for close support and battlefield interdiction—made some sort of appearance, with varying degrees of significance. This part examines the first four of these conflicts: the truly "small wars." Four of the remaining six will be examined in Part Three. To establish a reference applicable to both Part Two and Part Three, this study begins with a discussion on attack aircraft technology and doctrine to 1939.

Attack Aircraft Development in the Interwar Years

By the end of the First World War, the airplane had undeniably demonstrated its value as an element of combined arms warfare, revealing a flexibility of application, a degree of mobility, and a psychological impact (which, even today, is a considerable and potent attribute of aerial attack) that surpassed the most sanguine expectations of its prewar supporters. Unfortunately, the overall mystique of the airplane eventually generated some expectations that were, in the pre-nuclear age, extremely unrealistic. Most of these involved strategic air warfare, as enunciated by Italian air prophet Giulio Douhet.[1] In the United States, Douhet's doctrines, enunciated by partisans such as Billy Mitchell (a prophet in his own right), worked against the application of air power over the battlefield and, instead, emphasized strategic air warfare deep in an enemy's homeland, and development of defensive fighters and interceptors to counter any possible enemy bomber threat. (Much the same occurred in other nations as well.) Indeed, the rapid development of strategic bombing thought caused even attack aircraft partisans such as Mitchell himself to largely abandon their interest in attack operations and, instead, concentrate on the strategic bomber issue.[2] (This work, of course, is limited to an examination of battlefield air attack—essentially a tactical air power issue—and those readers desiring to examine the strategic air power question will find that a considerable body of excellent primary and secondary literature already exists.)

The time period from 1918 to the end of the 1930s was a particularly fruitful one for aircraft design, and major developments occurred in the evolution of aerodynamics, structures, and propulsion. As might be expected, the attack aircraft developed in the interwar years mirrored these technological advances. At first, they were little more than extrapolations of existing World War I technology. Gradually, however, designs appeared that made use of the growing technology base to generate entirely new capabilities. Generally speaking, thanks to the airframe, structures, and propulsion advances of the 1920s and 1930s, these new aircraft were increasingly able to fly farther and faster, were stronger than their predecessors (and thus more tolerant of battle damage), and had higher useful weights (and thus had greater payload—fuel and armament—potential). The following listing presents some selected characteristics of typical attack aircraft developed during the interwar years:[3]

Sopwith Salamander (Great Britain, 1918): Single-seat attack biplane; one 230-hp radial engine; two forward-firing machine guns; four 25-lb bombs; gross weight, 2510 pounds; max speed: 125 mph.

Curtiss A-3 (United States, 1927): Two-seat attack biplane; one 435-hp inline engine; four forward-firing machine guns plus two flexible rear guns; 200-lb bomb load; gross weight, 4378 pounds; max speed: 141 mph.

Polikarpov R-5 (Soviet Union, 1931): Two-seat attack biplane; one 680-hp inline engine; five forward-firing machine guns plus two flexible rear guns; 1102-lb bomb load; gross weight, approx. 6600 pounds; max speed: 142 mph.

Curtiss A-12 Shrike (United States, 1934): Two-seat attack monoplane; all-metal construction; one 670-hp radial engine; four forward-firing machine guns plus one flexible rear gun; 464-lb bomb load; gross weight, 5756 pounds; max speed: 177 mph.

PZL P-23B Karas B (Poland, 1936): Two-seat attack monoplane; all-metal construction; one 680-hp radial engine; one forward-firing machine gun plus one flexible dorsal gun and one flexible ventral gun; 1543-lb bomb load; gross weight, 7704 pounds; max speed: 198 mph.

Northrop A-17A (United States, 1937): Two-seat attack monoplane; all-metal construction with retractable landing gear; one 825-hp radial engine; four forward-firing machine guns and one flexible rear gun; 654-lb bomb load; gross weight, 7550 pounds; max speed: 220 mph.

Breda Ba 65bis (Italy, 1938): Two-seat attack monoplane; all-metal construction with retractable landing gear; one 1000-hp radial engine; four forward-firing machine guns and one rear gun in a rotating turret; 2205-lb bomb load; gross weight, 8014 pounds; max speed: 255 mph.

Junkers Ju 87B Stuka (Nazi Germany, 1938): Two-seat monoplane dive bomber; all-metal construction; one 900-hp inline engine; two forward-firing machine guns and one flexible rear gun; 1540-lb bomb load; gross weight, 9336 pounds; max speed: 217 mph.

Mitsubishi Ki-51b (Japan, 1939): Two-seat attack monoplane; all-metal construction; one 940-hp radial engine; two forward-firing machine guns and one flexible rear gun; 528-lb bomb load; gross weight, 6169 pounds; max speed: 263 mph.

Breguet 691 AB.2 (France, 1939): Two-seat attack monoplane; all-metal construction with retractable landing gear; two 700-hp radial engines; one forward-firing 20mm cannon, two forward-firing machine guns, one flexible dorsal gun, and one flexible ventral gun; 882-lb bomb load; gross weight, 11,023 pounds; max speed: 301 mph.

Martin Model 167F (United States, 1939): Three-seat attack monoplane; all-metal construction with retractable landing gear; two 1050-hp radial engines; four forward-firing machine guns and two rear-guns in a power-operated turret; 1760-lb bomb load; gross weight, 15,297 pounds; max speed: 304 mph.

Douglas A-20A Havoc (United States, 1939): Three-seat attack monoplane; all-metal construction with retractable landing gear; two 1600-hp radial engines; four forward-firing machine guns, one flexible dorsal gun, and one flexible ventral gun; 2400-lb bomb load; gross weight, 19,750 pounds; max speed 347 mph.

As can be seen, the attack aircraft generally was a two-seat single-engine aircraft up until the late 1930s, by which time twin-engine designs were becoming increasingly prevalent. Additionally, however, it will be noted that a new catagory of aircraft had appeared in the specialized attack family—the dive bomber, typified by the German Stuka.

Several significant points can be made about the aircraft in this listing. First, as the 1920s led into the 1930s, the performance disparity between the special-purpose attack aircraft and the fighter—at once its competitor for the ground-attack mission and its greatest adversary—grew ever wider. Then, the twin-engine attack airplane appeared. In part this stemmed from a widespread (and erroneous) belief in the efficacy of the "multipurpose" multiengine design for fighter, bomber, and reconnaissance duties—a belief soon shattered after 1939. But it basically stemmed from an attempt by designers to overcome the increasing performance disparity of the single-engine attack airplane by going to a twin-engine formula. It had several results. First, it blurred the distinction between the low-altitude attack airplane and the medium-to-high-altitude medium bomber (such as the North American B-25 Mitchell). Second, these attack "twins" were big airplanes for their time (an A-20A had a 61-ft wingspread), and they were large and relatively awkward targets for the rapid-firing antiaircraft cannon proliferating in service with ground forces. Very quickly, once war broke out, nations learned that their twin-engine attack aircraft, to survive, had to operate at higher altitudes, in a fashion akin to more conventional medium bombers attacking in formation, and in level flight.[4] Single-engine attack aircraft continued to be developed (such as the Soviet Ilyushin Il-2 Shturmovik), but they were increasingly the exception.[5]

One form of single-engine attack airplane deserving special mention is the dive bomber. Dive-bombing attacks had first appeared during the First World War, with modified bomb-carrying fighters diving on a target, the pilot aiming the entire aircraft as he would a rifle, and then dropping his bombs. The principal use of the dive bomber was, of course, as an antishipping aircraft, and it would excel in that role, as Pearl Harbor and Midway and a host of other naval engagements clearly attest. As a rule, dive bombers had rugged structures to withstand abrupt pullouts at low altitude and had special aerodynamic speed brakes to open into the airstream and slow the bomber during its dive to permit bomb release at low altitude. Greatly influenced by American dive-bomber development, Nazi Germany embarked upon the design of a category of "diving battleplanes" *(Sturzkampfflugzeug,* subsequently abbreviated to the notorious *Stuka)* that eventually spawned the Junkers Ju 87, a plane whose history is synonymous with the Nazi blitzkrieg. But—again as the Ju 87's own history indicates—the dive bomber was a dramatically overexaggerated threat, highly vulnerable to ground defenses and dependent on near-total control of the air to survive. Both Britain and the United States dabbled with dive bombers and used them in small numbers for ground attack in the Mediterranean, Southeast Asian, and South Pacific campaigns of World War II, almost always with disappointing results. The fighter-bomber soon demonstrated that it could undertake the dive bomber's mission and do it with greater success and survivability. By mid-war, certainly, air leaders of virtually all the combatant nations recognized that any gains in accuracy that might derive from the special-purpose dive bomber were offset by its inability to conduct other missions and, more importantly, to survive in the high-threat environment of the 1940s battlefields.[6]

In 1922, in an essay setting forth the requirements desirable in fighter aircraft, then-Major Carl Spaatz differentiated between four kinds of fighters: *offensive fighters,* characterized by structural strength, visibility, speed, and maneuverability; *defensive escort fighters,* usually heavily armed two-seaters; *night fighters;* and *attack fighters,* characterized by visibility, maneuverability, speed, and armor protection. (In actual practice, the two-seat heavily armed "defensive escort fighter" proved totally unsuccessful during World War II.) Technologically, there were great similarities in the requirements for what Spaatz characterized as offensive and attack fighters. Eventually, the fighter blended quite smoothly into the ground-attack role, becoming the fighter-bomber by the time of World War II.[7]

This occurred largely in spite of existing doctrine. In 1940, for example, Air Corps Field Manual FM 1-5 held that fighters were "not suitable" for ground attack "other than personnel or light material" except for temporary employment during emergencies.[8] This general inclination was not restricted to the United States. The classic fighters of the Second World War (all under development in the late 1930s and early 1940s) were all aircraft designed largely without thought to ground attack; in fact, a listing of the aircraft, their original rationale, and their ultimate use (Table 2) is most instructive.

The larger lesson to be learned from this is one that first appeared in the First World War, and one that has held true for fighter aircraft development since the Second World War: *virtually every fighter designed to undertake air-to-air combat (with the exception of a few highly specialized interceptors) has subsequently been modified to undertake ground-attack duties as well.* Put another way, during the interwar years, the attack aircraft concept gradually underwent a transformation; the attack airplane itself evolved into the specialized dive bomber on one hand, and merged with the light and medium bomber on the other. The fighter airplane gradually came to assume the duties of what had, to that time, been considered the traditional role of the "attack" airplane—striking at ground targets with bombs and machine-gun fire delivered from low-altitude terrain-hugging attacks.

Table 2. Development rationale and actual combat usage of World War II fighters

Type	Rationale	Combat Use
Hawker Hurricane	interceptor	fighter-bomber
Supermarine Spitfire	interceptor	fighter-bomber
Fiat C.R. 42	fighter	fighter-bomber
Reggiane Re.2000/2002	fighter	fighter-bomber
Mitsubishi A6M	fighter	fighter-bomber
Yakovlev Yak 7/9	fighter	fighter-bomber
Messerschmitt Bf 109	interceptor	fighter-bomber
Messerschmitt Bf 110	"destroyer"	fighter-bomber
Focke-Wulf FW 190	fighter	fighter-bomber
Lockheed P-38	interceptor	fighter-bomber
Bell P-39	interceptor	fighter-bomber
Curtiss P-40	fighter	fighter-bomber
Grumman F4F	fighter	fighter, antisub
Vought F4U	fighter	fighter-bomber
Grumman F6F	fighter	fighter-bomber
Westland Whirlwind	interceptor	fighter-bomber
Hawker Typhoon	interceptor	fighter-bomber
Republic P-47	interceptor	fighter-bomber
Nakajima Ki 84	interceptor	fighter-bomber
North American P-51	recce-fighter	fighter-bomber
Bristol Beaufighter	night-fighter	fighter-bomber
Hawker Tempest	fighter	fighter-bomber
Messerschmitt Me 262	interceptor	fighter-bomber

Ground-Attack Aviation and Military Thought

Despite the considerable debate over the precise mission and role of bomber, fighter, and attack aviation, military air and ground forces worldwide generally conceded that air attack against ground forces *could* have potentially decisive impact, particularly if *integrated* into the actions of friendly ground forces. To this end, the Royal Air Force and British Army developed complex and rudimentary cooperation procedures for aircraft working with troops, using a combination of radio and visual signals. In contrast to the U.S. Army Air Corps, the RAF believed fighters eminently suited for low attack work, though an official manual cautioned that "if . . . low-flying attacks are made against well-trained and well-equipped troops of high morale, severe casualties to attacking aircraft are likely to result," adding that such missions should not interfere with the primary mission of the fighter, namely to secure air superiority. The manual cautioned:

> Troops deployed are not suitable targets for fighters, nor should fighters be used to attack objectives which can be engaged by artillery or machine guns on the ground. Fighter squadrons must not be told to search for objectives; they should be given definite information and orders regarding the location of the targets which they are to attack. Such information will often best be obtained by aircraft engaged in tactical reconnaissance, and co-operation should therefore be arranged between army co-operation squadrons and fighter squadrons detailed for low-flying duties, in order to take immediate advantage of the information obtained. For this reason it may be desirable to place such fighter squadrons temporarily under the command of the army co-operation wing commander on the section of the front concerned.[9]

This cautionary note was one that the Army Air Corps was fully in accord with as the 1930s drew to a close; FM 1-5 bluntly stated:

> Support aviation is not employed against objectives which can be effectively engaged by available ground weapons within the time required. Aviation is poorly suited for direct attacks against small detachments or troops which are well-intrenched or disposed.[10]

But within this generally cautious framework, another Air Corps manual, FM 1-10, saw a wide range of missions possible for battlefield support of ground forces, including attacks on fortifications and "defensive organizations," strikes to isolate the battle area by attacking communications and transport, blocking the movement of enemy reserves, and attacks upon hostile mechanized forces.[11]

FM 1-10 reflected the reports coming from Asia and Europe in which modern military aviation forces were engaged in battle, and it emphasized most strongly the importance of command, control, and communications, particularly with friendly armored forces, using prearranged signals, pyrotechnical devices, panels, and—above all—direct radio communication between armor and air units. Two general texts issued by the Air Corps Tactical School at Maxwell Field in 1938 and 1939 argued forcefully that selective air strikes over a battle area could have a significant impact on the conduct of the ground war, using a mix of fragmentation bombs, machine-gun fire, and chemical weapons.[12] In a little over a year, the General Headquarters' maneuvers of 1941 would confirm that and add vital data points on the operations of air forces and ground forces, based on the mock-war of Louisiana and the Carolinas, including illumination of the problems of working together. As Christopher Gabel has written, however, the prevailing influence of strategic air warfare tended to work against developing a cohesive tactical air-ground doctrine. Despite undeniable interest in such a notion by Army ground commanders and by members of the attack and air support community within the Air Corps (and its successor the Army Air Forces—AAF), and despite the evidence of the GHQ maneuvers, such an effective partnership of air-ground forces did not emerge until in the midst of the Second World War itself, thanks largely to lessons learned in Tunisia and the inspiration of successful British air-ground support.[13]

It is interesting to note that the only American air service with combat experience in the interwar years—the U.S. Marine Corps—had a different view of the role of military aviation forces. The Marine perspective reflected the Corps's experiences in the "banana republic" wars of the 1920s and 1930s, as well as the notion of the Corps as basically a landing-force organization dependent upon aviation to substitute for a lack of comprehensive organic artillery, or to substitute for an absence of naval gunfire support. Clearly, the difference in the AAF and Marine view of air warfare—a difference that would be highlighted in the years after the Second World War, particularly around the time of Korea—had already developed in the years prior to Pearl Harbor. Consider, for example, this extract from the Marines' *Small Wars Manual* of 1940:

> *The primary mission of combat aviation in a small war is the direct support of the ground forces. This implies generally that all combat aviation will be used for ground attack* [emphasis added]. Air opposition will usually be nonexistent or weak, and friendly aviation should be able to operate against hostile ground troops at will. Fighting squadrons [i.e., fighter squadrons], if included in the force, may be employed as light bombers; while the bombing squadrons will find more use for their lighter bombs and offensive machine guns than they will for their major weapon—the heavy demolition bomb. Attack aviation, or its substitute, the dual-purpose scout, is the best type to cope with the targets likely to be encountered in small wars. Troop columns, pack trains, groups of river boats, occupied villages of flimsy construction, mountain strongholds, and hostile bivouac areas are all vulnerable to the weapons of the attack airplane—the light bomb and machine gun. Occasionally, targets of a more substantial nature may require the use of medium demolition bombs. As the type of campaign approaches the proportions of a major conflict, so will the employment of the different types of combat aviation approach that prescribed for major warfare. For the typical jungle country small war, the division of missions between the different types is not so clearly marked.[14]

The *Small Wars Manual* discussed for example, how a column of troops could be covered by a division of airplanes (a unit of six aircraft). Reconnaissance aircraft could spot route ambushes, and air strikes could "be coordinated with the ground attacks if communication facilities and the tactical situation permit, or they may be launched independently to prevent hostile interference with the march of the supported column."[15] In such operations, it was critically important that ground commanders mark the location of their positions by panels, also indicating the direction and estimated distance to an enemy position by panel signal. "The ground commander," the manual stated, "should also indicate, by whatever means is expedient, just when and where he wishes the fire of aviation to be concentrated. In short, he requests fire support in the same manner as he would from artillery." In reviewing the role of aviation in attacking hostile positions, the manual explained:

> Combat aviation may be used as a substitute for artillery in the organized attacks of hostile strongholds. As such it provides for the preliminary reduction of the hostile defenses by bombing, for the interdiction of lines of communication and supply, and for the direct close-in support of the attacking infantry by lying down [sic] a barrage of machine-gun bullets and fragmentation bombs on the enemy front lines. All these missions cannot of course be performed by one air unit; schedules of fire must be worked out, timed with the infantry advance, and executed by successive waves of aircraft. Details of this form of air support are worked out by the air commander, using such numbers and types of air units as are available and necessary. The ground commander must submit a definite plan if air attack is to be coordinated; otherwise the air commander on the spot must use his force as opportunity offers.[16]

In the course of the Second World War, all of these assumptions—those of the Royal Air Force, the AAF, the Marines, and the other of the world's air services—would be exhaustively put to the test. But that such doctrinal notions had already been derived, no matter how accurate or flawed they may have been, depended in great degree upon the actual combat experiences of the interwar years, from struggles in Europe, Africa, Asia, and South America.

6

Small Wars of the 1920s and 1930s

The Russo-Polish War of 1920

No sooner had shooting stopped on the Western Front than conflict erupted between the Allied powers and the emergent Bolshevik state of Soviet Russia. Then, resulting from the same revolutionary turmoil, war broke out between the Soviets and the newly revived Polish state. Meanwhile, Great Britain, France, and Spain found themselves embroiled in imperial conflicts in North Africa and the Middle East, and resorted to air warfare to help restore order and assist the operations of ground forces. In none of these conflicts can it be said that air application was decisive; but in most of them, the use of aircraft in support of ground forces added an important (and historically neglected) dimension to the fighting, and had significance for future military operations. Indeed, in some aspects, these conflicts offered a glimpse of what air power could be expected to accomplish—and fail to accomplish—in future wars, particularly those that would occur after 1945 in lands formerly controlled by the major European colonial powers.

Because of the Allies' experience with air operations over the Western Front, the American and British pilots who flew against the Bolsheviks with the intervention forces in Russia and with the so-called Kosciuszko Squadron in Poland applied many of the same tactics that had characterized air operations in the latter stages of the First World War. Because the Soviet air force itself was in but an embryonic state, the occasional ferocity that had characterized Western Front air warfare did not exist in the East; dogfights occurred (which the Soviets invariably seemed to lose), but the pilot's greatest enemies were not opposing airmen, but, rather, mechanical failure and ground fire. Pilot memoirs from this period tend to treat their experiences in Russia as almost a lark—an effort to retain the camaraderie and excitement of the wartime years.[17]

In this environment it was perhaps unrealistic to expect that either side would have a well-thought-out and comprehensive air doctrine, and, indeed, neither really did, beyond an apparent hazy recognition that, in some way, air operations could assist forces on the ground. The relatively permissive air combat environment caused both sides to get away with operations that simply could not have succeeded on the Western Front: operating slow and unarmed reconnaissance aircraft deep into enemy territory, for example, or slow and unmaneuverable cooperation aircraft over the battlefront. In sum, both the Allied and Bolshevik sides enjoyed complete freedom of movement over the battlefield, due in large measure to the small numbers of planes involved. (In April 1920, for example, Poland had 160 combat aircraft—130 of which were serviceable—while the Red air force had 210 in service, facing each other along a 500-mile front.) As discussed subsequently, this had a disastrous effect upon future Polish aircraft acquisition strategy.[18]

The activities of the famed Kosciuszko Squadron—a squadron made up of American volunteers, most of whom had extensive combat experience, as well as a small number of Polish officers—offers a good account of air-ground operations during the Russo-Polish war of 1920. By and large, air strikes were undertaken on a "pre-briefed" and armed reconnaissance basis, with no real air-ground communication and coordination. Aircraft struck at targets identified from aerial or ground (cavalry) reconnaissance, in response to requests from ground forces relayed by telephone or by messenger to an aerodrome nearest the front, or on the basis of the pilot's own observations as he roamed over the front. Ground fire was intense and surprisingly accurate, and the squadron experienced numerous forced landings but, fortunately, only one combat death. The squadron operated a mix of Austro-Hungarian-built Albatros D III fighters (a plane ill-suited to ground attack, lacking a robust structure—indeed, its lower wings had a built-in design flaw predisposing them to shed in flight), the Italian-built Ansaldo A.1 Balilla ("Hunter," a type known for its high speed but otherwise an undistinguished design, and the Breguet XIV (a very fine two-seat light bomber, as discussed earlier). In part to compensate for deficiencies in equipment and the nature of the threat environment, the Kosciuszko pilots resorted to imaginative stratagems; one pilot used terrain-masking while attacking a Bolshevik artillery battery, "dodging around trees and the domes and steeples of churches," so that antiaircraft gun crews could not track him and shoot him down. Using this tactic, and then popping up from behind cover to attack the guns of the battery, he was able to "silence" the guns—presumably by killing the crews, as he had previously dropped his bombs on other targets.[19]

In May 1920, a Bolshevik counteroffensive north and south of the Pripet Marshes triggered a nearly three-month retreat of Polish forces before the rapidly advancing cavalry of Semyon Budenny. During this retreat, the Kosciuszko Squadron had to move frequently from airfield to airfield, and conducted a series of rear-guard harassing actions against "Bolo" forces. Reflecting on these operations, a U.S. War Department summary stated:

> After locating [Budenny's advancing columns] the pilot would fly well out of range to a point about one half mile behind the rear of the column and then at a height of about 300 meters would fly back over . . . the wagon trains and the rear of the cavalry [where] two bombs would be dropped. After dropping the second bomb, the pilot would dive upon the troops and at a height of 100 meters open fire with both machine guns. Flying directly down the line by elevating and lowering the nose of the plane a very slight degree while at a height of 10 to 15 meters above the column, machine gun bullets could be sprayed through the entire length of the column, this causing casualties and creating great disorder and confusion. By this method one squadron was able to check the advance of 20,000 cavalry for several hours every day.[20]

It is difficult to state with certainty just how effective these attacks upon "the savage Bolshevik hordes" (as the Polish government dubbed them) actually were.[21] A memoir by Captain Merian C. Cooper, the founder of the Kosciuszko squadron, stated that following an attack by eight airplanes against Bolshevik forces, a wireless intercept picked up by Polish intelligence indicated that the "Bolo" commander estimated he had been attacked by no less than *thirty* airplanes.[22] Another wireless intercept from Budenny's headquarters to Moscow revealed that the cavalry commander wanted his own aviation force to offset that of the Poles, and allegedly asked "that the price originally placed upon the heads of the American pilots be doubled—twenty-five thousand gold rubles for each of their skulls, dead or alive."[23] In August 1920, Marshal Jozef Pilsudski launched a counteroffensive which broke through Bolshevik lines in three days, and by the end of the month, Communist forces were in disarray. In October, an armistice settled the Russo-Polish boundary, which lasted until Stalin invaded Poland in September 1939.

In the absence of any hard evidence, one must conclude that air assault during the Russo-Polish war contributed to the destruction of Bolshevik forces, and assisted the Poles in making an orderly withdrawal in mid-1920 until they were in a position to launch the Pilsudski counteroffensive in August. But beyond this, the evidence is simply not there to argue that battlefield air support in the Polish war tipped the scale of any particular battle or campaign. Clearly it helped, but it was not decisive, for even in its assistance, it could not be argued that it had the concentration of force and the level of effort that, say, characterized air support operations over the Western Front during the great offensives of 1918, or over Palestine near the end of the war. Interestingly, Poland apparently misread the experiences of the war, placing too great an emphasis upon developing aircraft for the army cooperation role. Initially, the *Lotnictwo Wojskowe* (Military Aviation) seemed to follow standard practice of separating bomber and fighter units, and attempted to develop a comprehensive air doctrine. Unfortunately, critical leadership changes halted any effort by Polish airmen to derive a comprehensive air doctrine and, instead, resulted (in the words of historian Jerzy Cynk) in "the *Lotnictwo Wojskowe* [being] put back firmly and squarely in its place as an insignificant army ancillary service. Pilsudski appreciated more its international prestige value than its operational usefulness, which in his view was limited to reconnaissance and liaison only."[24] This resulted in overemphasis upon developing *liniowe* (frontline support) aircraft which were designed to undertake a multitude of battlefield roles, such as artillery spotting, reconnaissance, and light attack. One of these designs, the PZL P-23 Karas, appeared in the mid-1930s and was generally useful, but totally obsolete by the time war broke out in 1939.[25]

Though overall it is much more difficult to state what impact the Russo-Polish air war had upon Soviet design practice, the results circumstantially speak for themselves; the Soviets emphasized development of ground-attack aircraft and by 1927 had formed no less than five specialized squadrons of *Shturmovaya Aviatsia* (attack aviation) for battlefield air assault; the appearance of the Polikarpov R-5 Shturmovik in 1931 gave the Soviet Air Force an aircraft as advanced as any in the world for ground support operations. Throughout the 1930s, ground-attack aviation played a major role in Soviet military aviation thought, and the vindication of that interest could be found in the record of Soviet attack aviation in the Second World War.

The British "Air Control" Experience

The Russo-Polish War of 1920 came closest in the years immediately after the First World War to the kind of mid- to high-intensity conflict commonly associated with military aviation operations. But another form of warfare that utilized aviation would become all too common in the years ahead: the low-intensity warfare of revolt, revolution, and terrorism. In this form of war, aviation would be employed quickly, and found—depending on circumstances—to be generally significant. In future wars of this sort, of course, counterinsurgency air operations would become commonplace; interestingly, many of the methods and notions of operational usage and control that governed subsequent air operations of this type first appeared in the small wars of the late 'teens and early 1920s, specifically, the wars of Great Britain, Spain, and France in Southwest Asia, the Middle East, and North Africa. Generally speaking, airmen came to refer to such operations in the 1920s and 1930s as "air control," though, in a strict sense, this was only a British term based on the use of the Royal Air Force in the Middle East and India. Military aircraft had been used in this role since before the First World War, but only after the Armistice were reasonably well-thought-out strategies and extended operations undertaken.

Britain's Royal Air Force was the first to undertake significant air-control activities; added to Britain's burden of maintaining the peace in its far-flung empire were new responsibilities imposed by its role in running mandate territories awarded after the collapse of the Ottoman Empire. In May 1919, war broke out when Afghan forces under the leadership of Amir Amanullah Khan invaded India; a rapid Anglo-Indian response resulted in this attack being turned back by mid-month, and a second (and potentially more serious) attack by Mohammed Nadir Kahn was likewise defeated. Amanullah Khan requested an armistice, and the Treaty of Rawalpindi in August 1919 brought the short war to an indecisive conclusion. During this brief war, British forces had been assisted by four squadrons of aircraft assigned to the North West Frontier Force and the Baluchistan Force. On the whole, the aircraft made little impact on the war's operations, except in the area of reconnaissance. They were elderly Royal Aircraft Factory B.E. 2c observation planes never really intended for offensive (bomb or machine-gunning) operations, and, indeed, had been obsolescent when war had broken out in 1914. Underpowered, they had extreme difficulty operating in the high-altitude terrain of the North West Frontier; nevertheless their pilots were game and did the best they could. For example, when the Anglo-Indian forces took Khargali Ridge on May 11, the RAF bombed and strafed the Afghans as they left the ridge, hotly pursued by British forces. The RAF also bombed key centers such as Dakka (now known as Dacca), Jalalabad, and Kabul, dropping "one ton and eight hundredweights of explosives . . . in a single day" on Jalalabad.[26] In response to heavy and accurate Afghan howitzer fire during the investment of Thal by Afghan forces, the RAF bombed the artillery positions, though this only gave temporary relief. The Afghan soldiers were determined fighters, and though new to air attack, shot back with small arms and rifles at low-flying aircraft; for example, during the attack on Chora fort, one RAF airplane was shot down and its crew taken prisoner. Another problem of air support operations manifested itself during the attack on Spin Baldak fort by Gurkha and Punjabi infantry; a bomb from a RAF plane fell among friendly troops, killing or wounding fifteen officers and men. In two cases, timely air reconnaissance and air intervention could have prevented disasters: on July 6, a convoy in a narrow defile near Babar was ambushed by a large party of Wazir tribesmen and severely mauled, as was another column in similar circumstances near Kapip over a week later.[27]

Overall, the General Staff Branch at Army Headquarters in India were less than enthusiastic but bluntly forthright when, in their final report on the Afghan war, they concluded:

> Little useful purpose can be served by drawing deductions from [aircraft] performances during this campaign. The machines with which the RAF were equipped when the Afghan War broke out were obsolete and worn out. Their climbing power was low, and this led to their being shot at from the hill tops as they passed along the valleys. Their morale effect was undoubtedly great, and the bombing of Dakka, Jalalabad and especially Kabul were factors which probably decided the Amir to sue for peace. The greatest credit is due to the officers of the Royal Air Force for the courage and skill which they displayed in performing their duties in these antiquated machines. They proved the value of the aeroplane in long-distance strategical reconnaissances, in bombing areas of concentration, supply depots and transport, but in short-distance tactical reconnaissances they were of no great value. The terrain was difficult and the tribesmen soon learnt how to break into small groups and to keep still when an aeroplane was overhead. The result of this was that bodies of the enemy were difficult to locate and negative information from the air had to be regarded with suspicion.[28]

The RAF went to war against a different kind of opponent in Somaliland in 1920, when they were used as part of an expedition to hunt down and destroy the "Mad Mullah," Mohammed bin Abdulla Hassan, a man noted for his violence and depravity, who had terrorized Somaliland for the better part of two decades. There was an undoubted "settling an old score" air about the expedition, for the Mullah and his fanatical followers had, on many occasions, frustrated other pursuers, inflicting heavy losses. After the 1919 expedition threatened to become bogged down in the countryside in the same manner that previous ones had, a flight of D.H. 9 two-seat bombers (known as Z Unit) joined a mixed force of British, Indian, and tribal troops and a Royal Navy contingent of three ships in an attempt to end the Mullah's menace once and for all. On January 19, eight D.H. 9's left Berbera, staging through Eil Dur Elan. On January 21, 1920, six left to raid the Mullah's headquarters at Medishe. The raid was a disappointment; one D.H. aborted with engine trouble, four attacked a Dervish fort manned by the Mullah's troops, and only one found the Mullah's *haroun*. That one made a distinct impression, however; during its approach some of the Mullah's followers informed him that the plane must be a chariot from Allah to take the Mullah up to heaven; another thought it was a messenger from the Turks to announce a Turkish victory in the war (news of the collapse of the Ottoman Empire obviously not having reached the Mullah). Then it dropped its bombs, the first of which killed the Mullah's chief councillor and singed the Mullah himself. Subsequent attacks killed or wounded forty of his followers, and over the next two days, the D.H. 9's undertook intensive bombing and strafing operations over the camp. When reconnaissance indicated that the Mullah had abandoned the position, the air commander announced that he would suspend all further independent air operations in favor of combined operations with the ground forces; this was the pattern, then, followed on subsequent attacks.[29]

As ground forces—including the 700-rifle Somaliland Camel Corps—chased the Mullah's demoralized followers down, Z Unit bombed and strafed the retreating Dervishes. The Mullah's personal fort at Tale was bombed with 112-lb bombs and incendiaries, but the Dervishes within retained their morale and kept up a brisk fire at the circling planes. Eventually, bereft of followers, the Mullah himself escaped into Abyssinian territory, setting up camp with "three or four hundred" refugees at Webbi Shebeli; the Abyssinians were less than pleased by his arrival, and threw his messengers into prison. The Mullah and most of the refugees died less than a month later from influenza and famine. Because of the novelty of aircraft attack in the frontiers of empire, the actions of Z Unit attracted a great deal of attention, and some alleged that it was due to Z Unit that the Mullah was finally brought down. In fact, while Z Unit played a most important role, the Mullah's days were, in all likelihood, numbered; the end of the First World War meant that sufficient resources could be applied to his suppression once and for all, and the size of the expedition against him is evidence itself of the resolve of British authorities in Somaliland to destroy him. Douglas Jardine, Secretary to the British Administration in Somaliland, took a negative view of the air operations, while the Viscount Milner, on the other hand, stated that "it is certain that, but for the hopes we based on the co-operation of the airmen, the campaign would never have been undertaken, and that they contributed greatly to its success."[30] Jardine noted that the Mullah expedition marked the first time that airplanes had been used in Africa as the main instrument of attack, and that it had attracted "the most profound interest" from those interested in both aerial bombardment and "the policing of Africa and Asia." Nevertheless he remained skeptical.[31]

British air control operations in the Middle East defy easy analysis or characterization, but overall, despite the enthusiastic support of partisans such as Hugh Trenchard and then-Secretary of State for the Colonies Winston S. Churchill, they must be judged to have been less successful than generally thought. Despite this, there were some interesting elements in these operations that had implications for future air-ground coordination and cooperation during attack. In mid-1921, Trenchard and Churchill secured approval to police the Middle East (particularly Iraq) with a force of eight squadrons: four bomber, one fighter, one fighter-reconnaissance, and two transport. These would cooperate with a ground force of RAF-manned armored cars and indigenous Arab army forces, a single British army brigade, and a pack artillery battery. In theory, the armored cars "worked in the shadow of the aircraft."[32]

In reality, such cooperation proved extremely difficult to achieve. Sir John Bagot Glubb, who served in Iraq throughout this period, roaming over what is now Saudi Arabia and the Gulf States through Iraq and Jordan, has written eloquently of his experiences as an Army officer detailed to the RAF. By the nature of air operations, aircrew tended not to recognize natives as friendly or unfriendly, since they did not live among the population. (Indeed, some pilot accounts infer that the aircrew really didn't care much about distinguishing friend from foe. The Air Staff believed all friendly natives should be cleared from an area so that remaining natives could be considered hostile—the exact opposite attitude of Glubb.) Arab raiders—particularly the feared Wahabi zealots—struck at their victims suddenly and only rarely were revealed by aerial reconnaissance. (Glubb eventually decided "to organize the Iraq tribes to defend themselves, using the RAF as a supporting arm rather than as the sole defensive weapon.") Bombing and strafing could scatter a raider party, but the members scattered so completely over the desert that it was difficult to achive any meaningful results against so scattered a target. The flying conditions themselves were inhospitable and limited aircraft operations. Coordination between the armored cars and the aircraft was poor at best and typically nonexistent. Finally, and of great significance, there was the problem of convincing the air component to send help to assist ground units. On many occasions—Glubb believed as often as half the time, or at least one case in three—reports of raids would be accurate enough to warrant RAF action, and the RAF would not be present. Such was particularly true of armored car operations, which Glubb believed to be "the only weapon which would have inflicted a really decisive defeat."[33] Every now and then, however, intelligence, communication, and control all came together, often aided by common sense and at times just plain good fortune. Then the system could work quite well. An example was the raid of Menanhi Ibn Ashwan into Shammar in December 1928. Glubb had arranged for an advance RAF base camp to be set up at Maghaizal, in the so-called Stony Desert. After receiving word of the raiding party from friendly tribesmen, Glubb arranged for RAF reconnaissance; on the second day of recon missions, and with an update on the likely location of the raiders from friendly shepherds, Glubb spotted a party of 100 camels, whose riders were hiding under the beasts. Amid heavy if inaccurate ground fire, the two D.H. 9's on the mission bombed and strafed the riders, leaving "only individual camelmen riding madly towards every point of the compass." The raiding party broke off its foray and returned to Mutair.[34]

A more interesting successful example occurred August 1924. By good fortune, an RAF driver near Amman learned from local Arabs that a Wahabi raiding party had reached the Transjordan. The Wahabi, fanatical Islamic fundamentalists, made it a practice to kill every male, adult and child, of tribes and communities that they raided, and thus the RAF could count on the strong support of the local populace in seeking out and destroying this party. An RAF recon aircraft from Amman located a force of 300 Wahabi near Madeba. Three RAF armored cars sortied from Ziza Station, an RAF base south of Amman, to scout for the raiders, who had already viciously attacked a community that morning, killing 130. They spotted hills "black with men"; in all, the Wahabi party numbered 5000, and had to cross two miles of open desert before getting to cover in an extensive network of hills that would hamper armored car operations. The armored cars lacked radio—it was fitted to them later—and thus were attempting to contact their headquarters by tapping into a railroad telegraph line when three D.H. 9's appeared, circled over the raiders, and began dropping bombs. The raiding party wheeled eastward, riding down the flat desert past the Ziza landing ground; it was a fatal mistake, for the armored cars now had a unique opportunity to come into play, and they mercilessly attacked the raiders with machine gun fire. The planes returned to Amman, were rearmed, and returned to the retreat, launching a second attack as the armored cars, ammunition exhausted, retired from the battle. A flight of Bristol Fighters from Palestine arrived over the scene, and continued to harass the remaining raiders with strafing attacks; during the Bristols' attack, one pilot and one gunner were wounded by groundfire. The air assault broke off as the raiders continued heading east; hundreds had been killed, and others wounded and taken prisoner. *The Royal Air Force Quarterly* considered the action at Ziza "an example of successful co-operation between air and mechanized land forces."[35]

One of the most influential of prewar British thinkers on air-ground cooperation and attack operations was then-Wing Commander (and subsequently Marshal of the Royal Air Force) John C. Slessor. Slessor was no great believer in the notion of the "dedicated" attack airplane designed exclusively for that purpose, and, instead, believed that

> assault aircraft must rely for protection on surprise, on high speed and manoeuvrability—the capacity to jink like a snipe when under fire, and to take advantage of trees, houses, and the folds of the ground. *For this reason the single-seater fighter was, and probably still is, the best type for low-flying action* [emphasis added].[36]

He believed that rather than requiring special-purpose designs, low-flying attack operations could be undertaken by aircraft normally used for other purposes, though he stated that "as a general rule . . . aircraft are not normally battle-field weapons."[37] By this he meant—in agreement with most other airmen worldwide—that aircraft were intended primarily for the attack of *rear* areas. He recognized that aircraft could play a critical role during the initial phase of an assault, "breaking the crust" (as he termed it) of defensive systems by operating in support of armored forces. But when he got his first chance to put some of his ideas into practice over the battlefield, it was not in the crucible of a general war, but rather in the episodic fighting of the Waziristan campaign of 1936–37 in the North West Frontier.[38]

As commander of an Army Co-operation Wing, Slessor was convinced that in low-intensity operations of the sort that characterized conflict in the North West Frontier, air and army forces had to work closely together. For example, unlike a conventional war where air recon could keep an army informed as to the enemy's location, air recon in low-intensity operations where a guerrilla force blended with the local population often was less satisfactory; indeed, the soldier on the ground, under fire, often had to communicate to the airplane above where the enemy actually was. Once air knew where an enemy was, its action could be decisive in helping break resistance. For this purpose, he believed the small fragmentation bomb (already the weapon of choice for attack personnel worldwide) was the ideal, and that bombing operations should be conducted closer than RAF-Army policy then dictated (namely 300 yards for a 20-lb antipersonnel bomb) even at the risk of friendly casualties. He argued it was better to take a small number of casualties from short bombing than a much larger number from enemy rifle and machine-gun fire. He organized what he termed the "Vickers-Bomb-Lewis" method of ground attack: a two-seat aircraft such as the Hawker Audax and Hardy, or the Westland Wapiti, would dive on enemy troops, strafing with its forward-firing Vickers gun, release its bombs, and during the vulnerable pull off the target, the rear gunner would spray the enemy with fire from his Lewis gun. Additionally, he and his staff recommended dividing maps into lettered grids, and arranging for aircraft to be "on-call" immediately for air-to-ground strikes via a "cab-rank" system whereby they would already be orbiting over the battlefield. Slessor assembled these and other thoughts on such operations in a small manual entitled *Close Support Tactics—Provisional*, and issued it to his air crews.[39]

During fighting at Dakai Kalai on December 22, 1936, an air action took place that tested air-ground cooperation to its fullest. Recon aircraft orbited the line of march of an infantry brigade as it approached a village; two Audax bombers were on strip alert, with two Wapiti as standby backup and two other Audax flying over the flanks of the column. A company of Punjabi troops came under intense and accurate fire, and the brigade commander radioed for air support "saying he considered it was a case for low-flying action." Confusion over the precise location of enemy and friendly units was compounded when some smoke candles failed to ignite. For safety's sake, the RAF opted to attack at a range undoubtedly greater than desired in front of the forward line of troops, and the recon pilot was directed to attack; he struck at the enemy using as a marker the bursting of friendly artillery fire. Two attack aircraft joined in the strikes, and supported a counterattack that recovered casualties from the enemy's first attack. Over the next three hours, six aircraft rotated over the battlefield, dropping a total of seventy-three bombs and firing approximately 1750 rounds of machine gun fire; when parties of enemy were seen they were bombed and strafed. On this note the fighting died away.[40]

To Slessor, this action

> proved the extreme difficulty of bringing down accurate fire from the air when the situation is not known at Column H.Q., and emphasizes the urgent necessity of some signal by which forward troops themselves can call for assistance, and indicate their positions. . . .
>
> The incident also proved the urgent need for a reliable R/T [radio] set by which clear orders can be conveyed quickly to pilots in the air.
>
> This was an occasion when bombing and machine-gun fire from the air on areas which *could* also be dealt with by artillery and machine guns from the ground was fully justified—namely an emergency, when every form of covering fire that could possibly be brought to bear was required to stave off a possible minor disaster. The air cannot claim on this occasion to have been more than a useful supplement to the excellent covering fire provided by the artillery. But as such I think it justified the not very considerable risks involved; it no doubt accounted for a share of the heavy enemy casualties (22 killed, 23 seriously wounded) in this engagement; and perhaps may claim some of the credit for the complete absence of casualties in the excellent counter-attack by 2 R. Sikhs.[41]

As time went on, newer technology replaced old, and as the Audax, Hardys, and Wapitis of the 1930s replaced the D.H. 9's and Bristol Fighters of the 1920s, they in turn gave way to monoplane successors such as the twin-engine Bristol Blenheim. The outbreak of the Second World War resulted in a decline in such "air control" activities, and the geopolitical changes following 1945 in the Middle East and Southwest Asia generally relegated air control to the past.

The Rif War in Morocco

Like Britain, Spain and France had their own experience in air control operations, notably in the troublesome war against Muhammad bin Abd al-Karin, the Moroccan Rif War of 1921–26. This romanticized conflict was, in fact, quite savage and bears some comparison–though there are also important differences–with the Algerian war of the late 1950s.

For years, al-Karim had fought the Spanish government in Morocco, and it flared into open rebellion in 1921, ultimately spilling over into the French protectorate, and resulting in a combined Franco-Spanish war against al-Karim's followers. While al-Karim seems to have had some nationalistic notions, his followers considered it merely a *jihad* (holy war), and thus it was (in historian Shannon Fleming's words) "a traditionalist manifestation of xenophobia and religious fanaticism."[42] The Rif soldiers emphasized surprise, maneuver, and concentration of forces. At first, Spain did not take the revolt very seriously, and Lt. Gen. Manuel Silvestre, commander of the Melillan sector, was confident that his forces would quickly crush the rebels. He underestimated the dedication and skill of the enemy, and overestimated the strength of his own forces, ignoring their relative lack of training, low morale, and corruption. Al-Karim's Riffi struck isolated Spanish forts strung out across Morocco and inflicted disturbing casualties. In one case, the Riffi assaulted a fort, killing 179 of 250 defenders. Subsequent attacks forced a retreat by Silvestre in July and August of 1921, a retreat which resulted in over 13,000 casualties, including Silvestre himself. Spain dealt relatively ineffectively with the rebellion, and it continued to fester; Spain began bombing operations by a few Breguet XIV bombers of the African Air Corps, but the campaign was not well thought-out, many of the bombs failed to detonate, and the Riffi disassembled them and made them into booby traps for use against the Spanish. In 1924, during fighting at Dar Aquba, al-Karim's men inflicted a staggering loss on the Spanish colonial forces, killing over 10,000 troops. That same year, a significant change took place in Spanish air tactics. The air-minded leader of Spain, Primo de Rivera, directed that the African Air Corps be concentrated so that it could undertake the systematic destruction of crops, markets, villages, and livestock; this purely Douhetian notion of air power marked the beginning of a concentrated bombing campaign against the Riffi and to elaborate on its effectiveness, the Spanish took to using poison-gas bombs as well as the traditional fragmentation and incendiary weapons. Over nearly the next five months, Spain dropped a total of 24,104 bombs on rebel communities and concentrations, but–and a lesson that other airpower enthusiasts would learn for themselves in the future–these Douhet-like tactics resulted in no appreciable decline in either enemy morale or ability to resist. It appears that after this experience in the spring and summer of 1924, the Spanish gradually shifted their air policy toward application of tactical airpower near and over the battlefield, using aircraft primarily for reconnaissance and bombing near the "front," and occasionally for bombing and strafing operations during an attack itself. Even here, however, the results were disappointing; in one case during fighting between Riffi and loyal colonial forces, the rebels counterattacked in the face of both a ground and air assault, routing the Spanish forces and killing their commander.[43]

Al-Karim's men, numbering in the few thousand, were increasingly facing supply shortages and outright starvation, and in a bid to change their situation, he invaded French Morocco in early 1925. Depending on one's viewpoint, this was either a desperate necessity or a foolish move. In any case, the combined forces of France and Spain now worked to defeat him, and though he scored a surprising early success against the French, it quickly became clear that al-Karim was overextended. Two amphibious operations during 1925 added immensely to his problems. First, on March 29–30, a Spanish task force invaded Alcazarsequir, imaginatively using combined artillery, bombing, and strafing of coastal defenses before the troops came ashore; once ashore, the soldiers quickly consolidated their hold and, supported by tactical bombing, moved inland on the offensive. Second, on September 8, a combined Franco-Spanish task force landed at Alhucemas, under bombardment from battleships, cruisers, destroyers, three squadrons of African Air Corps Breguets, and some Spanish Navy seaplanes; altogether, 100 aircraft supported the landing. Harried within an increasingly tightening perimeter, al-Karim's forces continued to fight on, but eventually resistance proved hopeless, and the rebellion collapsed in 1926 (though desultory fighting lasted into 1927), al-Karim going into French captivity.[44]

The air campaign in Morocco can hardly be considered to have been decisive, but when transformed from a poorly organized happenstance effort into an element of combined arms warfare, it made important contributions, particularly in reconnaissance. However, the French were the more perceptive and discerning in their use of air power during the campaign, and indeed, they employed it with a professionalism and insight into the nature of low-intensity warfare that was not to be found even in the contemporary Royal Air Force experience in Iraq and elsewhere.

To support its operations in Morocco, France established a mixed force of Breguet, Potez, and Farman aircraft used for bomber, reconnaissance, medical evacuation, liaison, and transport duties. Aircraft delivered messages and supplies to outposts and undertook photographic reconnaissance and observation. Artillery ranging proved less significant than battlefield observation due to the lack of long-range artillery in Morocco. Low-altitude operations drew enemy defensive fire, and ground fire did claim some aircraft and crews. But beyond this general experience, the greatest significance attached to French air operations in Morocco involved their integral ties into the workings of the traditional French colonial anti-insurgent force: the *Groupe Mobile*.[45]

As its name implied, the Groupe Mobile consisted of a mobile force of infantry, artillery, and cavalry. But now it was equipped with armored cars, mechanized transport, and small tanks, in addition to the traditional horses and camels that accompanied military operations in the Middle East. Further, it also could call on its own aviation resources, which were placed under the mobile group commander. Typically, a French "GM" employed in the Rif wars comprised from four to nine battalions, three to six artillery batteries, one or two *escadrons* of *spahis*, and, if possible, an additional cavalry *escadrille*; other combat forces (such as partisans) and support personnel accompanied the mobile group as it sought out the enemy. The emphasis in mobile group operations was on speed, maneuver, surprise, and concentration of forces. In effect, the mobile group functioned much like a naval task force at sea, using the desert as its ocean. And, like a naval task force, it could be arranged with great flexibility to confront whatever threat might be encountered.[46]

A summary examination of French tactical experience in Morocco placed great faith in the operations of the mobile group, and attached special significance to aviation operations in conjunction with the activities of the group, particularly when combined with attacks by tanks and armored cars. The author, a French Army colonel with extensive experience in Moroccan fighting, wrote: "The employment of aviation, tanks, and armored cars as a principle technique is noteworthy in as much as the enemy is powerless to oppose them by any similar means of defense."[47] (Interestingly, al-Karim had himself recognized Riffi helplessness in the face of air attack, and had actually purchased three aircraft in an effort to offset this; his pathetic "air force" was quickly destroyed by Spanish airmen.)

Generally, in the French Army's view, aviation had two major purposes: observation and "to intervene with its bombs and machine gun fire, particularly to force an [enemy] disengagement."[48] Additionally, it provided the mobile group with a means of "reconnaissance, harassment, and retaliation."[49] It assisted bombing and artillery strikes by radio-equipped observation aircraft that could call in, if necessary, air strikes or artillery shoots. Then, of course, beyond the range of artillery, it attacked with its own guns or bombs. In periods of relative calm, aircraft undertook prebriefed long-range strikes into enemy territory, attacking encampments and fortifications. In times of strain, the aircraft reprovisioned isolated outposts, dropping medicines and ammunition. Finally, when the mobile group was in direct contact with the enemy, aviation assisted over the battlefield with bombing and strafing attacks. "It is possible to strafe an adversary engaging friendly troops to force a disengagement," the manual warned, "but these attacks should not be undertaken by low-flying aircraft operating in isolation [i.e., without control and communications] so as to avoid errors [i.e., friendly casualties]. . . . These interventions are occasionally requested by the observation aircraft which, as a last resort in particularly urgent situations, can thus itself intervene with bombs and machine guns."[50]

Thus, the mobile group relied on orbiting observation aircraft covering the force, scouting ahead and over the flanks. These aircraft, in communication with the group headquarters and the aviation element (typically eight airplanes) thus functioned in some respects as forward air controllers (FAC), since they had authority to call in air and artillery strikes as needed. Their air observers were also trained artillery officers, and during battle,they transmitted information updates to the commander's staff. Armed themselves, these observation aircraft would attack the Riffi on their own if needed. Overall, the aviation element attached to the mobile group reported to the group commander (typically a major general), and the army arranged for an air officer to be assigned to the commander's staff to function as a tactical advisor and as a liaison officer between the aviation and the ground forces. Finally, after the conclusion of a battle, casualties were often "medevac'd" by special Hanriot and Breguet ambulance planes (the latter a derivation of the ubiquitous Breguet XIV bomber of World War I) off the battlefield direct to aid stations. As can be seen, then, the use of aviation by French forces in Morocco was perceptive, appropriate, and farsighted. It emphasized the combined arms nature of air-ground warfare, with the airplane incorporated as a coequal element with other combat arms. It was characterized neither by overoptimistic expectations about what aviation could provide nor by a tendency to regard aviation as a mere ancillary support service to ground forces. It foreshadowed the kind of counterinsurgency operations that became commonplace after the Second World War in such campaigns as Malaya, Indochina, and, particularly, Algeria.[51]

The Sandino War in Nicaragua

America's only significant experience with air combat operations in the interwar years occurred during the Sandino War in Nicaragua. Between their intervention in 1927 to assist the ill-trained and ill-equipped Nicaraguan military to their departure in 1933 (when Juan Bautista Sacasa was sworn in as President of Nicaragua), the Marine Corps lost a total of 136 Marines, 47 of whom died in battle or from wounds received in combat. The rebel forces, dubbed "Sandinistas" after their leader, Augusto Cesar Sandino, had been overcome in often bitter guerrilla warfare. In February 1933, as a result of a peace conference, Sandino agreed to submit to the authority of the Nicaraguan government and allow the eventual disarming of his followers, which occurred before the end of the month. A year later, following increasingly tense relations between the government and the former rebel chieftain, Sandino and many of his followers were seized and killed. The death of Sandino and the subsequent increasingly corrupt and repressive nature of the Nicaraguan government provided a powerful stimulus to continued unrest, which finally, in the last decade, resulted in the creation of the present Communist government in Nicaragua, which pays lip-service to Sandino as its spiritual founder, but follows its own totalitarian course.[52]

Marine aviation operations in Nicaragua were characterized by missions involving observation and reconnaissance, covering the activities of patrols, supporting assaults against Sandinista positions, message drops and pickups, aerial resupply, and medical evacuation. At the height of Marine operations in Nicaragua, the Corps had twenty-four airplanes, consisting of twelve Curtiss OC-2 Falcon and Vought O2U Corsair observation and attack biplanes, seven Loening OL amphibians for liaison, observation and light attack duties, and five Fokker TA

Trimotor transport aircraft. The Falcons and Corsairs replaced older De Havilland DH-4B's (the American-built version of the British D.H. 4) used by the Corps for observation and bombing duties when the Marines first arrived in Nicaragua in 1927.[53]

Undoubtedly, the most widely publicized "innovation" to come out of Nicaraguan air operations was the Marines' use of dive-bombing attacks against the Sandinistas. This form of attack, of course, was not new to aviation, having been pioneered by the British in close-support operations in 1918. In 1923, Marine Maj. Ross E. Rowell had been assigned to the Army Air Service for a tour of duty at Kelly Field, Texas. There he was exposed to Army interest in attack aviation, and the activities of the 3d Attack Group, which the AAS had established in 1921 as a specialized attack formation. The AAS had evaluated a number of experimental aircraft for ground attack, emphasizing heavy armor and armament, and was understandably disappointed when most of these designs proved to be definite "clunkers"; as a result, the AAS increasingly stressed—as did the Royal Air Force—modifying existing fighters and two-seat aircraft for ground-attack work. Along the way, influenced by the British experience in the First World War, the 3d Attack Group had experimented with dive bombing using the big DH-4B's, a hardly suitable aircraft, modified to carry A-3 bomb racks under the wings; fully loaded, the two-seat DH-4B could carry ten small bombs. The AAS quickly cooled to dive bombing, never to regain its enthusiasm, because most aviators believed a dive bomber would be too vulnerable to antiaircraft fire. In contrast, Rowell became convinced that dive-bombing attacks could be most useful in small guerrilla wars, and in antishipping operations. After a brief period with the Army, he returned to the Marines to command Marine Squadron VO-1M, a DH-4B unit. He quickly instructed them in dive-bombing techniques. The squadron's exhibitions resulted in the Navy adopting this method of bomb delivery for its own bomber units, and eventually spawned the first special-purpose dive bombers acquired by the Navy—the ancestors of the aircraft that would devastate the Japanese fleet in the Second World War. In February 1927, VO-1M received orders to take its DH-4B's to Nicaragua; by the end of the month, Rowell and his airmen were flying patrols over the dense countryside of that troubled nation.[54]

The Nicaraguan air service consisted of two Laird Swallow general aviation airplanes piloted by mercenary pilots who attacked Sandino's troops with crude and ineffective dynamite bombs. For their part—soon to be their undoing—the Sandinistas did not take airplanes seriously.

On July 17, 1927, 700 to 800 rebels attacked a small Marine outpost at Ocotal, with every expectation of exterminating its defenders (thirty-seven Marines and less than fifty Nicaraguan National Guardsmen). The attack came at 3 A.M., and was not discovered by Marine reconnaissance aircraft until 10:15 that morning, when gunfire and urgent panel signals on the ground gave them the news. Ocotal was 125 miles from Managua by air, and there was no Marine ground force within rescue distance; it was the air or nothing at all. Conditions conspired to make air operations difficult; the DH-4B's, laden with bombs, would have to take off from a 400-yard strip in hot noonday temperatures. Rowell's squadron carried full fuel and a limited bomb load, taking off shortly after noon "with mighty little runway to spare." The five DH-4B's dispatched formed up and headed north to Ocotal. As Rowell recalled,

> All the pilots had been trained in dive-bombing and that was the kind of attack that I planned to employ. As I made the approach on the town, we formed a bombing column. We were fired upon as we flew over the outposts along the [Rio Coco] river, but at 1500 feet we did not suffer any particular damage from rifle fire. I made one circle of the town in order to reconnoiter the situation and determine the focal point of attack. Fuel limitations required immediate action. The situation was made more unpleasant by a huge black front, full of lightning, wind and rain, approaching rapidly from the east. Light rain was already falling. I led off the attack and dived out of column from 1500 feet, pulling out at about 600. Later we ended up by diving in from 1000 and pulling out at about 300. Since the enemy had not been subjected to any form of bombing attack, other than the dynamite charges . . . they had no fear of us. They exposed themselves in such a manner that we were able to inflict damage which was out of proportion to what they might have suffered had they taken cover.[55]

As the De Havilland dove on the Sandinistas, the pilots strafed them with their forward-firing gun, helping suppress any defensive fire, and the rear gunners added their own strafing when the bombers were climbing away in their vulnerable pull off the target. This, of course, was an American equivalent to Slessor's "Vickers-Bomb-Lewis" method of attack employed subsequently in the Waziristan campaign. The attack, and the skilled defense of the Marines in the besieged town, cost the rebels over 100 casualties and perhaps twice that; it also gave them a bitter lesson in what air attack could accomplish. More importantly, the Marine garrison at Ocotal had become the first American unit that is known to have survived a ground assault by vastly superior forces thanks to aerial intervention.[56]

Only one other air strike achieved an equivalent distinction to that of the Ocotal operation; in June 1930, a band of 400 Sandinistas assembled on El Saraguazca mountain preparing to assault the strategic town of Jinotega. They attacked a National Guard unit which, by good fortune, was able to call upon a Marine strike force of six aircraft. The aircraft and Guardsmen fought off the attackers, who were able to slip away before another column of government reenforcements arrived. "We would have captured Jinotega, if it were not for the airplanes," one prisoner exclaimed gloomily.[57]

Beyond these spectacular strikes, much of the air operations work consisted of more routine reconnaissance and patrol activities; anti-aircraft fire increased in volume, and resulted in aircraft shot down; only two crewmen were killed, and these had landed successfully but were captured by Sandinistas, "tried" and shot. (It is perhaps a measure of the impact that air operations had against guerrilla insurgencies to note that in both the Nicaraguan campaign and in the Rif wars, prisoners were shot for aerial attacks; in the Rif war, for example, all Spanish officers held prisoner were shot in response to Spanish bombing.) Patrols over columns were particularly valuable: one column survived several days of intense enemy attacks due to timely warning of ambushers by observation planes coupled with prompt air attacks on bandit parties; seriously wounded Marines were air-evacuated from a hastily prepared landing strip only 600 feet long (the pilot subsequently received the Medal of Honor for his missions).[58]

Although Marine aircraft were equipped with radio, the prohibitive weight of early sets, coupled with the poor transmission characteristics amid the terrain of Nicaragua meant that most were removed from planes to permit carrying more fuel and weapons. "Communication with ground troops," then-Captain Matthew B. Ridgway, an Army observer in Nicaragua, reported, "has been almost entirely by Very Pistol, pick-up and drop messages, and panels."[59] Overall, despite the often violent weather, the mountains rising to approximately 7000 feet, the lack of airfields, and the primitive technology of some of the aircraft, Marine air operations in Nicaragua proved that in a guerrilla war of a different character than that in Morocco or Arabia, aircraft could make a substantial contribution to bringing an insurgency under control. Sixteen Marine aviators died in Nicaragua, two shot while Sandino's prisoners, and fourteen in accidents.[60]

The examination of military air power in the four cases discussed in this part indicates that while nascent, air power could occasionally make an important difference in what would now be termed low-intensity conflict. Lessons learned from these conflicts had subsequent profound impact. For example, British experience in "air control" operations influenced subsequent fighting in the Burma campaign during the Second World War and the Malayan campaign in the post–World War II years. The Marine experience in Nicaragua firmly wedded the Corps to the notion of precise air attack in support of ground forces engaged in close combat with an enemy. Of course, these wars typically pitted small numbers of well-equipped and well-armed military professionals against less well equipped, armed, and trained opponents who had a manpower advantage. Training, experience, and equipment (including aircraft) were all "force multipliers" that enabled the smaller, more professional forces to survive and win. In the larger conflicts of the interwar years—discussed in the next section—both sides tended to be more evenly matched. Likewise, the latter occurred at a time when the lethality of the airplane itself was rapidly increasing. But if battlefield aviation's promise still remained largely unfulfilled, its limitations were due not to inherent flaw, but rather to the emerging and still-adolescent nature of military aviation itself.

Royal Aircraft Factory F.E. 2b. During the First World War, the fighter airplane underwent continuous refinement and adaptation. By mid-1917, older fighter aircraft such as this Royal Aircraft Factory F.E. 2b had already proven obsolete for the rigors of air-to-air combat. Instead, air commanders began using them for "trench strafing" and "ground strafing" missions—what are now called close air support (CAS) and battlefield air interdiction (BAI). *Photo courtesy NASM.*

Sopwith F.1 Camel. Great Britain's Royal Flying Corps and Royal Naval Air Service (subsequently merged in 1918 to form the Royal Air Force) made extensive use of the small and extraordinarily agile Camel as a battlefield attacker. A true swing-role fighter, the Camel could undertake air superiority and ground-attack missions with equal ease and success, beginning a trend stretching to multipurpose fighters (such as the F-16) of the present day. *Photo courtesy NASM.*

Royal Aircraft Factory S.E. 5a. The S.E. 5a, though less agile than the Camel, was fast, strong, and robust: the F-4 Phantom II of its day. It saw extensive CAS/BAI service over the Western Front and in Palestine. *Photo courtesy NASM.*

Bristol F.2B Fighter. The two-seat Bristol Fighter, a surprisingly agile and maneuverable aircraft for its size, was a formidable ground attacker, and served well into the postwar years on RAF "air control" duties in the Middle East. *Photo courtesy NASM.*

American Breguet 14's during the St. Mihiel Offensive. The French-built Breguet 14B2 proved a mainstay as a battlefield interdictor during the Allied offensives of 1918; here a formation of American-flown Breguets from the 96th Aero Squadron are shown on a mission against German forces during the St. Mihiel offensive. The Breguet 14, like Great Britain's D.H. 4 and D.H. 9, went on to a long and distinguished postwar career in both military and civilian service. *Photo courtesy NASM.*

Fusing Bombs for a Battlefield Support Mission. The Breguet featured distinctive Michelin under-wing bomb racks capable of holding up to thirty-two 8-kg antipersonnel bombs. Here a ground crew sets the tail fuse on one of these bombs before affixing it to the rack. *Photo courtesy AFM.*

Halberstadt Cl II. Beginning in late 1917, the *Luftstreitkräfte* (Imperial German Air Service) made increasing use of the heavily armed two-seat *Schlachtflugzeug* (assault aircraft), such as this Halberstadt Cl II. Note the close placement of the pilot and gunner for good crew harmony, the flares for signaling, and the rack of antipersonnel stick grenades, which the gunner would drop during low-altitude attacks. Such aircraft were German equivalents to Britain's superlative Bristol Fighter. *Photo courtesy NASM.*

A Hannover Cl III Victimized by Allied Ground Fire. While these German ground-attack aircraft gave good service, they proved vulnerable—as did their Allied counterparts—to small arms and antiaircraft fire. Here is a Hannover Cl III, brought down by American ground fire in early October 1918. The pilot obviously made a good emergency landing; many *Schlasta* crews were not so fortunate. *Photo courtesy NASM.*

Junkers' J I "Furniture Van." In an effort to produce a new generation of more rugged and survivable warplanes, German manufacturer Hugo Junkers experimented with all-metal structures. His research led to the Junkers J I *Möbelwagen* (Furniture van) of late 1917, a heavily armored ground attacker that saw widespread service with the *Schlachtfliegertruppen* in 1918. The prototype of this singularly ugly (but quite effective) warplane is shown with co-designer Otto Reuter. *Photo courtesy NASM.*

Sopwith Camel "Trench Fighter." In an attempt to improve the survivability of their own battlefield attackers, British designers experimented with heavily armed and armored "trench fighters" for CAS missions. Here is an experimental Sopwith Camel modified for such duties; it had two downward-firing machine guns mounted in the fuselage, as well as extensive armor plating. Performance penalties were so great, however, that the RAF opted instead for a totally new ground-attack fighter, the Salamander. *Photo courtesy NASM.*

Sopwith Salamander. The Sopwith Salamander represented the epitome of British ground-attack fighter development in the First World War. Appearing on the Western Front just too late to see combat service, the Salamander would have been extensively employed for CAS/BAI missions had the war continued into 1919. *Photo courtesy British Aerospace Aircraft Group.*

Junkers Cl I. The sleek Cl I, an all-metal internally braced assault monoplane far removed from the jarring angularity of the earlier J I biplane, constituted the apotheosis of German ground-attack aircraft technology during the Great War. Just entering service when the war ended, this revolutionary warplane—approximately a decade ahead of its time—saw combat in the postwar period during fighting in the Baltic states against the Bolsheviks. *Photo courtesy AFM.*

Engineering Division GAX. In contrast to the well-thought-out Junkers Cl I, the U.S. Army's experimental GAX ground-attack triplane, designed by the Engineering Division at McCook Field, Ohio, was an extraordinarily grotesque design. Slow, noisy, and an invitingly large target, the postwar GAX was to battlefield aviation what the Barling bomber was to strategic bombardment: an aeronautical abomination unsuitable for military service. *Photo courtesy History Office, Aeronautical Systems Division, Wright-Patterson AFB.*

Patrolling the Border from the Air. Typifying more practical approaches to postwar battlefield air support requirements were aircraft such as this formation of De Havilland DH-4B's flying on a tactical exercise near the Mexican border during a period of tension caused by the threat of cross-border bandit raids. Dive-bombing experiments by the DH-4-equipped 3d Attack Group greatly influenced Marine and naval aviation and, by extension, the Luftwaffe of Hitler's era. *Photo courtesy AFM.*

The 3d Attack Group Exercising its A-3's. As newer technology replaced old, Great War-era De Havillands and their like were replaced by more suitable higher-performance designs such as the Curtiss A-3 Falcon. Here the 3d Attack Group practices low-level attack tactics; note the shadows of the aircraft on the ground. *Photo courtesy AFM.*

Potez Type 25. Interwar cooperation and ground-attack designs saw extensive service in British and French air control operations, and in various wars in Central and South America. The French-designed Potez Type 25 was one such aircraft that appeared in conflicts from the Rif War to the "Green Hell" of the Gran Chaco. *Photo courtesy NASM.*

Westland Wapiti. The British Westland Wapiti, a hefty multipurpose biplane, proved well suited to the demanding high-altitude conditions of air control warfare on the North West Frontier. Aircraft such as the Wapiti, and Hawker's Audax and Hardy, were worthy successors to the Bristols and De Havillands of the First World War. *Photo courtesy NASM.*

Vought O2U Corsair. Flying elderly DH-4B's, Marine airmen went to war in Nicaragua against Sandinista guerrillas. Eventually the earlier DH-4's were replaced by the excellent Vought O2U Corsair, a rugged and dependable biplane that flew a variety of counterinsurgency missions ranging from CAS and BAI to reconnaissance and casualty evacuation. Around the globe, the Chinese Central Government operated O2U's in various "Bandit Extermination Campaigns," including attacks against Mao's Communists during their famed "Long March" to the west of China. *Photo courtesy NASM.*

Caproni Ca 101. During the Abyssinian War, Mussolini's Regia Aeronautica had an opportunity to try out the teachings of Guilio Douhet in operations against Abyssinian tribesmen. The Caproni Ca 101, an early trimotor monoplane bomber, flew strategic bombardment missions against Abyssinian towns and population centers. Such missions proved far less useful than actual battlefield air support operations, for the Italians quickly learned that the bombing did not materially break the will of the tribesmen to resist—something that CAS/BAI operations against troops did achieve. Here a line-up of Caproni bombers are shown being refueled and rearmed for another mission. *Photo courtesy NASM.*

Savoia-Marchetti S.M. 81 Pipistrello. The rugged and dependable S.M. 81 Pipistrello (Bat), a more modern trimotor bomber and transport, flew both "strategic" and tactical bombing missions during the Abyssinian war. Neither this aircraft, nor the various Caproni bombers, were immune to ground fire; nine were shot down during the brief war. Here three from the XXVI° Gruppo are pictured over the Abyssinian countryside near Macalle. *Photo courtesy NASM.*

Meridionali Ro 1. More useful to the day-to-day activities of the Italian ground forces in Abyssinia were tactical aircraft such as the Meridionali Ro 1 observation plane. On occasion, its crews would themselves attack Abyssinian forces, particularly in the assault and pursuit phase of battle. *Photo courtesy AFM.*

Meridionali Ro 37. The Ro 37, a more modern two-seat reconnaissance and attack biplane, flew in both the Abyssinian war and Spanish Civil War. By the late 1930s, however, the value of such multipurpose biplane aircraft was diminishing rapidly, as the proliferation of antiaircraft cannon and high-performance monoplane fighters rendered them increasingly anachronistic and vulnerable. *Photo courtesy NASM.*

Polikarpov I-16. The Soviet Union furnished large quantities of fighter, bomber, and attack aircraft to Spanish Republican (Loyalist) forces during the Spanish Civil War. Here a flight of Polikarpov I-16 Type 10 Moscas (Flies) of the 3a Escuadrilla de Moscas prepares for takeoff. The stubby I-16 proved valuable both as a dogfighter and ground-attack airplane. *Photo courtesy NASM.*

Polikarpov R-5. The Loyalists made extensive use of the Polikarpov R-5, a two-seat light attack biplane dubbed the Rasante (Leveler). Aircraft of this type played a major role in the Loyalist victory at Guadalajara in 1937. *Photo courtesy NASM.*

Heinkel He 51. Nazi Germany assisted the Spanish Nationalists by providing a wide range of aircraft, including Heinkel He 51 fighters. The He 51 quickly proved a disappointment as a dogfighter, and, instead, was largely relegated to ground attack duties by the Legion Condor and Spanish Cadena (chain) squadrons. Here are three early He 51A's cruising over Germany in 1934, before the Luftwaffe came out in the open. *Photo courtesy NASM.*

"Cockroaches" on Patrol. Fascist Italy quickly assisted Francisco Franco's forces by establishing the Aviacion d'el Tercio, equipped with a variety of combat airplanes including the nimble Fiat C.R. 32. The Fiat, an excellent fighter possessing superlative handling characteristics, was flown by most Italian and Spanish fighter squadrons and used on both air-to-air and air-to-ground missions. Here a flight from the XVI° Gruppo "La Cucaracha" (Cockroaches) is shown over Spain in 1938. *Photo courtesy NASM.*

Henschel Hs 123A. The stocky little Henschel Hs 123A, developed as a dive bomber, flew in Spain primarily on low-level ground attack, proving an excellent close support airplane. Shown is a Spanish Nationalist Hs 123A from the Arma de Aviacion's Grupo 24, based at Tablada in early 1939. *Photo courtesy AFM.*

Junkers Ju 87B Stuka. Junker's angular Ju 87 Stuka dive bomber flew in Spain with the Jolanthe Staffel of the Legion Condor's Kampfgruppe 88, earning the nickname "Sharpshooter" because of its ability to precisely attack ground targets. Here a flight of Ju 87B-1's is shown over the Ebro front in late 1938. *Photo courtesy NASM.*

Fiat Type I's over Japan. The Sino-Japanese war furnished Japan with an opportunity to evaluate its military forces in combat, including a wide range of combat airplanes purchased from abroad or built indigenously. Here are two Italian-built Fiat Type I bombers (equivalent to the Italian B.R. 20) of the Japanese Army Air Force shown over Japan in 1938. The Fiats were eventually used for both terror bombing of civilian population centers and attacks on Chinese and Soviet military forces, proving vulnerable to enemy ground and air defenses. *Photo courtesy AFM.*

Mitsubishi Ki-51b. Responding to an increased need for tactical air support of Japanese ground forces, the JAAF developed the Mitsubishi Ki-51 two-seat attack monoplane, and it entered service in China shortly before the attack on Pearl Harbor. The Ki-51, despite pleasant flying characteristics, exemplified the limitations of the overspecialized ground-attack airplane. During the Second World War, like other Japanese bombers, it proved woefully vulnerable to marauding Allied fighters. *Photo courtesy AFM.*

Breguet Attack Bombers. Even France's small and speedy Breguet 690-693 family of twin-engine attack bombers were not immune to loss; for example, five of six were lost during a low attack on Tongres, and the survivor had to be junked. The American-built Curtiss Hawk 75 fighters in the background would have been better able to survive (they did very well against the Luftwaffe's Messerschmitt Bf 109E), but the Armée de l'Air made little serious attempt to employ its fighters as fighter-bombers during the brief and bitter battle of France. *Photo courtesy NASM.*

Fairey Battles. Light flak and fighters also smothered attacks against the Meuse bridges by the Royal Air Force's Fairey Battles, a single-engine light bomber of but modest performance. Battle crews displayed extraordinary courage in pressing home their attacks in the face of opposition so intense that captured airmen were subsequently lectured by their Nazi captors on the futility of such assaults. *Photo courtesy NASM.*

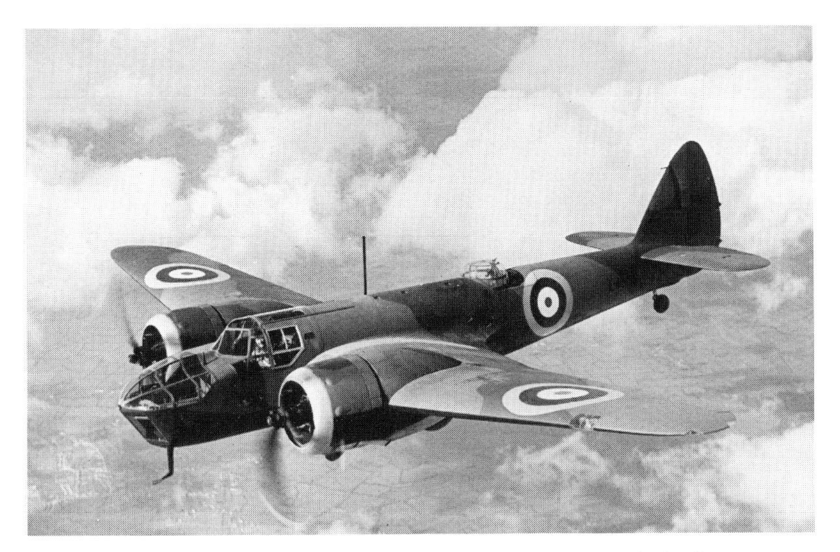

Bristol Blenheim. The twin-engine and more modern Bristol Blenheim, considered a "fast" bomber, fared little better than its Battle predecessor and its French compatriots. Battle and Blenheim losses were staggering. In one day, on May 14, 1940, 40 out of 71 Battles and Blenheims sent against the Nazi bridgehead at Sedan were shot down; individual unit losses varied between 25 and 100 percent of the force dispatched. *Photo courtesy NASM.*

Hawker Hartbee. The Allies did markedly better during the long Western Desert campaign against German and Italian forces in North Africa, developing a highly effective air-ground support system. An important precedent, however, was the British and South African campaign against Italian forces in East Africa. Flying Hawker Hartbee light attack aircraft, the South Africans seriously disrupted Italian troop movements and cooperated closely with friendly ground forces. The elegant Hartbee, an outgrowth of the earlier Audax, represented the epitome of British "cooperation" biplane design. *Photo courtesy NASM.*

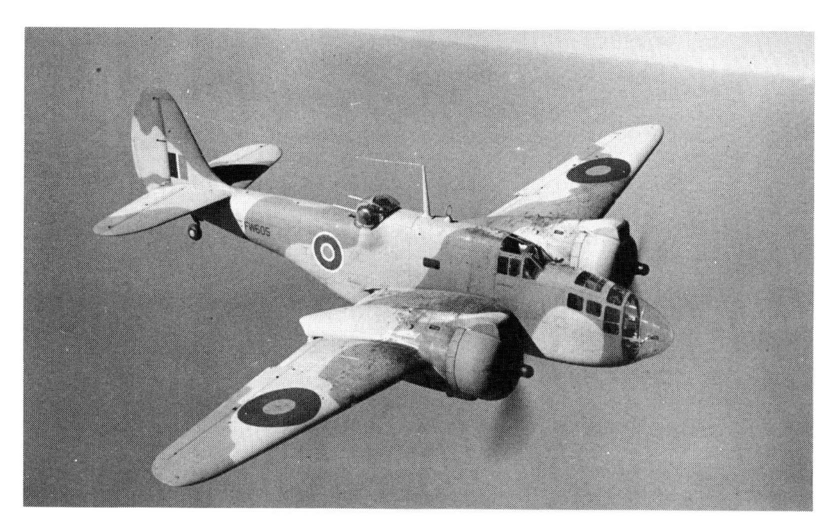

Martin Baltimore Mk. V. The Mediterranean air war constituted a valuable training experience for Allied tactical air power. Attack aircraft such as the twin-engine American-built Martin Baltimore (shown here) and the Douglas Boston took heavy losses when operated unescorted on deep strikes or when flown in low-altitude attacks against well-protected German formations. They were much more useful when flown like medium bombers, attacking from altitude and with adequate fighter protection, as part of Air Vice Marshal Arthur "Mary" Coningham's well-thought-out Allied air strategy. *Photo courtesy NASM.*

Hawker Hurricane Mk. IIC. Much more useful were sorties by British fighter-bombers, such as the Hawker Hurricane. Armed with four 20mm cannon and equipped to carry two 250-lb bombs (such as those shown here), the Hurricane Mk. IIC proved a formidable battlefield support airplane. Dubbed the "Hurribomber," the Hurricane was a true swing-role fighter that echoed the Camel or S.E. 5a of the Great War. *Photo courtesy NASM.*

Hawker Hurricane Mk. IID. In 1942, during the battle of Bir Hacheim, Coningham unveiled yet another air support development: the Hawker Hurricane Mk. IID, armed with two 40mm Vickers guns and intended specifically to destroy German tanks. Scoring some immediate and impressive successes—such as one mission where four IID's destroyed five tanks, five trucks, and an antitank gun—the 40mm Hurricane anticipated the later Junkers Ju 87G, Henschel Hs 129, Yak 9T, and Ilyushin Il-2m3 Mod that battled opposing armor on the Eastern Front from 1943 onward. *Photo courtesy NASM.*

Defanged and Declawed Tigers. Even the heaviest German tanks, such as the Mk. VI Tiger, were not immune to attack from the Hurribomber. Here two lie harmlessly beside a Tunisian road, being inspected by American troops. *Photo courtesy MHI.*

Woods Were No Cover. . . Here American A-20 light bombers blasted a brood of Tigers nestled in a Tunisian grove. The disaster that befell Rommel's forces in North Africa was but a foretaste of what he would experience a year later in Normandy. *Photo courtesy Edwards Collection.*

A Shattered Riposte. Much as the Allies had attempted without success to counter Nazi air-land warfare in 1940, so, too, did the Germans attempt to regain some measure of initiative in North Africa. Here the remains of a Henschel Hs 129B attack airplane, flown by Schlachtgeschwader 2 from Tunis, serve as a convenient backdrop for an A-20 pilot's souvenir photographs of a squadron mate. The slab-sided and heavily armored Hs 129—note the heavy construction and thick cockpit framing—proved mechanically unreliable, and, in any case, was fielded in too few numbers to make any difference. This experience would be repeated on a grander scale on the Eastern Front. *Photo courtesy Edwards Collection.*

Planning for War. During the early days of the war, prewar American notions of low-level attack still seemed sound. Here two A-20's "attack" a formation of M3 Stuart light tanks at the Army's Mojave Desert Training Center during exercises in 1942. Note, however, that each tank is tracking the bombers with a .30-caliber machine gun; large attack aircraft such as the A-20 were extremely vulnerable to such concentrated ground fire, as the French and British learned at the hands of Nazi flak in France; only the much faster and agile fighter-bomber could get away with such attacks. After the debacle of American air support at Kasserine, AAF battlefield air support was thoroughly reorganized to reflect the experience and lessons learned by the RAF in North Africa. *Photo courtesy MHI.*

Bell P-400 Airacobra. America's first experience with battlefield CAS in World War II came with the Guadalcanal campaign in 1942. During that brutal battle, Army, Navy, and Marine pilots devised local solutions to air support problems that served as precedents for subsequent Pacific practice. Here a Bell P-400 Airacobra—an export model of the P-39 retained by the AAF and operated from Henderson Field by the 67th Fighter Squadron—is readied for a support sortie. Unsuitable for dogfighting at high altitudes with lightweight and nimble Japanese fighters, the Airacobra nevertheless proved an excellent low-altitude ground-attack fighter. *Photo courtesy AFM.*

Douglas A-24 Dauntless. Far less satisfactory was the Army version of the famed Navy Douglas SBD dive bomber, the A-24. During the New Guinea campaign, Dauntlesses proved vulnerable to Japanese fighters and were quickly relegated to noncombat liaison and training duties. Dive bombers, regardless of the nation that employed them, were simply too limited to satisfactorily confront both the ground-to-air and air-to-air threats likely to be encountered over a battlefield. *Photo courtesy NASM.*

Whirraways Whirring Away. During the New Guinea campaign, Australian airmen pioneered in using airborne strike coordinators to mark ground targets, direct battlefield support missions and, in general, act much like later Forward Air Controllers (FAC) of Italy, Korea, and Vietnam. Initially they employed the Commonwealth CA-1 Whirraway, an Australian-built derivation from an American trainer. Here a formation of three Whirraways from 21 Squadron, Royal Australian Air Force, are seen on a prewar training flight. *Photo courtesy NASM.*

Commonwealth CA-13 Boomerang. Later, the RAAF employed a more suitable light strike and target-marking airplane, the Commonwealth Boomerang. With fighterlike armament and performance—something the Whirraway lacked—the Boomerang could act like a "Fast FAC" in much more intensive combat environments. Here Flt. Lt. A. W. Clarke shows off *Sinbad II,* a Boomerang from 5 Squadron, RAAF, in March 1944. *Photo courtesy NASM.*

North American A-36A Invader. During the Sicilian and Italian campaign, the AAF employed the disappointing A-36 Invader, a dive bomber derived from the Allison-powered P-51A Mustang. The A-36 gained an unenviable reputation for attacking friendly troops, often, in fact, when those troops were locked in desperate combat with Nazi forces. Eventually, most Invaders were operated with their dive brakes wired shut, in a manner similar to other Allied fighter-bombers. Experience with the A-36 confirmed for the AAF that the dive bomber had nothing to offer over the fighter-bomber as a battlefield support system. *Photo courtesy AFM.*

Martin B-26 Marauders. Interestingly, when flown by highly skilled crews under conditions of relatively static ground warfare, the Martin B-26 Marauders of the AAF's 42d Bomb Wing did yeoman service as battlefield support aircraft, particularly in the Anzio and Cassino campaigns. Here a formation of B-26's attack German positions; note the top turret gunners maintaining watch "up-sun" to check for any lurking Nazi fighters that might be in the area. Bombing from altitude minimized losses from light and medium antiaircraft fire. *Photo courtesy AFM.*

Grasshopper Patrol. In June 1944, the XII Tactical Air Command began employing Piper L-5 Grasshoppers—previously used only for liaison and artillery-spotting duties—as airborne forward air controllers. Flown by fighter-bomber pilots assigned to FAC duty, carrying an Army observer trained to differentiate between American and German armor, and equipped with Army-compatible radios, these FAC's, known by the call-sign "Horsefly," proved invaluable for strike direction and coordination. Here an L-5 flits low over the Italian countryside in 1944. *Photo courtesy MHI.*

BATTLEFIELD AIR SUPPORT: FROM THE LIBERATION OF FRANCE TO THE SHATTERING OF THE WEHRMACHT IN THE EAST

Thunderbolt Takeoff. Successfully invading France in 1944 depended on achieving air supremacy over Hitler's Luftwaffe. The massive Republic P-47D Thunderbolt (two of which are shown here taking off on a bomber escort mission) played a leading role in shattering the Luftwaffe, and then went on to become an outstanding battlefield support aircraft during the fighting across Europe. *Photo courtesy NASM.*

Looking for Trouble: North American P-51 Mustangs. The Merlin-powered P-51 Mustang, particularly the "bubble top" P-51D, was the finest propeller-driven air-superiority fighter of the Second World War. Mustangs continued the American ascendancy over the Luftwaffe first established by the P-47, but they proved far less satisfactory for battlefield support. Particularly vulnerable to ground fire because of its liquid-cooled engine, the Mustang's basic unsuitability for CAS/BAI missions would be reaffirmed by the Korean War, at, unfortunately, even greater cost. *Photo courtesy NASM.*

Marauding the Reich. With the Luftwaffe busy fighting for its life, the Ninth Air Force's medium bombers were free to embark on a complex Allied air campaign aimed at crippling German ground forces in France. Here a formation of 322d Bomb Group B-26's are off on a raid. The aircraft in the foreground is the most famous B-26 of all: *Flak Bait,* shown on its 200th mission (this plane is now in the collections of the National Air and Space Museum). *Photo courtesy NASM.*

Not a Good Day . . . If German fighters were largely no longer a problem, German flak certainly was. Here a Ninth Air Force Douglas A-20J Havoc erupts in flames from a direct 88mm hit, underscoring how vulnerable even reasonably fast and agile bombers such as the A-20 were to German ground defenses. *Photo courtesy AFM.*

The Case of the Disappearing Jabos. The Luftwaffe's own Jagdbomber (Fighter-bomber), the Focke-Wulf FW 190, was conspicuously absent from the battle for France. So overwhelming was Allied air superiority in Italy that S.G. 4, a Schlachtgeschwader equipped with the 190, was unable to prosecute battlefield attack missions. Its fighters were transferred not to Normandy (where the Allies by now possessed air supremacy, and where Rommel needed all the help he could get), but to Russia. They were joined there by the III Gruppe of S.G. 4, which *had* been in France, but had been forced to abandon Normandy for the *safer* climes of the Eastern Front!

Hawker's Typhoon: An Imposing Tank-buster. In the summer of 1944, the skies over Northern France belonged to Allied fighter-bombers, such as the Hawker Typhoon. Shown is an Mk. IB armed with four 20mm cannon and eight 3-in rockets, each having a 60-lb armor-piercing warhead. More vulnerable than the air-cooled Thunderbolt to ground fire, the liquid-cooled Typhoon was, nevertheless, a most formidable swing-role fighter. Like Republic's "Jug," the "Tiffie" had proven more than a match for the Bf 109 and FW 190; now, from D-day onward, it would make its reputation as a destroyer of Nazi armor and motorized transport. *Photo courtesy NASM.*

Available as Needed . . . This Republic P-47D Thunderbolt is shown with two 500-lb bombs and an external fuel tank, a typical offensive load carried in the 1944 campaign across France. The combination of bombs and eight .50-caliber machine guns, together with a rugged structure, high speed, and good maneuverability, made the P-47D the Allies' premier fighter-bomber. *Photo courtesy NASM.*

Clearing the Path . . . The Thunderbolt was particularly successful flying armored column cover (ACC) missions; P-47's of Brig. Gen. Elwood Quesada's IX Tactical Air Command supported Gen. Omar Bradley's 12th Army Group tankers with on-scene reconnaissance and strike missions that greatly facilitated the American ground assault. Fighter-bomber pilots rode in specially equipped M4 Sherman tanks having aircraft radios, coordinating Thunderbolt support from dawn to dusk. *Photo courtesy NASM.*

Night Was Not Entirely a Sanctuary . . . Though CAS/BAI operations essentially stopped at nightfall—a problem since the Second World War, as well—modified Allied nightfighters such as the British De Havilland Mosquito and the American Northrop P-61B Black Widow (shown here) were used for night BAI missions since they were not really required to confront the Luftwaffe. Often these night intruders would attack under the light of flares dropped from other aircraft—a risky business, but one followed in Korea and Vietnam as well. *Photo courtesy AFM.*

A Grim Payoff . . . During the battle of Mortain, Typhoons devastated German tank and mechanized columns attempting to reach the French coast. Here are the results of one such attack. *Photo courtesy MHI.*

Couloir de la Mort: Falaise "Corridor of Death." Allied fighter-bombers mercilessly attacked Wehrmacht units attempting to retreat through the Falaise-Argentan "gap" to the German frontier; here is one such scene, repeated literally hundreds of times in mid-August 1944. *Photo courtesy NASM.*

Air Contra Armor. Here an Mk. V Panther lies shattered after being attacked by American fighter-bombers during the drive across France. *Photo courtesy MHI.*

Air Contra Motorized Transport. The heavily armed Thunderbolt was even more devastating against lightly armored motorized transport. Shown is a German road convoy caught by P-47's near Kaiserslautern late in the war and totally destroyed. *Photo courtesy AFM.*

Once More to the Attack . . . The ubiquitous Stuka gained a new lease on life with the unleashing of Hitler's offensive against the Soviet Union. For much of the Nazi-Soviet war, the permissive conditions characterizing Eastern Front air warfare enabled the Stuka to survive and contribute. Here a Ju 87D armed with small bombs takes off on a battlefield support mission. *Photo courtesy NASM.*

One T-34 That Didn't Make Berlin . . . Chief prey of Ju 87's and other tank-hunters was the Soviets' formidable T-34 tank—the finest all-around armored fighting vehicle of the Second World War. Shown is an early model T-34 with a 76mm cannon, destroyed by a direct bomb hit—a most unusual occurrence. Later T-34's had more powerful 85mm guns. *Photo courtesy NASM.*

If Bombs Don't Work, then Try Cannon . . . Due to their greater accuracy, guns proved far deadlier against tanks than bombs. Accordingly, following operational experimentation, the Luftwaffe introduced the Ju 87G, carrying two 37mm cannon permanently affixed under the outer wing panels. "Gun Stukas" were incapable of dive bombing, and possessed even less maneuverability than the basic Ju 87. Nevertheless, in the hands of such expert Schlachtflieger as Hans-Ulrich Rudel, they proved useful destroyers of Soviet armor. *Photo courtesy AFM.*

Henschel Hs 129B. The air war on the Eastern Front was characterized by massive air-land engagements such as Kursk in 1943 and the Soviet offensive of 1944 that expelled the Nazis from Russia. Antiarmor aircraft played a major role on both sides. Shown is the Henschel Hs 129B, which first flew in the Tunisian campaign of 1943. Plagued by unreliable engines, the Hs 129 nevertheless did useful service as an antiarmor aircraft, particularly after being modified to carry a variety of large aerial cannon. Thus equipped, it earned the nickname *"Büchsenöffner"* ("Can opener"). *Photo courtesy NASM.*

"Bombing-up" a Jabo. Germany made extensive use of fighter-bombers in Russia, beginning with the Messerschmitt Bf 109E in 1941–42. Here a German ground crew wrestles with a 500-kg bomb amid the soft earth of the Russian steppes, preparing to load it on a Bf 109 Emil belonging to S.G. 1. *Photo courtesy NASM.*

The Search for Better Weapons. The introduction of the powerful FW 190 Jabo gave the Schlachtgeschwadern a potent ground-attack swing-role fighter. To further the capabilities of German ground-attack forces, the Luftwaffe developed a number of specialized weapons (though, interestingly, they largely ignored rockets). Here an FW 190 is shown carrying a mutiple ejector rack with four 50-kg antipersonnel bombs, one of many weapons options available from 1943 onward. *Photo courtesy AFM.*

Readying a Cluster Bomb. Both the Luftwaffe and the VVS made extensive use of cluster bombs and small bomblet dispensers to attack troops and "soft" targets such as motorized transport. Here is the *Abwurfbehälter,* a bomb dispenser having a hinged shell that would burst open over a battlefield, scattering a deadly pattern of bomblets. One such weapon is shown being readied for attachment to a *Schlacht* FW 190. The best-known of Soviet cluster bombs was the infamous "Molotov Breadbasket," though the later PTAB antiarmor munition did the greatest damage on the battlefield. *Photo courtesy NASM.*

Winter Ops. Weather played a dominant role in Eastern Front air warfare, and the Russians always had an ally in "General Winter." Here an FW 190F is poised for takeoff, carrying a single cluster bomb munition on its centerline bomb-rack. The running ground crewman may be racing for a "last chance" check—or he may be trying to give the pilot updated targeting information before takeoff. Eastern Front warfare was so fluid in 1943–45 that battlefield situations frequently changed even as strike flights flew toward their targets. *Photo courtesy NASM.*

Bombing Up a Seagull. After being nearly destroyed in Hitler's opening assault against Russia, the VVS staged a remarkable comeback from defeat. In the war's opening months, Soviet ground attack was largely undertaken by older fighters such as the Polikarpov I-153 Chaika. Here a ground crewman secures a small bomb to the wing rack of an I-153 standing strip alert at an improvised airfield. *Photo courtesy NASM.*

Fast, but Not Fast Enough. Considered the most formidable high-speed medium bomber of the mid-1930s, the Tupolev SB, like its French and British counterparts, proved woefully vulnerable to German fighters and flak during the opening stages of the Nazi blitz into Russia. Many were lost in battlefield air attacks in 1941–42, causing the VVS to place greater emphasis on more advanced designs such as the Il-2, Pe 2, and fighter-bombers. *Photo courtesy NASM.*

Petlyakov's Peshka. Sometimes called the "Russian Mosquito" (a reference to Britain's outstanding warplane), the Petlyakov Pe 2 combined the ruggedness of a dive bomber with the speed and agility of a fighter and the payload of a light bomber. The result was an excellent battlefield support airplane known affectionately to its crews as the "Peshka." The Pe 2, while not so heavily armored as the Ilyusha (the Il-2), relied on its higher performance for enhanced survivability against air-to-air and ground-to-air threats. This immaculate mid-war example belonged to a VVS Guards regiment. *Photo courtesy NASM.*

Ilyushin's Flying Tank. The heavily armored and armed Ilyushin Il-2 Shturmovik was a devastatingly effective battlefield support airplane, carrying a variety of cannon, bombs, and rockets. In late 1941, Josef Stalin stated that the "Il-2 is as needed by [the] Red Army as air or bread," demanding greater production. Largely invulnerable to German antiaircraft fire because of its thick armor and high speed "on the deck," the Il-2 proved adaptable to many roles, earning the nickname *der Schwarze Tod* (the Black Death) from the Wehrmacht. Here is the Il-2m3 Mod, an antitank version mounting two 37mm cannon, which appeared just in time for the battle of Kursk in 1943. *Photo courtesy NASM.*

Bell's "Little Shaver." Complementing Soviet designs were large quantities of American and British Lend-Lease aircraft, such as the Douglas A-20 Havoc and the Bell P-39 Airacobra. Shown are six late-production P-39Q's with four-bladed propellers and long-range fuel tanks awaiting delivery to the VVS. Well suited to the free-wheeling tactical air war of the Eastern Front, the P-39's proved excellent low-altitude fighters and ground-support airplanes, earning the nickname of "Little Shaver" from their pilots. *Photo courtesy NASM.*

A Mosquito's Deadly Bite. The gutting of postwar Air Force and Navy-Marine tactical air power after 1945 came back to haunt the United States in the opening months of the Korean War. Numerous lessons of battlefield support had to be relearned and recast to fit the requirements of a "limited" war. One rediscovery was the airborne Forward Air Controller (FAC). Modified North American T-6G Texan trainers, called Mosquitoes, carried smoke rockets and radios permitting them to mark targets and direct airborne strike flights. Here a Mosquito has just fired a white phosphorous marking rocket at a Communist troop position. *Photo courtesy AFM.*

Return of a Vet. The Mustang, now designated F-51, soldiered on in Korea, but at heavy cost in lost airplanes and captured or killed pilots. Here an F-51D on a support mission taxis through typical Korean conditions armed with bombs and four 5-in high-velocity rockets. *Photo courtesy AFM.*

Going Downtown. The air war in Vietnam conjures visions of the intensive campaign against MiGs, surface-to-air missiles (SAMs), and guns waged by American airmen over North Vietnam. Here a strike flight of Republic F-105D Thunderchiefs—better known as Thuds—refuel from a Boeing KC-135 Stratotanker before ingressing to Route Package 6, the heart of North Vietnam's air defense network. But not often appreciated is that these aircraft, and numerous others, were frequently required to undertake "in country" air support tasks as well, flying CAS and BAI missions. *Photo courtesy AFM.*

A Time for Retreads. Like Korea before it, Vietnam was a time when older aircraft reappeared in combat service. For example, the Douglas B-26B Invader—the A-26 of the Second World War—flew CAS and BAI missions both for the French in Indochina and for the South Vietnamese and Americans a decade later. Shown is a B-26B photographed while escorting a Fairchild C-123 transport over the Mekong delta. A disastrous series of wing fatigue failures eventually spelled the end to the B-26's combat career. *Photo courtesy AFM.*

Bird Dogs Sniffing About. As if once wasn't enough, the need for FACs to help coordinate and control CAS/BAI missions had to be relearned yet again for Vietnam, using Cessna O-1 Bird Dog spotters acquired from the U.S. Army. Here are two of these 100-mph machines, armed with marking rockets and equipped with additional communication gear. Such worked well until the Communists began fielding increasingly heavier antiaircraft defenses. Being a FAC required uncommon courage and devotion to duty. *Photo courtesy AFM.*

North American–Rockwell OV-10A Bronco. The OV-10A Bronco represented the result of several years of joint service studies on the requirements for specialized counterinsurgency (COIN) aircraft, coupled with the fascination of senior Department of Defense officials with guerilla war and acquisition commonality. But in many ways it was a mélange of conflicting requirements—part reconnaissance aircraft, part FAC, part CAS/BAI strike aircraft, part transport, and all compromised. Though a much better and more survivable FAC than the O-1 that it supplemented (though it did not fully replace), the OV-10 was underpowered and overall a disappointment, like another McNamara-era project, the TFX. *Photo courtesy AFM.*

Up Close and Personal . . . Ed Heinemann's magnificent Douglas A-1 Skyraider—the AD "Able Dog" of the Korean War—returned to combat in Vietnam as a counterinsurgency and air rescue support airplane. Capable of carrying prodigious weapons loads, the heavily armored and exceptionally rugged A-1, known as the Spad by some and Sandy by others, was responsible in numerous cases for saving small hamlets and outposts from being overrun by the Viet Cong. Eventually the introduction of the SA-7 shoulder-launched SAM and the increasing proliferation of light flak rendered it vulnerable to ground fire, though some continued in service until the collapse of South Vietnam in 1975. Here an A-1H is shown dropping 500-lb Mk. 82 bombs over Communist positions. *Photo courtesy AFM.*

A Corsair Comes Ashore . . . Interest in upgrading the Air Force's strike forces led to acquisition of the Navy Ling-Temco-Vought A-7 Corsair II light attack airplane, inspired by the earlier Vought F-8 Crusader fighter. The A-7D shown here had near-fighterlike performance, excellent payload characteristics, and a superb bombing system. The A-7 served well in Vietnam, and went on to a long postwar career. Even now an advanced version with a new engine and avionics is under study for future CAS/BAI mission requirements. *Photo courtesy History Office, Edwards AFB.*

Bell AH-1G Cobra. Vietnam marked the practical employment of armed helicopter "gunships." The specially designed AH-1G Cobra proved an effective and devastating aerial fire support platform, escorting troop helicopters, covering landing zones, and, finally, attacking Soviet-built tanks with TOW antitank missiles during the 1972 North Vietnamese spring invasion. *Photo courtesy NASM.*

Bell AH-1S Cobra: a Deadlier Derivative. Following the Vietnam war, Bell continuously refined the Cobra. Here is the AH-1S, intended as a Warsaw Pact tank-killer, carrying a mix of missiles, rockets, and cannon. Note the exhaust diffuser and infrared jammer to suppress heat-seeking missiles of the type that proved so deadly in the latter stages of the Vietnam war—and which, more recently, took a high toll of Soviet aircraft during the war in Afghanistan. *Photo courtesy NASM.*

A Different Kind of Bird of Prey . . . The recognized vulnerability of conventional military airfields to attack inspired Sir Sidney Camm of Hawker to develop an experimental vertical/short-takeoff-and-landing technology demonstrator, the P.1127. Years of refinement led to the British Aerospace Harrier, a combat-proven V/STOL strike fighter. The Harrier offered a new capability of quick-reaction battlefield air support, a capability acknowledged by the U.S. Marine Corps, which ordered an American version (the McDonnell-Douglas AV-8A), as well as a substantially redesigned and upgraded successor, the contemporary AV-8B Harrier II. *Photo courtesy NASM.*

Warthogs on Review. The search for alternatives to reopening the Douglas A-1 Skyraider production line in the 1960s spawned the AX design competition which resulted in the production of the Fairchild A-10A Thunderbolt II, known to its pilots as the Warthog. As the rationale for the helicopter gunship changed from counterinsurgency jungle warfare to Warsaw Pact tank-killer, so, too, did that of the A-10. Born of the need for a high-payload bomb-dropper for Southeast Asia, the A-10 was, instead, procured as a 30mm cannon-toting tank buster; its development inspired the Soviets to proceed with their own "A-10ski," the Sukhoi Su 25 Frogfoot, which proved vulnerable to shoulder-launched Stinger SAMs in Afghanistan. Questions over the A-10's survivability by the 1990s have triggered an intense interest in new and more capable CAS/BAI aircraft. Shown are three of the A-10 development aircraft, flying near the Air Force Flight Test Center. *Photo courtesy History Office, Edwards AFB.*

PART THREE

Abyssinia, Spain, and War in Asia

During the 1920s and 1930s, a significant number of larger conflicts took place around the world that involved some or all of the following: employment of sizable forces, operations on a broad geographic scale, and big-power involvement. Six wars and "incidents" fall into this category (four of which are examined in this study): the Gran Chaco war (1932–35), the Abyssinian (Ethiopian) war (1935–36), the Spanish Civil War (1936–39), the Sino-Japanese war (1937–41), the Changkufeng/Khasan incident (1938), and the Nomonhan/Khalkhin Gol incident (1939). In all of these, aviation played a role. In some cases, such as Abyssinia, Spain, China, and the Nomonhan episode, that role was extensive and influential. In others, such as Gran Chaco and Changkufeng, its use was largely inconsequential. As might be expected, the more extensive the use, the more significant were accompanying lessons in terms of doctrine, theory of employment, and expectations for future war. As is already recognized, some of these conflicts—notably Spain and the Sino-Japanese war—served as testing grounds for new military technology and ideas, in much the same way that subsequent surrogate warfare in the Middle East has served as a proving ground for Soviet and Western doctrine and equipment. In virtually all of these conflicts, the military equipment utilized—particularly aircraft—represented the existing norm or standard then employed in service; these were not "bush wars" typified by use of outdated or immature technology; even in the Gran Chaco and Abyssinian conflicts, the aircraft tended to be reasonable representatives of then-current design practice. By the end of the 1930s, the aircraft appearing in conflicts such as the Spanish Civil War and the latter stages of the Sino-Japanese war included large numbers of designs that were just entering service and which would, in fully developed form, serve throughout the Second World War.

One of the most bitterly fought of these "larger" wars was the Gran Chaco war (1932–35) fought between Bolivia and Paraguay over a desolate tract of land called the Chaco and commonly referred to as the "Green Hell." Air power played a largely meaningless role in the Chaco war, though air support operations (by both sides) were an occasional feature of Chaco fighting. In most of the decisive battles of the Chaco war, however, thirst proved the real difference between victory and defeat.[1] Neither Bolivia nor Paraguay was really prepared for war, though the bellicose rumblings of the Bolivian high command gave an impression of military professionalism that was, in fact, sorely lacking.[2] Paraguay, for its part, could count on the distinguished generalship of José Felix Estigarribia, a graduate of French military schools; he quickly got the better of his opposite number, the German émigré general Hans Kundt, who served as chief of the Bolivian general staff.[3]

At the war's outset, Bolivia (with between 60 and 80 aircraft) outnumbered Paraguay nearly ten to one in the air. Neither side really had a clear conception of what aircraft could mean in military strategy, though the Bolivian air service was headed by a particularly fine airman, Bernardino Bilbao Rioja.[4] Not surprisingly, Bolivia dominated the skies, but even so, found it necessary to rely on the services of foreign mercenary airmen.[5] Bolivian "air superiority" meant little in the Chaco fighting, for in contrast to the ineptly led Bolivians, the Paraguayans fought stubbornly and, indeed, brilliantly.[6] Bolivian airmen increasingly encountered dense antiaircraft fire from machine guns and even 20mm Oerlikon cannon, though actual losses were low.[7] Historian David Zook has accurately concluded that the air campaign was ineffectual, noting that "poor employment of air power" undoubtedly contributed to the eventual Bolivian defeat in June 1935.[8] Overall operational control and the fundamental comprehension of joint air-land operations were seriously deficient (on both sides); basic communication was limited to panels and hand signals; and there were no procedures for coordination and managing air operations over the battlefield. At best, the level of air support furnished during attacks was deficient even by the standards of the First World War, and there are no apparent cases where air attacks had profoundly destructive results. Thus, while aviation did play a role in the Chaco war, it is more profitable to examine three conflicts where it had far more serious consequences: in Abyssinia, Spain, and Asia.

7

The Abyssinian War

Background to War

In 1896, while trying to reassert its control over Abyssinia (Ethiopia), Italy suffered a disastrous defeat at Adowa, in which (in the words of Brigadiers Peter Young and Michael Calvert) "the dead were more fortunate than the prisoners."[9] For nearly four decades, memory of that humiliating and savage defeat lingered in Rome, and finally, for that and other more geopolitical reasons, Benito Mussolini, the Fascist *Il Duce* of Italy, determined to reestablish the Italian protectorate over Ethiopia by a two-pronged assault from Eritrea (on the shore of the Red Sea) southward, and from Italian Somaliland (on Africa's Indian Ocean and Gulf of Aden coastlines) northwest into Ethiopian territory. Mussolini reached his decision in the autumn of 1933, and communicated his wishes to Marshal Emilio De Bono, a vigorous near-septuagenarian with extensive colonial experience. From that moment on, the Abyssinian invasion plan nearly occupied De Bono's full time and attention, particularly since Mussolini had directed that Ethiopia be subdued by the end of 1936. Despite his years, De Bono was no Hans Kundt, and he paid particular attention to the needs of the Italian air service, the *Regia Aeronautica.*[10]

Italy had been one of the pioneering nations in military aviation, and was, of course, the nation of Douhet. Italian aviators had established a distinguished combat record in the First World War and in the years following the war. During the 1920s and 1930s, Italian aircraft technology generally kept pace with that of other leading nations. Italian air racing designs showed a profound appreciation for high-speed flight, and Italian long-range flying boats made pioneering flights across the Atlantic. Italian military aircraft technology equalled that of other nations, and the country established a reputation for quality high-performance and highly efficient military machines. In part, this was due to Mussolini himself who, like Josef Stalin, appreciated the public and psychological impact of advanced aeronautical technology in conveying the alleged progressivism of a totalitarian state; both men, in fact, had sons who became military aviators.

Despite the more exaggerated pronouncements of Douhet, Italy's military aviators showed a pragmatic appreciation of coupling air and land action. In 1931, Italy established its first attack aviation elements (as distinct from bomber and fighter aviation), the so-called *gruppi d'assalto*. In the mid-1930s, influenced by the notions of General Amedeo Meccozzi, the Regia Aeronautica embarked on a development program to produce specialized multipurpose military aircraft suitable for combined fighter, light bomber, and assault missions. Unfortunately, Italy emphasized the bomber side of the equation, generating aircraft such as the Breda 64 and 65 family that tended to be slow and relatively unmaneuverable, and thus disappointments when general war broke out between Italy and the Allies in 1940. During the Spanish Civil War these aircraft served reasonably well; they were not available for service during Italy's Abyssinian war in 1935–36. Instead, Italy made use of large numbers of older aircraft such as the Caproni Ca 101 trimotor bomber and its derivatives, the single-engine (and hence underpowered) Ca 111, and the trimotor Ca 133; the trimotor Savoia-Marchetti S.M. 81 Pipistrello ("Bat") bomber; a few Fiat C.R. 20 fighters; and larger quantities of the Meridionali Ro 1 observation plane (a license-built version of the Fokker C V) and the Ro 37 two-seat reconnaissance biplane (which was extensively used for ground-attack duties both in Ethiopia and subsequently in Spain).[11]

In contrast to Italy, which approached the Abyssinian war as a European power already committed to waging a mechanized war integrating the combat effort of air and land forces, Ethiopia was backward. The Ethiopians could count on only thirteen airplanes, none of which was up to the latest standard in aircraft design, and its air force consisted of mercenary pilots dismissed by De Bono as "amateurs." It is difficult to ascertain whether there was a brief "air superiority" war; Italian memoirs by De Bono and Marshal Pietro Badoglio make no mention of such combat, and neither do the accumulated attaché reports of American observers; indeed, De Bono emphatically states that at no time did he even see an Ethiopian airplane. One popular (but suspect) account states that "one after another the [Ethiopian] aircraft were shot down by Italian pursuit planes."[12] In any case, Italy clearly never had to worry about interference as its air force went about its business. The individual Ethiopian soldier was personally courageous, shrewd, and ferocious; often ineptly led and thrown away in foolishly planned attacks, he proved nevertheless to be a dangerous opponent and capable of operating modern infantry weapons with skill. The popular image of a modern European power rapaciously devastating submissive and ignorant tribesmen is thus not merely inaccurate; it is likewise an insult to both the Ethiopians who furnished an intense level of defense and the Italians who coped surprisingly well with the strenuous demands placed upon their armed forces by this distant and complex campaign.

In his planning for war, De Bono recognized that the immense distances involved in the Ethiopian campaign made it highly desirable to use aircraft (because of their great mobility) as much as possible for reconnaissance and liaison tasks. At the same time he realized that the distances and the primitive nature of support facilities would tax air operations to their maximum. Thus he emphasized airfield construction and arranging for logistical support of the aircraft in place or soon to arrive. For the most part, aircraft arrived by ship via the Suez Canal; after being unloaded, six could be assembled by technicians every forty-eight hours. Above all, De Bono wished it to be clearly understood that the air elements would not operate independently, and he successfully lobbied to have the air commander placed under his overall command, reflecting testily in his memoirs that "Unity of command is indispensible. *One man only* must exercise command."[13]

The Air Campaign

On October 3, 1935, Italy invaded Ethiopia, triggering a bitter seven-month campaign culminating in the capture of Addis Ababa and the annexation of Ethiopia within the Mussolini empire. During the war, Italian air power was used extensively during attack and pursuit operations, and proved (in the words of American intelligence officers) "tremendously effective," particularly when the Ethiopians employed mass attack tactics. In contrast to the general success of aircraft, tanks proved a serious disappointment in the intensely fluid conditions of Abyssinian warfare; they could not maneuver rapidly enough, were hampered by the rugged terrain, and were frequently overwhelmed by infantrymen (who, in one notable case, used heavy rocks to bend the tanks' machine guns, and then killed the crews as they evacuated their damaged vehicles). But the air operations were not without some problems. Commanders quickly learned that the crevice- and cave-strewn terrain of Ethiopia meant that air strikes often failed to dislodge or kill defenders, and that difficult ground-clearing operations by infantry were still a necessity. Italy turned to chemical warfare—namely dropping mustard gas bombs and using flamethrowers—to compensate for this problem. Officially Italy denied gas bombings, aside from a few special cases. An American intelligence officer accompanying Ethiopian forces stated, however, that Italian use of mustard gas was "the most effective" of all Italian weapons used in the campaign because it so thoroughly broke Ethiopian resistance. It is uncertain how widespread its use was; Italian commanders denied virtually all use, though Ethiopians claimed it was quite common. The truth is likely in the middle. An American intelligence officer serving with the Italian forces stated that gas bomb use had not been widespread, and that its overall effectiveness was "very slight." Perhaps his statement reflects unrealistic expectations on the part of the Italians who utilized it; certainly the Ethiopians themselves—and foreigners attached to Ethiopian units, including one of his fellow intelligence officers—felt differently.[14] As in any air campaign, command and control problems appeared that had to be worked out.

Perhaps the most notable of these occurred in November 1935, when Italian forces under the command of General Luigi Frusci advanced on the town of Gorrahei. Though De Bono was pleased with the degree of control that he exercised over the air arm, subordinate commanders found that they could not adequately coordinate their needs with the air command when they asked for air support. Frusci had planned a combined infantry-air assault on Gorrahei with a view to utterly destroying the Ethiopian forces there. Instead, as a postwar American intelligence summary stated,

> the air force carried out its bombing operations prematurely and prior to the arrival of the ground columns within striking distance. The severe air bombardment killed a dozen or so Ethiopians, wounded the commander, and caused an evacuation of the town and withdrawal, before the Italian ground forces could make their presence felt. This was hailed in some quarters as a decisive triumph for the air—the first battle won by an air force alone. Actually it destroyed the opportunity for a decisive victory and probable elimination of the Ethiopian forces involved. The Italian ground commander charged with the operation had

no control over the air attack. It was to have been coordinated by G.H.Q. and is a rather outstanding instance of failure in cooperation.[15]

As Slessor found out in Waziristan, and as the Marines had learned in Nicaragua, dispersed bodies of troops were not easily located even by aerial reconnaissance when they practiced camouflage and deception techniques and took advantage of available ground coverage. Repeatedly during the Ethiopian war, Italian ground commanders found that they could not always draw upon aerial reconnaissance information for precise intelligence on enemy locations and strengths. Additionally, the rapid pace of the Italian advance taxed the ability of the airmen to maintain support operations; they quickly found that they were operating at maximum range from the few available airfields, and had to rapidly arrange for appropriate logistical support to continue furnishing services to the ground forces. This situation would have been very serious had it not been for the steps that De Bono had taken in the months prior to the invasion to arrange for expansion and improvement of the airfields available to the Italians in East Africa. While airborne control and observation of artillery firing was helpful, the lack of detailed maps hampered precise targeting of enemy positions and batteries for artillery fire.[16]

Still, over time, operations became smoother and working relationships solidified, and the quality of Italian air support to ground forces constantly improved. In one important area, troop resupply, air played a prominent role. During the Tembien offensive in February 1936, the Italian III Corps received five tons of food, water, and munitions as it moved across the Asta Plain, an arid area devoid of roads and wells. Toward the end of the Abyssinian campaign, Italian airmen had perfected resupply methods, delivering 385 tons in one twenty-one-day period. Air-ground liaison generally functioned well, and Italian commanders relied heavily on radio communications, sometimes so much so that they saturated their network; insofar as reconnaissance and observation could assist the ground forces, ground columns were constantly kept aware of the latest tactical situation as observed from above. It was in aerial attack, of course, that Italian air power made its presence most forcefully felt.[17]

Italian attacks against Ethiopian forces consisted of two basic missions: level bombing sorties from an altitude of 1000 to 3000 feet using the various trimotor bombers available, and low-level attack missions using other machines. While the former were conducted on a pre-briefed basis and generally along Douhet-like lines, the latter were usually much more tactically oriented, consisting of pre-briefed strikes in support of assaults, and then what might be considered "armed reconnaissance," target of opportunity, and emergency response actions. Italy carefully avoided bombing operations against targets of major cultural and political significance (such as Addis Ababa) that might have attracted even more unfavorable publicity than the invasion effort had already received. Nevertheless, between October 1935 and the end of the war in May 1936, Italian airmen flew 872 bombardment missions against towns, fortifications, caravans, and troop assembly points. When Mussolini relieved De Bono for disagreeing with certain policy actions (though the two men remained close, and Il Duce even wrote the introduction to De Bono's subsequent memoirs), it triggered a period of inactivity until De Bono's replacement, Marshal Pietro Badoglio, could launch his own offensive that would finish the war. During this lull, bombing substituted for land warfare in keeping pressure on Ethiopian forces.[18]

Very quickly, the Italians learned first-hand that Douhet's notion of bombing a populace into submission did not work; though Ethiopians undoubtedly feared bombing, it did not materially break their will to resist. Bombing of troops, however, was a very different matter, and intelligence reports repeatedly emphasized its effectiveness, one such document stating that

> the long range bombardment of troops was very effective, and Italian aviation was invaluable to the ground forces in this respect. A very strong Ethiopian column of several thousand men was discovered by the air en route from Gondar north toward Dabat on December 4, 1935. It was repeatedly attacked on December 5 and 6 by the 4th and 27th Bombardment Groups with 30 planes. In spite of violent Ethiopian reaction during which all planes were hit by rifle or machine gun fire, the column was completely disorganized, men and animals killed and supplies destroyed. Many similar attacks took place in the rear areas with valuable results in furthering the Italian plan of operations.[19]

Air strikes proved particularly significant when they preceded or accompanied ground assault. In these missions, Italian airmen employed larger aircraft (such as the Ca 101 or the S.M. 81) against fortifications, but generally made use of smaller single-and two-seaters, carrying light bombs and relying on strafing attacks. Air supplemented, but did not supplant, artillery in supporting infantry assault, not surprising given the small bomb tonnage dropped. For example, though Italian aviators dropped a total of 25,700 bombs in one six-day period during the battle of Enderta in February 1936, this only represented 192 total tons of bombs, as the majority were small 10- to 15-lb antipersonnel weapons preferred for attack-type missions. On a daily basis, it was roughly equivalent to the weight of fire from a single battery of 155mm howitzers. Repeatedly, however, air strikes added a vital impetus to attack, and they proved mercilessly effective in pursuit of shattered Ethiopian forces. During the initial assault into Ethiopia, close support aircraft attacked Ethiopian forces ahead of advancing Italian columns, notably at the Mareb River on October 3, 1935. Marshal Badoglio's memoirs repeatedly refer to the devastating effect of air attack on fleeing Ethiopian troops, and copies of his operational orders included as appendices to this work reveal that he depended on air support extensively. The battle of Enderta seemed to him a particularly good example of the cooperation of air and land forces during an attack. He relied on attack aircraft to bomb and strafe ahead of advancing troops, and once the battle had been decided in Italy's favor, he released the aircraft to attack the retreating Ethiopian forces; in one of these missions, a plane strafed the column of the Ethiopian commander, Ras Mulugeta, killing him. In all, Ethiopia lost approximately 6000 troops—and twice that number wounded—in the Enderta battle, against total Italian casualties of just over 800 Italian and native troops. Enderta was by no means an isolated example; during the climactic battle of Lake Ascianghi, attack aircraft struck from altitudes as low as thirty feet, inflicting thousands of casualties with high-explosive and gas bombs, and machine-gun fire. Reflecting on the Abyssinian experience, Badoglio wrote subsequently that

> *l'aviazione* was present in all the phases of the war and functioned throughout each battle. In the absence of enemy aviation, it dominated the sky. It is the combat arm of the future and will be increasingly important in many [combat] areas. However great its role, it must act in coordination with the army. Neither can ever act again on its own to make war.[20]

Despite this generally impressive performance, there were some significant indications of the growing vulnerability of aircraft to ground defenses. Again in contrast to their popular image as spear-carriers, the Ethiopian forces were equipped with rifles and automatic weapons (including Oerlikon cannon), though not in great quantity. A postwar American intelligence analysis estimated that Italy lost a total of fifty airplanes due to all causes during the war, many from the demanding environmental conditions (such as terrain, high-altitude operations, and weather), but sixteen from ground fire. These sixteen consisted of three

Ca 101, two Ca 111, two Ca 133, and two S.M. 81 bombers; one C.R. 20 fighter; five Ro 1 observation aircraft; and one Ro 37 recce/strike aircraft. Personnel losses amounted to 78 crewmen killed in action, and 148 wounded. The study concluded that

> on certain missions of ground strafing all planes participating were reported struck by fire from the ground. 259 planes were reported hit by antiaircraft fire, which means small arms fire or the fire of the Oerlikon 20mm gun.
>
> . . . During the ground strafing in the Mai Mescic valley following the battle of Amba Aradam [Enderta], all planes (S-81's and Ca-101's) participating were hit by fire from the ground, one of them 19 times. One plane was lost. During the battle of Lake Ascianghi 25 planes were hit and one plane brought down. During the battle of Birgot, 7 planes were struck, 2 pilots wounded and 2 Ro-37's were forced down behind their own lines.
>
> The Italian pilots operated with great daring at low altitudes, which they probably could not afford to do against a well-armed and well-trained opponent, and it seems that, almost invariably, the majority of the planes so operating were struck by small arms fire from the ground. This is particularly significant in view of the class of opposition: insufficiently equipped with effective types of weapons and untrained in proper methods of defense.[21]

Overall, the Abyssinian war offered an indication of how modern military forces could work together on the battlefield in joint operations involving coordinated air and land action. Beyond this, it did give yet another example of the vulnerability of ground forces that could not call upon their own air elements to prevent an opponent from undertaking unrestricted air attacks against them. It also showed that even a relatively unsophisticated opponent could be expected to inflict casualties on low-flying attack aircraft, particularly if those aircraft were relatively large, unmaneuverable, and slow (less than 200 mph). From the perspective of 1940, an Army Air Corps officer tasked by the Army War College with evaluating the effectiveness of new weapons enthusiastically concluded that in the Abyssinian war, "the influence of air power could be classed as decisive."[22] In fact, even before the war began, the outcome—pending no foreign intervention—was never in doubt. In reality, as was recognized by intelligence officers at the time, the Regia Aeronautica's activities were not in and of themselves decisive; as an element of combined air-land warfare, however, the Italian air force's "aid was invaluable to an *early* and successful conclusion of the campaign [emphasis in original]."[23]

8

The Spanish Civil War

Myths, Realities, and the Nature of the Air War

Perhaps no conflict of the twentieth century has so polarized individuals as has the Spanish Civil War (1936–39). It became overnight virtually a political litmus test between the left and the right, and, to a large degree, it has remained such since that time. The Spanish Civil War has been characterized as a conflict between democracy and fascism, a conflict won by the forces of Hitler, Mussolini, and Franco, at the expense of Spanish democracy and the Spanish people. It is remembered primarily as a callous proving ground for the Second World War, a huge open-air laboratory in which Hitler's Panzer forces and the adolescent Luftwaffe practiced the techniques of blitzkrieg: "lightning war" that would overwhelm Poland, the Low Countries, and France in 1939–40. In truth, of course, the war was much more complex. The Spanish Civil War did not destroy Spanish democracy; rather, democracy had been destroyed long before some of Spain's military rose up in revolt in Morocco. Those who speak in glowing terms of the Spanish Republic would do well to keep in mind the brutal political repression and the wholesale murders of clergy (and sacking of churches) that accompanied the regime. The onset of civil war merely accelerated antidemocratic tendencies present in both the left and the right, in the increasingly leftist and anticlerical government of the Republic and the burgeoning right-wing (though *not* fascist) movement of Francisco Franco y Bahamonde. By mid-1937, any pretext about fighting for democracy was gone, save for propaganda pronouncements. By that time, the war pitted a right-wing coalition (the Nationalists) aided by Italy and Nazi Germany against a coalition of left-wing forces (the Loyalists) representing Republicans and Communists (with a sampling of anarchists, socialists, trade-unionists, and Trotskyites subsequently purged by more doctrinaire Stalinist elements) actively assisted by the Soviet Union and France's leftist Popular Front government. Seen in this context, the Spanish Civil War becomes far more than a particularly vicious and savage fratricidal conflict. There was, virtually from the outset, a substantial European involvement that went well beyond Germany and Italy, and which resulted in the Loyalists receiving military aid and equipment from many European nations, including Czechoslovakia and Holland. Portugal and Poland likewise assisted the Nationalists. Mercenaries operated on both sides, but primarily with the Loyalists, who also benefited from the Comintern's establishment of the International Brigades, composed of Communists and other leftists, and far fewer numbers of nonaligned idealists.

In the years since the Spanish Civil War, the conflict's political and social dimensions have been increasingly romanticized and stereotyped by those writers who have had an emotional stake in the story; there are few works—typified notably by Hugh Thomas's impressive *The Spanish Civil War*—that have attempted to go beyond the commonly accepted one-dimensional portrayal of the war as a struggle between "democracy" and "fascism."[24] Likewise, the war's military dimension has suffered equally from simplistic interpretation. Germany and Italy did not enter the Spanish Civil War with a coherent strategy or doctrine that they wished to "test"; neither, for that matter, did the Soviet Union. At first the military efforts of these countries were aimed primarily at furnishing equipment and support personnel. Very soon, as the Spanish military proved incapable of mastering some of the newer technology—particularly advanced aircraft—quickly enough, these nations changed their policies to permit using their personnel in Spain in direct combat roles. Hitler's decision to furnish aid seems to have been less ideological—in the sense of supporting a fellow anti-Communist totalitarian—than due to a desire to gain access to Spain's vast iron ore deposits. The "blitzkrieg" doctrine could hardly be tested in Spain because it was only then gestating within Germany, though German tank advocates recognized that Spain could likely serve as another "Aldershot" (a reference to a British proving grounds). Likewise, as German advisors in Spain were to discover, the Spanish military character, which emphasized macho head-on encounters, was not responsive to the notion of a doctrine which emphasized maneuver, deception, and flank assault.[25] Thus, when Spanish forces and their outside allies clashed, it was not in the sense of a battle between established military doctrines and philosophies; rather, it was in the uncertain nature of real war—a war in which ideas were generated out of necessity, tried, and found wanting or successful.

The Spanish Civil War involved complex air operations on a scale unmatched until the latter stages of the Sino-Japanese and Japanese-Russian clashes in the Far East, and the outbreak of the Second World War itself. During the fighting in Spain, air power was utilized in a number of significant ways. Initially, for example, the Nationalist revolt was kept alive by a massive (for its time) airlift of Moroccan troops from North Africa to Spain. Reconnaissance, observation, and bombardment all played significant roles in attacks, as did low-level strikes by fighters and bombers operating in a close air support and battlefield air interdiction role. One must be careful in characterizing lessons from the Spanish air war, however, because Spain was a classic example of a war in which a generational clash occurred within technologies. Further, it was a constrained war and smaller in scope than a more general conflict, and thus lessons learned from it were not necessarily valid for a larger war between opposing nations.

For example, bomber advocates could point to the relative invulnerability of bombers to fighter interception as supporting the current "bomber will always get through" school of thought. In fact, such was true when modern twin-engine monoplane bombers such as the Soviet-built Tupolev SB or German Dornier Do 17 encountered older biplane fighters such as the Fiat C.R. 32 or Polikarpov I-15. But when such aircraft encountered contemporary monoplane fighters such as the Messerschmitt Bf 109 or the Polikarpov I-16, it was quite another story. As early as February 1937, the American military attaché in Valencia was writing that "the peacetime theory of the complete invulnerability of the modern type of bombardment airplane no longer holds. The increased speeds and modern armament of both the bombardment and pursuit plane have worked in favor of the pursuit. . . . The flying fortress died in Spain."[26] (Failure to heed this lesson resulted in many more Flying Fortresses dying over places such as Schweinfurt and Regensburg in 1943–44.)

Likewise, it is commonly accepted that close air support and army cooperation came directly from the lessons in Spain. In fact, to a great degree, these were lessons from the Luftstreitkräfte's experience with the Schlachtstaffeln of 1917–18 that were merely rediscovered and elaborated upon, as well as from the influence of Soviet experience in this area. The Spanish Civil War reintroduced the notion of aerial "dogfighting" between fighters, a notion that had been considered passé once fighter air speeds rose above the 200-mph mark. Throughout Spanish air operations, fighters operated in both an air superiority and air-to-ground role. Again following a traditional pattern, when a fighter proved unsuitable for air-to-air combat (as with the Heinkel He 51) it was relegated primarily to air-to-ground attack. As in the Rif War or in Italy's more recent Abyssinian experiences, the use of bombing to generate a Douhet-like crushing of morale proved elusive. Tactical bombing proved quite valuable to both sides, as did the direct application of bombing and strafing in support of ground forces—in short, low-level air attack. This was particularly true in the significant battles that marked the progress of the war: the invasion of Majorca (August 1936), the drive on Madrid (September–November 1936), Guadalajara (March 1937), the Northern Campaign (March–July 1937), Teruel (December 1937–February 1938), and the Ebro (July–October 1938).[27]

The Beginnings of Spanish Air Support

As might be expected, these conflicts reveal that over time the use of air power, particularly for air support operations, became more sophisticated. To a degree this mirrored the state of available aviation forces. At the time of the revolt in July 1936, "aeroplanes," as Hugh Thomas has written, ". . . seemed to be the key to victory."[28] The Loyalists largely retained control of the air force, but it was in any case a second-rate service built around Breguet XIX two-seat bombers and Nieuport-Delage NiD 52 fighters, both aging biplanes. Pierre Cot, the aviation minister of Leon Blum's leftist Front Populaire government in France, arranged for the transfer of more modern Potez 540 bombers and Dewoitine 372 fighters from French stocks to the Loyalist air force, and made other provisions for the Loyalist government, including permitting Loyalist fighters to refuel in France during cross-countries from Catalonia to the Basque Atlantic coast (under the guise of "navigational errors"), letting Loyalist pilots train at French flight schools using French instructors, planes, and government facilities, and setting up Air Pyrennées, a government-sponsored airline to support the Loyalist cause.[29] Italy and Germany were even quicker in offering support to Franco when he asked for assistance in transporting the Spanish Foreign

Legion and Moroccan troops across the Strait of Gibralter; the first Italian aircraft arrived at the end of July (amid some embarrassing crashes), and the first German transports and a few Heinkel He 51 fighters arrived early in August. Shortly thereafter, Germany and Italy began the building up of forces that would lead to the establishment of the so-called *Legion Condor* and *l'aviazione legionaria,* the former built around a force of Junkers Ju 52/3m bomber-transports (later replaced by Heinkel He 111 and Dornier Do 17 aircraft) and Heinkel He 51 fighters (subsequently replaced by Messerschmitt Bf 109's), and the latter built around Savoia-Marchetti S.M. 81 and Savoia-Marchetti S.M. 79 trimotor bombers, Fiat B.R. 20 bombers, Fiat C.R. 32 fighters, and Meridionali Ro 37 reconnaissance/assault aircraft. In October 1937, the first Soviet contingents of personnel and planes arrived in Spain, under the direction of "General Douglas" (the alias of Jacob Schmutchkievich; eventually the future Soviet Marshals Zhukov, Konev, Malinovsky, Rokossovskiy, and Nedelin—the latter killed in an incredible Soviet launch pad explosion in the early 1960s—all served in Spain. They had better luck than the unfortunate Schmutchkievich, purged by Stalin in 1940). The Soviet aircraft were Polikarpov I-15 (and eventually I-152) biplanes, I-16 monoplanes, Tupolev SB bombers, and Polikarpov R-5 reconnaissance and Shturmovik aircraft. They arrived in time for service during the latter stages of the first Nationalist assault on Madrid. For its part, the Roosevelt administration steered—however unwelcome—a strictly neutralist policy, though some American aircraft did arrive in Spain for service with the Loyalists, via individual acts of subterfuge.[30]

Thus at the outset the Nationalists had difficulty securing air superiority, even before Soviet support arrived. Once that support did reach Spain, Soviet and foreign pilots flying the I-15 had no difficulty establishing supremacy over the He 51, to the discomfiture of Legion Condor and Spanish pilots. The Fiat C.R. 32 proved a match for the I-15, but not for the I-16, and it was not until the Messerschmitt Bf 109 arrived in Spain in early 1937 that the Nationalists could assert air superiority over the Loyalists, though even the early "109" had difficulty with the by-now mature I-16. In bombers it was a different story: by the end of the northern (Bilbao) offensive in 1937, the Nationalists had a clear supremacy in bombers. After the Soviets effectively gave up support in 1938, command of the air unquestionably passed to the Nationalists for good.[31]

As with almost any war, the first attempts at employing air power were halting and relatively immature, but still, on the whole, quite effective. On August 16, 1936, an initial landing force of 2500 Loyalists, subsequently augmented by 10,000 more later that day, made an amphibious landing near Porto Crispo, on the Nationalist-held island of Majorca. Indecisiveness caused this force to linger on the beachhead and in the town of Porto Crispo, and allowed the defenders time so that some Italian fighters and three bombers could begin concentrated air attacks against the invaders. Largely by air assault alone, the defenders were able to pin down the invaders until troop reenforcements arrived from Africa. The Italian fighters prevented Loyalist bombers from interfering over the battle area, and undertook their own strafing and bombing attacks. The invading force, estimated at the time as only 8000 troops (less than two-thirds its true size) was devastated and forced to abandon the island. As the American attaché in Paris subsequently reported, the force, ". . . well supported and munitioned, after an initial success, suffered disaster due to Italian bombing. For two days aviation harassed the landing force, inflicting heavy losses and finally forced a precipitate withdrawal."[32] Here, then, was an interesting case of battlefield air support operating in the defensive during an amphibious operation, and succeeding magnificently. Majorca remained firmly in the Franco camp for the rest of the war.

Initially, the revolt had split Spain in two, with the Loyalists in firm control of the eastern portion of Spain along the Mediterranean, and the northern portion on the Biscay coast. Very quickly the Nationalists, having established a strong foothold in the south of Spain, drove northeastward toward Madrid. Because of the professionalism of their troops, the Nationalists were able to make good progress on their drive, but eventually stiffened resistance, assisted by the presence of the International Brigades and the intervention of Soviet air and ground forces stopped Franco's forces on the edge of Madrid, leading to a war of stalemate marked by sporadic attempts to restore a war of movement that lasted until Madrid fell in 1939. During the drive, from September into November, Nationalist attack aircraft and bombers were active in attempting to break down the resistance of Loyalist ground forces. Early operations were relatively primitive; the Junkers Ju 52/3m bomber-transports operated by both the Legion Condor and the Nationalist air service lacked sophisticated bomb racks, and crewmen actually would open a hatch in the belly of these trimotors to drop 10- and 50-kg bombs by hand! Subsequently, when racks were fitted, things became a bit more professional, and bombs up to 250 kg could be carried. By late 1936, as José Larios, a Ju bombardier soon to become a leading Nationalist fighter pilot recalled, "The enemy was now in full retreat toward Madrid. We were bombing and strafing convoys and road junctions intensely, and we ran into heavy flak most days."[33] Eventually, vulnerability of the Ju to flak and, especially, the influx of Soviet fighters severely curtailed its usefulness. For the time being, however, accompanied by other Nationalist aircraft, the Ju was a key element in Nationalist air support.

The intervention of Soviet air and land forces had a decisive impact on this first drive upon Madrid; in early October, the Soviets launched a major tank drive against Nationalist cavalry, operating their tanks according to the principles outlined by Germany's Heinz Guderian, whose writings had, ironically, greatly influenced soon-to-be-purged Soviet Marshal Mikhail Tukhachevsky. In response, Germany stepped up its assistance to Franco, and in early November the Legion Condor, now a full-fledged military expeditionary force, took camp at Seville to support the fighting around Madrid. The Legion consisted of both air and ground elements, including fighters, bombers, reconnaissance, maritime patrol, antiaircraft, antitank, and armored forces; the ground component of the Legion generally did not receive the attention that was its due, but many of the lessons it learned—such as the efficacy of light and heavy flak cannon against troops and tanks—would be replayed with spectacular results in 1939–41. Interestingly, the Legion had not yet employed a genuine ground-attack aircraft, such as the Henschel Hs 123 biplane or the Junkers Ju 87 Stuka; that would come later, in the following year. Until that time, particularly after the arrival of the Bf 109 for air-to-air operations, the Legion made do with the Heinkel He 51 for air-to-ground attack. Italy's l'aviazione legionaria did have an attack element virtually from the outset, namely the XXII Gruppo consisting of two squadrons of Ro 37bis recce and assault aircraft, and later added another Gruppo with two squadrons of B.R. 20 bombers and a squadron of Breda 65 attack aircraft used in the battlefield assault role, as well as using the legionaria's Fiat fighters in an air-to-ground attack role on occasion.[34]

Perhaps not unexpectedly, given their early interest in such operations, the Soviet forces in Spain quickly began using their air element in a ground-attack role. At first, even the twin-engine Tupolev SB flew low-level battlefield attack sorties, assaulting Nationalist trench positions with bombs and machine-gun fire, but after several were lost, the Russians operated them in a more conventional mid- to high-level bombing role; the vulnerability of the large twin-engine attack airplane during low-level operations was a lesson that would itself be relearned by virtually all the powers in the Second World War to come. Polikarpov's armored two-place R-Z attack biplane (a variant of the R-5 recce aircraft dubbed the Rasante) proved more suitable, armed with eight machine guns and carrying small antipersonnel bombs, typically twenty 10-kg bombs. Flights of three in V formation would approach a target at altitudes of less than 100 feet, pop up to 500 feet over the target, roll-in and drop the bombs, and exit the target area at low level while the rear gunners suppressed enemy fire. Strafing operations were undertaken on a "lo-lo-lo" basis, maintaining low altitude throughout the mission. Accompanying fighters—I-15's and I-16's—would prevent enemy aircraft from "bouncing" the attack planes, and, if no enemy were present, would themselves join the air-to-ground attack, particularly acting to suppress enemy air defenses. Significantly, after studying the German, Italian, and Russian "volunteer" forces flying in Spain, the American attaché in Valencia wrote early in 1937 that "the Russians are the only ones to have developed any special attack aviation as known in our Air Corps."[35] Indeed, an examination of Soviet operations in Spain indicates that, when the Spanish war broke out and for at least its first year, the Soviets appear to have had a much stronger desire to prosecute low-level close air support and battlefield interdiction-type missions, and that this reflected a much more mature and doctrinally more secure interpretation of the battlefield support role that aircraft could play. Italy was next, followed by the Legion Condor. The Germans appear to have first become interested in low-level high-speed ground attack only as an expedient following the discovery that the He 51 was unsuitable for air-to-air combat. Only later did the potentialities for such attack become so apparent as to warrant increased emphasis. What is unclear is the degree to which the Germans were influenced by the Italians, by the hard experience of observing Russian Shturmovik operations (particularly at Guadalajara), and by recollection of Germany's legacy of *Schlachtflieger* operations in the Great War. In any case, as will be discussed subsequently, the German experience in Spain did *not* immediately result in a radical reshaping of the Luftwaffe to emphasize CAS and BAI-type missions. The same, however, was not true for the Soviet air force, which placed even greater emphasis upon Shturmovik development.[36]

During fighting around Madrid, the Nationalists increasingly turned their attention to supporting attacking and defending forces by air action. In late November, for example, concentrated Nationalist air attack helped defend the key supply center of Talavera against a strong Loyalist attack. As Larios subsequently recalled:

> At daybreak our bomber and fighter squadrons were in the air hammering away at the enemy positions. The Savoias and Romeos [Ro 37's] struck first. We flew in next and thoroughly plastered the heights from low level. We made six runs. I managed to place a 250-kilo bomb exactly in the center of a group of farm buildings sheltering one of the enemy batteries which was making a thorough nuisance of itself. After dropping all our bombs, we flew down to ground level and harrassed the enemy transports with our machine guns. Our fighter squadrons did an excellent job strafing and dropping small 2½-kilo bombs. A continual relay was kept up from dawn to dusk with such good results that the Reds became thoroughly demoralized and fled, leaving their artillery behind.[37]

There was a profound effect upon morale by these attacks, though as troops became more experienced the morale impact declined. Nevertheless, at first it was quite severe, and one attaché report noted:

> A member of the International Brigade who had seen considerable service on the Madrid front said that frequently their lines were straffed [*sic*] by enemy pursuit immediately prior to an infantry attack. . . . The morale effect was unbelievably severe and even though many casualties resulted the worst effect was the lowering of the will to resist. He added that at such a time the only way to again raise the spirits of the men was to have friendly aircraft come into action. The French [volunteer pilots] have told me that on many occasions they have received hurried requests to appear over such-and-such sector for no other reason than to reassure the ground forces. The morale effect of the presence of aircraft either friendly or otherwise cannot be denied.[38]

Even at the battle of Jarama in February 1937, when the Loyalists held unquestioned air superiority and were, in fact, able to prevent Nationalist bombers from intervening over the battlefield, what Nationalist attacks that did occur discomforted Loyalist troops, particularly those of the inexperienced Abraham Lincoln Battalion, whose members feared air assault more than artillery fire, and who missed many meals because air attack disrupted supply columns. Because Jarama ended in a stalemate, both sides claimed victory; in fact, it drained much of the force of the Loyalists—and the Loyalist cause could not afford draining stalemates.[39]

The Battle of Guadalajara

Exactly what well-coordinated air action could accomplish over a battlefield was graphically demonstrated in March 1937 during the battle of

Guadalajara, when Loyalist air attack devastated an Italian armored and mechanized column threatening Madrid from the north; like Majorca, it was one of the most significant air-to-ground operations of the entire civil war, and, arguably, one of the most significant of all time.[40]

The Nationalist assault on Guadalajara (a provincial capital northeast of Madrid) consisted of 50,000 troops divided between Spanish and Italian leadership. The more formidable were 30,000 Italian troops in four divisions (the Black Shirts, Black Flames, Black Arrows, and Littorio Division) on the left of the advance, commanded by Gen. Mario Roatta. Despite the formidable names, these were units deficient in training and expertise, and with some serious morale problems of their own. The force included 250 tanks and 180 mobile artillery pieces, had an average of seventy trucks per battalion, and could call upon fifty fighters and twelve recce aircraft. In fact, weather curtailed Nationalist air sorties, and when the weather lifted and they could appear, the evident air superiority of the Loyalists kept Nationalist airmen busy fighting for their lives. The Loyalists were concentrated on airfields in and around the battle area, particularly at Alcalá de Henares, Albacete, Azuqueca, and Barajas, and the evident good drainage of these fields—allegedly hard-surfaced—gave the Loyalists an advantage over the Nationalists, who, being farther away to the north, had to operate from dirt fields that flooded and were located in high terrain frequently swathed in fog and low cloud.

A chronology of the two-week battle emphasizing the bitter fighting between Italian and Loyalist forces (which was of greater significance than the activities of the Spanish Nationalist forces versus the Loyalists) gives some idea of the role of air attack; (the aircraft totals are from intelligence reports and undoubtedly reflect in many cases multiple sorties by the same planes):

March 8: Offensive opens with an early morning assault by the combined Italo-Spanish force. Black Flames break through and begin advancing toward Brihuega. Bad weather conditions effectively limit both Nationalist and Loyalist air operations; eight SB's attack advancing Italian forces without significant result. Column advances nine miles from line of departure.

March 9: Black Flames and Black Arrows continue advance; weather limits air operations, no air strikes against column. Nationalists ineffectively attack airfield at Alcalá de Henares. Loyalist reenforcements establish defensive positions along Brihuega-Torija road.

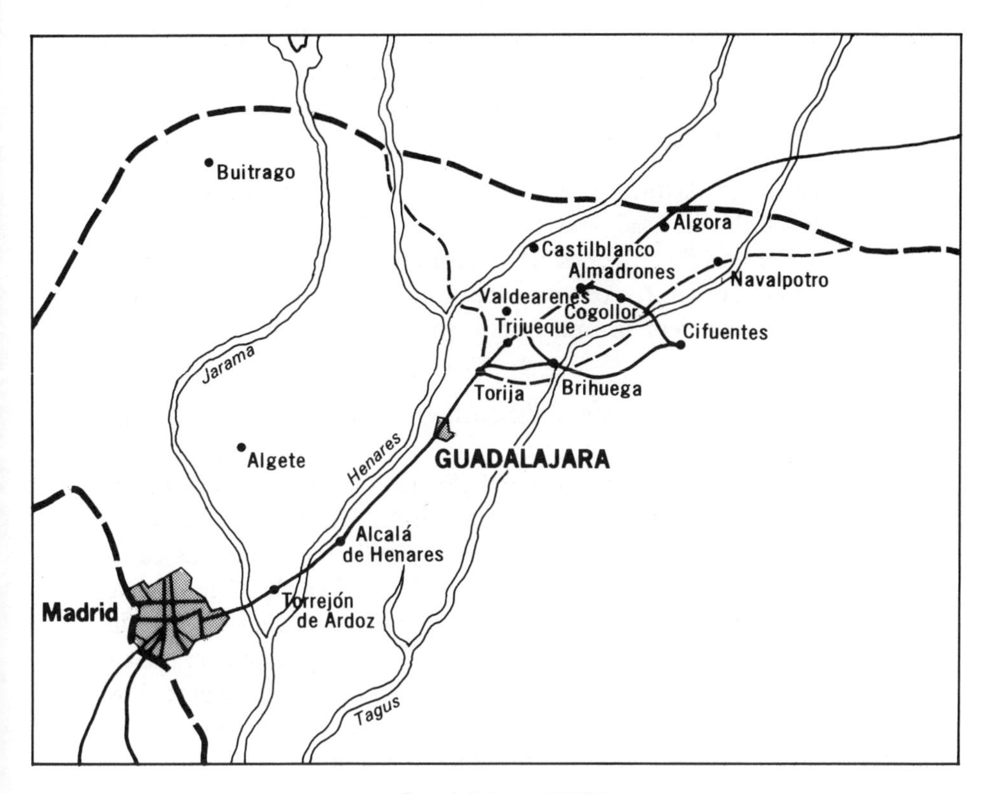

Guadalajara, 1937

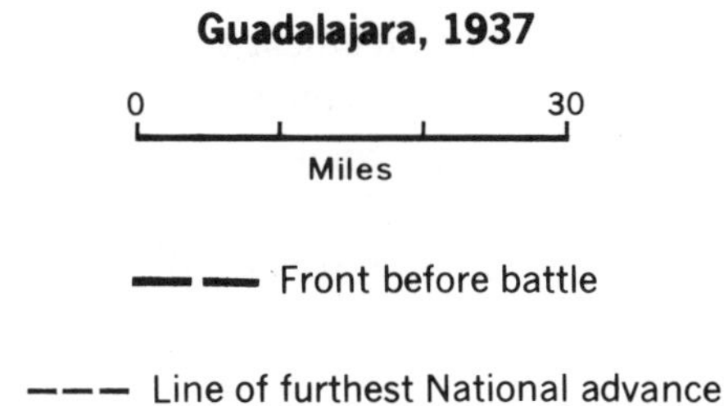

March 10: Black Flames and Black Arrows seize Brihuega. Littorio Division follows as a reserve. Later in day Black Flames clash with Loyalist's Garibaldi Battalion. Weather worsens, preventing aerial intervention by either Nationalists or Loyalists.

March 11: Black Arrows seize Trijueque. Loyalist forces continue to hold Torija-Brihuega road, and hold Trijueque-Torija road. I-15's from Alcalá de Henares fly an armed reconnaissance sortie amid rain squalls and low cloud, and make a one-pass attack on Italian forces, ascertaining that they are deficient in antiaircraft and hampered by flooded and boggy terrain.

March 12: Weather improves sufficiently to permit Loyalist operations against Italians, but not Nationalist air intervention. Intense air attacks against Italian columns by 140 aircraft; fighters strafe the Black Shirts, killing regimental commander Alberto Liuzzi. Thirty R-Z attack biplanes strafe and bomb the Black Arrows, inflicting numerous casualties, including Roatta's chief of staff, Col. Emilio Faldella. The strafing I-15's are joined by I-16's not needed for counter-air operations. Total of 492 bombs dropped and 200,000 rounds fired. Immediately after the air strike, Soviet Gen. Dimitri Pavlov launches a tank attack supported by infantry that smashes Italian positions. Trijueque retaken by Loyalists, and Garibaldi Battalion finally defeats Black Flames.

March 13: Roatta attempts reenforcement of front by Black Shirts and Littorio Division, but they are beaten off by Loyalists. Weather lifts and Nationalist aircraft intervene over the battlefield, but are driven off by Loyalist fighters. Loyalist air attacks continue, including a strike by twenty-eight R-Z's against Italian forces on the Almadrones-Brihuega road.

March 14: Pavlov pushes forward, but lack of infantry limits his advance. Italians attempt disengagement to regroup. Continued Loyalist air attack, but Nationalist aircraft force Loyalist fighters to remain in primarily an air superiority role.

March 15: Loyalists take Valdearenes. Nationalist and Loyalists grapple for air superiority inconclusively. Loyalist attacks against retreating Italian columns continue.

March 16: Twenty-five Tupolev SB bombers attack Italian positions in and around Brihuega, dropping 770 bombs ranging from 50 kg to 250 kg. Thirty I-15's undertake a low level attack with 120 bombs and machine-gun fire. Other I-15's prevent Nationalist fighters from intercepting the SB's and I-15's. Two bombing attacks are undertaken against the Nationalist center of Sigüenza to the north.

March 17: Weather prevents Loyalist and Nationalist air support activities. Lull in ground fighting.

March 18: Loyalists go on the offensive on the Guadalajara front. Three air strikes commence at 1:30 p.m., consisting of level and low attack bombers, and strafing fighters. At 2:00 p.m., two Loyalist divisions supported by seventy tanks attack around Brihuega, attempting to encircle trapped Italians. At 8:00 p.m., the Italian forces are saved from total annihilation by a timely evacuation order.

March 19: Deteriorating weather; Loyalists undertake air reconnaissance only.

March 20: Intensive Loyalist air attack lasting 2 hours against retreating Italian columns at Almadrones, Algora, and Navalpotro by 125 aircraft consisting of 75 R-Z attack aircraft and SB bombers and 50 I-15 and I-16 fighters. I-16's drive off Nationalist aircraft attempting to intervene. Twelve SB's raid Sigüenza rail station. Italian losses judged by attachés to be "tremendous."

March 21: Deteriorating weather ends all air activity. Italian forces continue to retreat and regroup; Loyalists lack strength to exploit pursuit further; battle ends. Italian losses are approximately 500 killed, 2000 wounded, and 500 POW, plus extensive losses of equipment. (One estimate placed the losses at 1000 vehicles, 100 machine guns, 25 field pieces, and "large quantities" of ammunition.) Loyalist losses are about the same, but with few taken POW, and negligible equipment losses.

The battle of Guadalajara was the only actual battle that the Loyalists clearly won in the entire civil war. Legion Condor communications personnel attached to the Italian columns were appalled at the rapid collapse of the Italian forces. From the outset the Nationalists had, in effect, planned their own defeat. They took no account of the impact weather would have on their operations; weather, of course, prevented their air service from participating to any meaningful degree in the battle, and it also restricted the Italian forces to operating on the road system, since the surrounding terrain was too soft to permit operation of mechanized forces. Further, many of the roads passed through narrow defiles that set up perfect killing grounds for the attack aircraft, as at

Wadi el Far'a. They had no provisions for antiaircraft, and though some Loyalist pilots and planes were lost to machine-gun fire, this reflected bad luck rather than predictable hazard. The linkage of Schmutchkievich's air actions at Guadalajara with Pavlov's tank assaults were examples of a Soviet awareness of air-land synergy that would become far more commonplace during fighting on the Russian Front during the Second World War.

Nevertheless, the Loyalists made their share of errors, primarily in the opening stages of the battle. Lack of effective air reconnaissance prior to the Italian assault had resulted in the assault catching defenders unaware. Fortunately, deteriorating weather helped offset the attacker's advantage. Subsequently, when the ground situation had stabilized and the defenders swung over to the attack, poor coordination between tanks and infantry prevented Pavlov from taking fullest advantage of the shock he had delivered to Italian positions, and possibly establishing a breakthrough as far as Sigüenza. Eventually, the weather acted to save the Italians from further destruction from the air. But no matter: the Loyalist defense at Guadalajara had saved Madrid. The Loyalists may have taught the lesson at Guadalajara, but the Nationalist students proved ultimately more able than their instructors.

The Maturation of Spanish Air Support: Bilbao, Brunete, Teruel, and the Ebro

That the Nationalists quickly surpassed the Loyalists in the effectiveness and organization of ground attack was largely due to the activities of the Legion Condor and the l'aviazione legionaria, particularly the former. The Legion Condor had an extremely capable chief of staff, Oberst Wolfram von Richthofen, a cousin of the famed "Red Baron" of First World War fame. Von Richthofen subsequently had as great an influence on the Luftwaffe's ground-attack forces as his fighter pilot cousin had on the Luftstreitkräfte's operations in the Great War. He believed strongly in using air attack closely coordinated with infantry and artillery action, and put his notions to good use in the Nationalist's northern offensive in March to September 1937. For its part, the Italian air service in Spain emphasized air-to-ground operations at least as much as it had air-to-air operations. Then, some of the better Spanish Nationalist pilots made their own innovations that became accepted tactical doctrine.

The vulnerability of the Heinkel He 51 to predatory I-15's and I-16's had caused the Germans and the Spanish who used the type to rely upon it for air-to-ground attack rather than air superiority sorties, in much the same way that the Royal Flying Corps had used the De Havilland D.H. 5 in World War I, or that the Bell P-39 Airacobra would be used in the Second World War. At the beginning of 1937, the fighter component of the German legion, Jagdgruppe 88, consisted of four squadrons equipped with nine to twelve He 51's apiece. The third, 3/J/88, functioned exclusively as a ground-attack unit, becoming the prototype of the reborn *schlacht* squadrons and wings that the Luftwaffe would begin to establish, starting in May 1938. During operations against the so-called "Iron Ring" of Bilbao, the attack He 51's undertook bombing and strafing as close as 150 feet in front of advancing troops, flying as many as seven missions per day. Typical attack tactics consisted of flying in three-plane V formations and dropping bombs from 500 feet. These missions were of a pre-briefed nature, as the planes lacked radio. Pilots also undertook armed recce sorties over the front looking for suitable targets of opportunity. A six-hour bombardment by seventy bombers and forty ground-attack fighters "ruptured" the Iron Ring at Bilbao, shattering the morale of its defenders.

When the Loyalists launched the ill-fated Brunete offensive in July 1937 in an attempt to prevent disaster in the north, they undertook an intensive three-hour air attack upon Nationalist positions, virtually destroying the Nationalists' first-line defenses. Loyalist fighters and bombers attacked artillery, broke up attempted Nationalist counterattacks, attacked truck columns attempting to reach the front, and supplemented deficiencies in Loyalist artillery strength. For its part, the Nationalist response was equally devastating. The He 51's suppressed enemy air defenses so that lumbering Ju 52 trimotors could get through and bomb enemy ground forces. The Legion Condor, now operating Bf 109's as well, used its Messerschmitts to grapple with the I-15's and I-16's, shooting down no less than twenty-one aircraft in a single day. Flak was intensive, and the low-flying He 51's were often hit, with several being lost. The Heinkels dropped a mix of small fragmentation bombs and fused external fuel drop tanks (anticipating the development of napalm during the Second World War) and followed up bombing attacks with ground strafing. Though well-dug-in troops had little to fear, troops out in the open were a different matter entirely. (Earlier, in one notable case during the northern campaign, Legion Condor aircraft intensely attacked Loyalist positions on Mount Monchetegui near

Ochandiano, dropping sixty tons of high-explosive and incendiary bombs over the heavily wooded area in two minutes, following up with another twenty tons of bombs. The Loyalist defenders fled in disarray, helping open the path to Bilbao.) For their part, the Italians flew no less than 2700 hours of tactical sorties during the Brunete battle, mostly on close air support and battlefield air interdiction missions. The bitter battle of Brunete ended in a substantial Loyalist loss, for even though the ground battle had seesawed back and forth (proving again that the Italian ground forces were not as competent as their aerial comrades), it had been so draining and costly to the Loyalists so as not to constitute even a Pyrrhic victory. Ahead lay Teruel and even greater disappointment for the Loyalist cause.[41]

Franco's victory in the northern campaign enabled the Nationalists to concentrate their entire effort on the remaining Republican territory bordering the Mediterranean, and marked the beginning of a war of attrition that lasted from mid-1937 through the fall of 1938. During this time three major offensives occurred: Teruel, the Aragon offensive, and the Ebro. Teruel, the "Gettysburg" of the Spanish Civil War, began in December 1937 when the Loyalists, amid foul and bitter winter weather, launched an offensive against the town of Teruel, at the head of a Nationalist salient within Republican territory. They quickly surrounded it, and after much bitter fighting in weather too bad for Nationalist air power to intervene, its defenders surrendered in January 1938 (many were subsequently murdered by their captors). The retaking of Teruel took another month, until early February. Teruel was another costly, draining loss for the Republicans, who could ill-afford the kind of troop-wasting endurance warfare that they pursued. From this point onward, the Republic was on the skids, its leaders—still bickering with one another—hoping desperately for foreign intervention that could not possibly come.[42]

Teruel was noteworthy for its application of Nationalist air power over the battlefield, and presaged the kind of intensive air operations that would be seen in the Ebro fighting before the end of the year. German air doctrine of the 1930s held that the dive bomber was an element of strategic air warfare, for deep attacks requiring precision strike. In late 1936, the Legion Condor's von Richthofen arranged for the deployment of a small evaluation batch of Henschel Hs 123 biplane dive bombers, but his interest was more in employing these aircraft as close support attack aircraft using low-level bombing and strafing tactics. The Hs 123's proved extraordinarily useful in this role, so much so that the

Luftwaffe, largely as a result of von Richthofen's urging, subsequently employed the Hs 123 virtually exclusively as a *Schlachtflugzeug* rather than a dive bomber (Stuka). A year after the Hs 123's arrived in Spain, von Richthofen arranged for the importation of a test and evaluation *kette* (three aircraft) of Junkers' new Ju 87 Stuka, and these arrived in time for employment during the Teruel fighting, where they were employed as dive bombers for precise attacks on troops and strong points, earning the nickname "sharpshooter."

Again, ground-attack control procedures were rudimentary, consisting (if at all) in pre-briefings and occasional ground panels. Not unexpectedly, under such conditions aircraft sometimes struck friendly positions. During fighting at Alfambra, which climaxed the struggle for Teruel, German airmen accidently attacked a Spanish division, killing or wounding 400 men, over ten times the total the division had lost in ground fighting. Operations were conducted at an extremely low level; Spanish Nationalist pilot José Larios repeatedly notes that it was not at all uncommon for attack aircraft to return punctured from the shrapnel of their own bombs, and, indeed, the Legion Condor lost a Dornier 17 peppered by its own bomb blast. Likewise, operations were undertaken at an intense pace; on January 19, for example, 400 Nationalist aircraft flew shuttle sorties all day between their bases and the front, dropping 100 tons of bombs. During the Alfambra fighting, the Nationalists decisively demonstrated how the air war had swung in their favor; Bf 109's prevented Loyalist aircraft from intervening over the battlefield, while He 51's and the three Ju 87's worked together. During fighting in Teruel itself, the He 51's strafed at low level while the Ju 87's, armed with 250-kg bombs, attacked fortifications. Losses were high, particularly among He 51 units (Larios has referred to strike flights being "smothered" in flak), for the Loyalists were increasingly armed with light cannon such as 20mm and 37mm or 40mm weapons, as well as machine guns. Additionally, Loyalist troops increasingly displayed good discipline while under air attack, taking cover and firing back at the attackers with small arms.

Losses likely would have been higher during Nationalist air-to-ground attacks as undertaken at Teruel had it not been for a form of air attack dubbed the *Cadena* ("chain"). Although it is difficult to say so precisely, the chain attack is credited to Joaquin Garcia Morato, a Nationalist Fiat pilot, who allegedly devised it using Fiats armed with 2.5-kg bombs, although it is more commonly associated with the German and Spanish Heinkel He 51 formations. A chain attack would be undertaken by a squadron of aircraft against a ground target, usually as part of a combined air-ground assault. Typically, as Nationalist troops and armored forces were poised for assault, a formation of level bombers would attack Loyalist trench and artillery positions. As the bombers pulled off target, a formation of Heinkels would begin a diving attack, armed typically with six 10-kg fragmentation bombs, and following the leader in line-astern formation. As the first Heinkel attacked, the second was on the verge of attack, and the third was already beginning its own run. As the leader pulled off target, the defenders had little opportunity to shoot at him during his vulnerable climb back to altitude, as the second, then the third, and then the remaining aircraft in turn were already attacking their position. Eventually, the attack would assume that of an inclined circle, with aircraft spaced evenly around the circle, some pulling off the target, some attacking it, and some preparing to attack. When the chain concluded its attack, the infantry and armored forces would immediately begin their own, usually covered by a small force of attack airplanes to strike at any positions offering resistance to the ground forces. Chain units had a high degree of esprit, and had their own distinctive insignia; the chain attack—which had appeared haphazardly on the Western Front during the First World War, though that previous history seems not to have influenced its "discovery" in Spain—first appeared during the northern offensive in 1937 and quickly became a standard element of Nationalist air attack practice.

After the Nationalist success at Teruel, the Franco forces began a general Aragon offensive in which ground-attack aircraft cooperated most effectively with advancing ground forces; one letter from an American attached to the Abraham Lincoln Brigade stated, "The sky is black with their planes bombing and strafing every minute"; another Lincoln veteran recalled that the Loyalists railed against their own air service: " 'Where are our _____ planes?' we cursed. 'Why the hell aren't they around when we need them?' "[43] It was at the Ebro, however, that the full fury of the Nationalists' mature air support aviation fell upon the Loyalists.

The Ebro has gone down in some Loyalist romantic lore as a victory; in fact, it was the penultimate Nationalist triumph, for after the Ebro, all that remained was the brief Catalonian campaign that resulted in the collapse of the Loyalist cause. The battle of the Ebro, which lasted from late July through mid-November 1938, resembles in some respects the Battle of the Bulge in December 1944. In both cases a foe on the defensive and on the verge of collapse mounted a brief and surprisingly successful offensive, an offensive ultimately stopped amid bitter fighting, and then reversed by sharp counterattack accompanied by extensive air-to-ground employment of air power. The fighting along the Ebro took place at the height of summer, and at first, thanks to good planning, not a little luck, and a nighttime assault across the Ebro River, the Loyalists made good progress. In early August their attack was blunted, and the Nationalists now turned to reducing the Loyalist bulge in their lines by concentrated ground action coupled with intensive bombing by horizontal bombers, dive bombers, and low-level attack aircraft. The actual Nationalist counterattack came at the end of October. Two weeks later, pushed back across the Ebro, the Loyalist army had virtually ceased to exist; 75 percent of the International Brigades that had entered combat had been lost, together with prodigious quantities of arms and supplies.[44]

During the initial stages of the battle, when the Loyalists were on the offensive, German, Italian, and Spanish air elements had to substitute for deficiencies in artillery, or for artillery that had been overrun and lost. Later, of course, when the Nationalists went on the offensive, the air elements worked as part of a well-coordinated, combined-arms (air-artillery-armor-infantry) offensive. Loyalist General Enrique Lister conceded that the Nationalist air attacks were "terrifying and deadly," as have the memoirs of surviving members of the International Brigades.[45] In the period from August 15 to August 19, for example, more than 200 Nationalist aircraft constantly bombed and strafed Loyalist positions, dropping an average of *10,000* bombs per day. Former Loyalist officer Ferdinand Miksche noted that the Ebro fighting represented "the highest point of co-operation of air and land forces," and bomb tonnage closely matched artillery tonnage, in some cases exceeding it. For example, the aggregate of air and artillery bomb and shell tonnage dropped/fired during six actions from early August through early October revealed (according to Loyalist General Staff intelligence estimates) that 9000 tons of artillery rounds had been fired and 8127 tons of aircraft bombs had been dropped. In sum, then, air was more than a supplement to artillery—it was clearly a full-fledged partner: something that it had not been before.[46] At no time was this more evident than in the first few days of the Loyalist offensive, when Nationalist air power had to secure a decisive advantage over a rapidly unfolding and rolling offensive. Larios has given an indication of the pace of these operations:

> A continuous shuttle service going back and forth to the Ebro was kept up by forty Savoia 79's, thirty He lll's, eight Dorniers [Do 17's], thirty Junkers 52's,

nine Bredas [Ba 65], twenty Savoia 81's, seven *Cadena* squadrons, and Morato's [Fiat C.R. 32] fighter group, hammering away relentlessly at the enemy. We were hard at work several days giving top cover to the Junkers and *Cadenas* over the Fayon sector, where the Reds were expected to drive a wedge.

When our bombers and fighter bombers had completed their missions, we would penetrate deep into Red territory to keep the Red wings off our lines. The sky over the Ebro would become dark with bursting shells, jolting and peppering our machines as we swept back and forth.[47]

With the conclusion of the Ebro battle, it was only a matter of time before the war ended. Resistance continued into 1939, and during the Catalonian campaign the Nationalist forces continued their rampage toward victory. By this time the Soviets were out of Spain, thanks to changes in Stalin's policies that would eventually spawn the infamous nonaggression pact on the eve of the Second World War. Indeed, by this time many of the Soviets who had played so decisive a role in enabling the Loyalist cause to survive as long as it had were already dead, victims of Stalin's violent xenophobic purges. By the evening of March 31, 1939, the Nationalist victory was complete, and the Spanish war came to an end, leaving a legacy of bitterness and mythic distortion not completely eradicated to this day. But Europe would only have a brief respite before war would break out anew, a war with infinitely greater violence and consequences than the savage fratricidal conflict that had ravaged the Iberian peninsula between 1936 and 1939.

9

The Spanish Legacy

Lessons Read and Misread

Writing just after the outbreak of the Second World War, Spanish war veteran Ferdinand Miksche perceptively stated:

> The air force has become the hammer of modern warfare on land. Employed in close co-ordination with tanks, motorized infantry, and other ground forces it can bring battle to swift development. The great mobility of aircraft enables the attacker, because he holds the initiative, to seize almost at a flash the mastery of the air over the battlefield. Aviation gives modern land battle a third dimension: height. Forces no longer fight for surfaces, limited to length and breadth; modern battle is the fight for cubic space.[48]

Miksche generously illustrated his argument with examples from the Spanish Civil War, which he had experienced first-hand. But Miksche was an unusually perceptive commentator, and his views of war were certainly not shared by all military experts, particularly since the confusion often inherent in the Spanish Civil War generated a number of lessons misunderstood. For example, in the wake of the Spanish Civil War, almost as much confusion existed over the role of aviation and armored forces in battle as had existed before it began. Tom Wintringham, another popular military commentator who had served with the International Brigades, wrote:

> By 1938 . . . air power had become capable of taking over a large share of the functions of artillery. In particular it would take over the function of supporting tanks and motorized infantry. As a flying artillery, planes could do what guns used to do in Napoleon's hands: blast open a breach in the enemy's position for

decisive manoeuvre. From 1938 it was clear that the dive-bomber was the best type of plane for this decisive function.[49]

Yet French Gen. Maurice Duval, interpreting "lessons" learned in Spain, saw no such decisive role for air power; to him, air attack was at best only a supplement to artillery. Writing of the Cadena form of attack, he stated that

> it is not in doubt whether the cannon is the more superior to the airplane for the execution of attack preparation. It places its projectives on the objective with an exactitude unknown to "the chain," and at much less risk. "The chain" cannot substitute for nor justify an insufficiency of artillery.[50]

It is this aspect of the Spanish Civil War that is most interesting: intelligent, well-informed critics could reach often diametrically opposite conclusions, or, interpreting data correctly, nevertheless reach a flawed conclusion. Wintringham, for example, was largely correct in his analysis of air power by 1938. But the dive bomber, except under very special circumstances—namely under the mantle of protective air cover or under conditions where enemy air opposition was not a factor—proved far from ideal for battlefield support. For his part, Duval was essentially correct in recognizing that air power, on its own, could not win battles. But in combination with other arms—infantry, artillery, and armor—it generated a tremendous synergy for both offensive and defensive warfare (though, as Miksche pointed out, it had particular value for the attacker). It might be said that critics took away from the Spanish war what they wished to believe, and they searched its lessons carefully to selectively acquire supporting data for their own particular viewpoint. Certainly, for example, the vulnerability of the unescorted bomber was a lesson that all the European nations—and the United States as well—should have clearly learned. Oddly, Italy alone among the Spanish combatants devoted negligible attention to the war, paying for this folly in the war to come. France, Germany, and Russia paid greater attention, looking for clues to the future.

France

France seriously misread some of the lessons of the Spanish Civil War, notably the value of integrated air-land operations in maneuver warfare. Then, there were political problems within France that mitigated against the country developing a realistic defense strategy. Pierre Cot,

Leon Blum's aviation minister (and a man fond of quoting Robispierre, arguably one of the fathers of modern terrorism), recognized the value of attack aircraft, particularly in an antiarmor role, and sought the advice of Soviet military aviation experts to create the first French parachute airborne assault and ground-attack units. He wanted a balanced air force in which attack aircraft would have represented 16 percent of the total combat force, but the collapse of the Front Populaire government brought this plan to an end. In any case, as historian Robert Allan Doughty has shown, the seeds of French defeat in 1940 were rooted deeply in the soil of doctrinal malaise. Born under a dubious premise—that the modern battlefield was so lethal as to give a tremendous advantage to the defender—French doctrine emphasized methodical battle and firepower dominance, and looked to new weapons such as the airplane and the tank merely to help achieve these ends. It failed to recognize the tremendous advantages of mobility, mass, and maneuver that these weapons gave the attacker. General Duval's "lessons" from Spain supported these views probably as a result of unconscious selection, emphasizing the supposed vulnerability of tanks and the inadequacies of aircraft. While specific lessons may have had greater or lesser validity, their incorporation into an already flawed framework was what generated a prescription for disaster. Without a comprehensive doctrine perceptively noting the changes in warfare brought about by new technology and integrating air elements into a genuine combined-arms warfighting scheme—as the Germans and Russians, for example, were increasingly refining—France simply could not win, and Cot's plan for buying attack aircraft could not serve as an adequate substitute for this lack of doctrine. Rather, it was mere purchasing of technology.[51]

Germany

Writing after the defeat of Germany, former Luftwaffe General Karl Drum stated:

> From the point of view of the "blitzkrieg," it is clear that the concept of a special close-support force, with its organizational and operational ramifications, was one of the most important ever evolved by the German *Luftwaffe*. . . .
>
> It was *Generalmajor Freiherr* [Wolfram] von Richthofen who was responsible for systematically gathering and evaluating experience in close-support missions in Spain and for seeing that it was utilized by *Luftwaffe* leaders in Germany. . . .

Of all the experience gained by the Condor Legion in Spain, it was that pertaining to the methods of tactical air employment which was most significant and most far-reaching in its effects.[52]

Undoubtedly, this is accurate. But as Drum points out, Germany's development of ground-attack awareness—essentially a blending of the collective memory of the Schlachtflieger of the Great War with the force of circumstances in Spain (such as the unsuitability of the He 51 for air-to-air combat) did not come without some price being paid. The creation of special ground-attack organizations came at the expense of the fighter corps, which gave up fighter pilots for training as attack pilots instead. The development of the dive bomber and its employment in large numbers (336 by the time of the invasion of Poland) drew away pilots from the bomber corps, and resulted in a special-purpose airplane not suitable for other forms of bombing. Drum provocatively suggests that the resources put into the development of attack and dive-bombing aircraft and forces might have been better spent (in light of the Luftwaffe's failures in 1939–45) in generating more fighters and bombers and fighter and bomber units. Such, he believes, might have made a critical difference during the Battle of Britain (1940), whereas the availability of attack and Stuka forces really was not critical for the success of German forces in 1939–40 on the battlefields of Europe.

When confronted with high losses among the He 51's in Spain from the increasingly effective Loyalist antiaircraft fire, the lesson Germany drew for the future was flawed in two regards: analysts emphasized developing armored close support aircraft at the expense of speed and agility, and they overemphasized the effectiveness of antiaircraft fire against bombers. Both errors mitigated against developing swing-role fighters that could undertake dual air-to-air missions such as bomber interception and air-to-ground attack. So it was that the Luftwaffe placed too great an emphasis on developing bombers capable of undertaking dive bombing (such as the Junkers Ju 88 and even the four-engine Heinkel He 177) and on developing heavily armored ground-attack aircraft (such as the Henschel Hs 129) that could not survive the kind of air-to-air threats likely to be encountered over both the Eastern and Western fronts in 1943–45. German ground-attack and dive-bomber units played a virtually negligible role in the air war against the Allies advancing through Western Europe in 1944–45 precisely because they did not have a protective umbrella of fighters to shield them from Allied interception and destruction.[53]

In one respect, however, Germany learned a very important lesson from Spain, and that was in the area of command and control. Following the battle of Brunete (during the northern offensive), von Richthofen established a control network as well as a supportive climate and environment conducive to good ground support. He arranged for air commanders to become intimately familiar with the front before them, created procedures for the issuance of orders integrating both ground and air needs via liaison between army headquarters and the Legion Condor's own headquarters, assigned air liaison officers to the front and to army headquarters, and established a tradition of a senior air representative being on the ground at the focal point of ground operations during an actual offensive or battle. Accompanying this was development of direct communication procedures by radio or land lines between air and ground forces, establishment of an identification system to differentiate enemy from friendly lines using cloth and light signals, directional searchlights, and blinker lights (the latter two for use at night), and establishment of control procedures for bombers being used in close support (chiefly by establishing a command post at the front staffed by air liaison officers who could observe the battle and direct the air strike themselves).[54]

It might be expected that having achieved all this, the Luftwaffe immediately restructured itself to address the needs of armies in the field. Such, in fact, did not take place. The von Richthofen initiatives favoring *direct* support of ground forces were by no means universally accepted, and, indeed, until the outbreak of the Second World War, the direct vs. indirect support notion remained in basic conflict. In 1935, the *Luftwaffen Dienstvorschrift nr 16* (the Luftwaffe's operational handbook) did not emphasize *schlacht* operations at all, but rather, stated that in a battle's critical phase, all air elements—fighters and bombers—would work together, though primarily for the *morale* benefit of friendly troops, rather than for any decisive military impact that they might achieve. Von Richthofen's influence was such, however, that in May 1938, Germany established five *Schlachtfliegergruppen* equipped with a mix of Hs 123's, Arado Ar 66's, Heinkel He 46's, and He 51's. Ironically, less than six months later, four of these five reequipped with the Ju 87 Stuka dive bomber, becoming dive-bomber units and, as such, part of the "strategic" rather than "tactical" arm of the Luftwaffe! Thanks to von Richthofen's persistence, the Luftwaffe established a *Nahkampfverband z.b.V.* (Close battle association for special duties) assigned primary responsibility for supporting ground forces through both direct (Hs 123) and indirect (Ju 87) operations. Eventually, this evolved into VIII Fliegerkorps which, under von Richthofen's leadership, played a leading role in devastating France in 1940.[55]

The Soviet Union

Aside from Germany, the Soviet Union learned the most from the fighting in Spain, and the Soviets did so more in the sense of confirming strongly held beliefs that aviation could be of importance during direct support of ground forces. Soviet observers, who published widely in military journals, noted that by June 1938, fully 80 percent of the Loyalist air operations were devoted to some form of battlefield support. (If anything, by this time the Nationalist figure was probably 85 percent or higher.) They recognized that light bombers could not rely merely on interlocking fields of fire for mutual support, but that fighters had to take an active role in escorting bombers and in sweeping the skies ahead of them. Further, they recognized that air-ground coordination was critical, particularly if casualties from friendly bombing were to be avoided. Air attack had a shattering effect upon morale, undisciplined troops often fleeing the battle by the time of a second pass by attack aircraft. Thus, commanders had to put a premium on disciplining troops to stand up under air attack and fire back at the attackers. When troops did fire back, aircraft took losses, hence any future ground-attack aircraft should ideally be armored to some degree, or at least damage tolerant. Ideally, commentators believed that future attack aircraft should be fast enough to free fighters to concentrate on the air superiority battle. In offensive operations, attack aircraft would be attached to an army corps, and would cooperate with infantry, artillery, and armored forces during an attack. In particular, aircraft were seen as ideal for suppressing defenses so that infantry and armor could overrun them, and to do so, an army corps needed to be able to call upon 100 to 120 Shturmovik aircraft to assist it "in effecting a tactical penetration without which the offensive will fall short of its objective."[56]

Thus, Soviet air doctrine increasingly emphasized massing and concentrating air elements to participate in the land battle. Heavily emphasizing offensive operations, Soviet doctrine saw defense as only a transitory stage to prepare the path for a decisive counteroffensive; thus, defense in and of itself could not be decisive in a war-winning sense. The lessons from Spain arrived in the Soviet Union in time for two major events: the Third Five-Year Plan (beginning in 1937), which placed renewed emphasis on force modernization via new military systems, and the purges Stalin instituted within Soviet military and civilian leadership. The former assisted the Soviet Union in its preparation for eventual war; the latter undoubtedly did not. By 1941, the Soviet military was being reequipped at a rapid rate with new technology exemplified by the T-34 and KV-1 tanks; the Yak, MiG, and LaGG family of fighters; the Ilyushin Il-2 Shturmovik (arguably the finest "dedicated" attack aircraft of the war) and Petlyakov Pe 2 dive and attack bomber; and new and more capable battlefield artillery. Soviet military spokesmen recognized the interrelatedness and unity of air-land warfare in a way that calls to mind the "Airland Battle" notion of the present day. In 1939, for example, commenting on aviation in modern war, A. N. Lapchinsky wrote, "In order to conduct maneuver war, to win the air-land battles, which begin in the air and end on the ground, one must concentrate all air forces at a given time, on a given front."[57] The Soviet invasion of Finland and, in turn, the German invasion of Russia revealed grave deficiencies in Soviet strategy, training, equipment, and leadership. Yet, if one looks at Soviet professional military thought in the late 1930s and early 1940s, one sees within it the core of the Red Army's—and Red Air Force's—victory on the Eastern Front in 1943–45.

10

War in Asia

Writing in 1940, one Army Air Corps officer assigned to the Army War College stated that "the principal lessons to be learned from the air operations in connection with the China Incident are: that Chinese towns burn briskly when bombed, and the numerous attacks launched from the air by the Japanese do not appear to have had any appreciable influence towards bringing the conflict to a close."[58] As is typical of most such off-the-cuff remarks from conflicts of the interwar years, the reality of air operational experience in Asia was more complex and significant. Indeed, Brig. Gen. Henry H. "Hap" Arnold had alluded to this in a lecture at the Army War College in October 1937, when he remarked that

> Japan has shown unmistakably that she knows how to employ her air force. She has demonstrated this by long-range operations. . . . In general she has demonstrated her soundness in air strategy in the selection of her objectives. She has not assigned her air force to operate against front-line trenches as have the Spaniards, but she has employed it against rail centers, manufacturing areas, and industrial establishments, and against Chinese aerodromes. . . . The employment of the Japanese air force is directly in line with the most up-to-date teachings of our own Air Corps Tactical School and with the doctrines of our own GHQ Air Force. That is significant. It is important and must never be forgotten. There is abroad in the world a first-rate air power which knows how to use its air strength.[59]

Allowing for his overemphasis on strategic warfare (and his apparent dismissal of the significance of battlefield air power), Arnold's cautionary warning—together with similar messages from attachés and military observers concerning Japan's growing air and naval strength—

strikes an ironic note today, given the shock of Pearl Harbor in 1941. As with the war in Spain, the conflicts in Asia during the interwar years held a variety of lessons for participants and observers, but also offered equal opportunities for misreading and misinterpretation.

There were three major conflicts in Asia during this time period, and in reality they were all linked to the general Asiatic war of 1941–45. The first was the Sino-Japanese War; next came a savage little "incident" at Changkufeng, followed by a bigger "incident" at Nomonhan. While the first of these was, on its face, a war between China and Japan, it involved (as had Spain) numerous foreign entanglements. The latter two were between Japan and the Soviet Union, and would forever alter the nature of the Soviet-Japanese relationship. Like Spain, the wars in Asia involved cross-generational technological conflict between aircraft of the biplane era and the newly emergent monoplane ascendency. Likewise, they involved large foreign forces—specifically Soviet, but with mercenaries and foreign volunteers from a wide range of other nations, particularly the United States. But in other ways the wars in Asia—particularly the Sino-Japanese war that broke out in 1937—were quite different in character from the Spanish Civil War. Whereas the Spanish war (as Arnold noted) had been distinguished by tactical air operations over the battlefront, the Sino-Japanese war demonstrated strategic air operations over vast distances, including raids from Japan and Formosa against China, raids from China against Japan and Formosa, and extensive naval air operations against China by Japanese aircraft carriers off the coast. The Soviets were especially impressed by the length of China war air operations, with some Chinese missions ranging as far as 1600 km from the front, while the Japanese sortied as far as 1200 km from the front.[60]

The Sino-Japanese War

The Sino-Japanese War erupted directly from the famous Marco Polo Bridge incident in which Japanese and Chinese troops clashed near Peiping (subsequently Peking, and now Beijing) on July 7, 1937. Japan followed this immediately by an invasion of North China, and accompanied this invasion by intensive air activity. By joint agreement of the so-called Central Authorities (the combined Army and Navy General Staffs and the War and Navy Ministries, later superseded by the creation of Imperial General Headquarters before the end of the year), the Army and Navy divided responsibilities for air operations. The Japanese Army

Air Force (JAAF) assumed responsibility for destroying Chinese air power in North China, and the Japanese Naval Air Force (JNAF) had responsibility for air superiority operations in South and Central China. Over time, the JAAF and JNAF used this basic guideline to influence their other air operations as well, such as control of strategic air operations such as bombing missions. Initially, however, the so-called Provisional Air Group sent from Japan to accompany the invasion forces had as its mission a more tactical orientation, including "close cooperation with land operations," with the proviso that "international relations will be taken into consideration when targets are being chosen for ground strafing and bombing."[61] Total JAAF and JNAF air strength supporting the invasion initially consisted of 451 aircraft (including 169 fighters, 152 bombers, 103 recce aircraft, and 27 torpedo planes), though other aircraft were available from the so-called Kwantung Army. Initially, these Japanese aircraft consisted of a mix of aging and modern types; very quickly, however, the Japanese employed the best available technology that they had, in the form of the Navy's Mitsubishi A5M fighter, the Army's Nakajima Ki-27 fighter, the Navy's Mitsubishi G3M bomber, the Army's Type I bomber (a Japanese version of the Fiat B.R. 20) and the later Mitsubishi Ki-21 bomber, and the Army's Mitsubishi Ki-30 light bomber. Eventually, before the attack on Pearl Harbor and the Japanese assault on Southeast Asia triggered the eruption of a general Pacific war, Japan employed even later generation aircraft such as the Mitsubishi A6M fighter (the famed Zero), and the Army's Mitsubishi Ki-51 attack aircraft, a derivative of the Ki-30 bomber. Japan did not have an attack aviation element per se, and it is thus interesting to note that the JAAF began development of the Ki-51 at the end of 1937, after the first lessons of the China war were commencing.[62]

In contrast to Japan's well-organized, highly trained, and tightly controlled military air services equipped with aircraft of indigenous manufacture, the air force of China was much more decentralized and equipped with a nightmarish mix of suitable and unsuitable aircraft imported from a variety of nations. Indeed, it is difficult to speak of even "one" Chinese air force, as many warlords had their own private air arms. Basically, what eventually became the Chinese Air Force (CAF) began with a Soviet mission to the Kuomintang in 1924–25. Because of the nature of internal Chinese politics, the Soviets fell out of—and eventually back into—favor; the clash of Japanese and Chinese forces in Manchuria in 1931–33 led to creation of an American mission to China headed by Col. John J. Jouett, who was directly responsible for establishment of the CAF as a separate military force in 1932. At the same time, an Italian mission arrived to help train pilots and create a Chinese aircraft industry. China now turned to the United States and Europe for aircraft, placing sizable orders for a wide range of aircraft built by American, Italian, British, German, and French manufacturers. Some of these were quite suitable, but others were more a tribute to the skills of manufacturer's salesmen.[63]

Chiang Kai-shek's air forces had already experienced combat in operations during the so-called Bandit Extermination Campaigns, as well as in major and minor insurgencies. In most of these, government pilots flew light bombing and attack sorties against insurgents, including the Communist forces of Mao Tse-tung (Mao Zhedong) during the latter's "Long March" to the west of China. These operations, and the rising standard of Chinese pilot training, gave China a basic (but rudimentary) skill level at air warfare. Nevertheless, the CAF remained basically ill-equipped. At the time of the Marco Polo Bridge incident, the CAF had a listed strength of 600 aircraft, including over 300 of modern design. In fact, its true strength was only 230 aircraft, with only 91 of these being "modern" types. The Chinese Air Force was composed of ten groups of between two and four squadrons each, with the groups broken down according to mission, such as fighter, bomber, reconnaissance, or attack. Unlike the JAAF and JNAF, then, the CAF did have an attack group organized along American lines, and it was equipped initially with the aging two-seat Vought Corsair biplane. China soon replaced the Corsair with twenty new Curtiss Export Shrikes, a more powerful variant of the Curtiss A-12 used by the Army Air Corps. China's most numerous fighters were the Curtiss Hawk II and Hawk III biplanes, the former having a fixed landing gear, and the latter having a retractable one. Both were agile but increasingly obsolescent. One squadron of Boeing Model 281's—an export model of the famed P-26A Peashooter—was in service, but even this monoplane was increasingly outdated by newer Japanese designs such as the Ki-27 and the A5M. There were a variety of bombers in service, of which the most modern were a few Heinkel He 111's, some Northrop 2E Gammas, and a few Martin 139W's; the Heinkels—six in total—were undoubtedly the best of this strange mix, and (together with the Martin 139's, an export model of the famed B-10 bomber) constituted the "long-range" element of the CAF, as well as its only twin-engine force.[64]

Thus, when the war broke out, China was inferior both qualitatively and quantitatively to its Japanese opponent. Yet as the war went on, the

Chinese proved—from the Japanese viewpoint—exasperatingly difficult to defeat. The Japanese government repeatedly announced the imminent demise of the CAF, but each time it managed to keep on going, and—more importantly—continued to extract a toll of Japanese aircraft. Such was the influence of Chinese air combat successes that as late as the eve of the Pacific war, Japan *still* could not claim to have effectively gained air superiority over the Chinese Air Force. It is obvious, then, that the CAF proved a resolute and determined opponent. It is important to remember this, for much of the post–World War II literature has accepted a conventional Japanese point of view that implies that Japan quickly destroyed the Chinese Air Force and thus was able to operate with a free hand. In fact, Japanese airmen fought less well than they believe they did, and the Chinese—admittedly with an important assist from the Soviets, beginning in late September 1937—fought better than they are credited with. The situation resembles that of the Russian Front in the Second World War, where the defender (China/Russia) refuses to crumble in the face of the attacker (Japan/Germany).[65]

In September 1937, the Soviet Union sent its first aircraft and support personnel (including combat pilots) to China; the aircraft consisted of Polikarpov I-15bis and I-16 fighters and Tupolev SB bombers—essentially the same kind of equipment then being flown by Soviet and Loyalist pilots in Spain. The Soviet pilots proved the equals of the Japanese, and gradually as Soviet influence increased, the Chinese Air Force became predominantly a Soviet-equipped force, with fewer and fewer (as attrition rose) American aircraft in service. (It must be kept in mind that even though Claire Chennault was in China as an advisor, his American Volunteer Group—the Flying Tigers—did not yet exist, and would not in point of fact enter combat until after Pearl Harbor.) Ultimately the Soviets delivered 885 aircraft to China, of which 489 were fighters and 322 were bombers. Japan would find it ultimately impossible to decisively bring the Chinese to battle and inflict a defeat that would spell the end of the Chinese Air Force—and her inability to do so merely mirrored her inability to defeat China's ground forces in a similar manner. The I-15's and I-16's prevented Japanese bombers from operating with impunity, even with attendant escort fighters. With the new SB's and the few Martin 139's remaining, Chinese airmen (and their Soviet colleagues) struck from the mainland to Formosa and from the mainland to Japan itself where the Chinese, in a political masterstroke, dropped not bombs but leaflets.[66]

Generally speaking, despite Hap Arnold's interpretation of the fighting in China, it is evident that the Japanese were, in fact, interested in applying air power over the battlefield and in support of land and amphibious operations. Attaché reports from Japan and China include numerous references to air-ground cooperation, as does the Shiba history of Japanese air operations in China, completed for the U.S. Army after the Second World War, and contemporary Japanese and foreign accounts. In August 1937, as if to underscore this interest, a JAAF staff officer wrote, "There appears the need for direct cooperation of the attack aviation with ground forces and the need for light scout planes."[67]

During the battle of Taiyuan, in September–November 1937, Japanese light bombers struck at Chinese troop concentrations, supported infantry attacks by bombing ahead of the infantry, and harassed retreating Chinese columns; the official Chinese history of the war admits that "heavy enemy bombing, followed by attacks of the enemy's mechanized units, made it extremely difficult for our forces to maintain a foothold."[68] The Taiyuan defeat was but a prelude to others that followed, notably the fall of Shanghai itself.

In the Shanghai operation, which involved a complex amphibious assault as well as decisive fighting on land, Japanese naval aircraft undertook recce operations over the area, strafed and bombed ground positions, maintained a combat air patrol against incursions by Chinese fighters and bombers, and undertook short-range bombing near Shanghai.[69] Operations such as Taiyuan and Shanghai established a pattern of air-ground cooperation that continued throughout Japan's war in China. Typically, reconnaissance aircraft would observe Chinese positions and flanks prior to an attack; as Japanese assault forces approached the Chinese forward line of troops, bombers would strike at strongpoints in Chinese positions, and also attack adjacent Chinese positions to fix the Chinese forces in place and prevent lateral movement. Formations of as many as thirty medium bombers would attack Chinese artillery positions, and fighters would strafe and bomb Chinese defensive positions with machine-gun fire and small fragmentation bombs, often attacking in short dives and with formations of three aircraft. These attacks would continue as Japanese infantry and armor forces advanced; armor generally operated in support of infantry, but occasionally operated as far as five kilometers in advance of friendly troops. It is evident, then, that even though Japan lacked "dedicated" ground-attack air forces, her military appreciated the need for cooperation and integration of effort to win on the battlefield.[70]

The Chinese Air Force was less successful when it tried to emulate the activities of the highly trained and experienced Japanese airmen, but Chinese airmen persisted, and their efforts could never be discounted or entirely dismissed as unproductive. Early Chinese air operations around the Shanghai area were marked by persistence and courage, if lack of skill. Raids by Northrop Gamma bombers against Japanese shipping—particularly the cruiser *Idzumo*—resulted in numerous friendly casualties from misplaced bombs and no hits on the ship. Fighter operations went much better, and the Japanese bombers took staggering losses from Chinese fighters—reputedly no less than fifty-four bombers destroyed in the first three Japanese raids—until Japanese fighters began escorting the vulnerable bombers. (Intelligence sources subsequently estimated that between 80 and 85 percent of all Japanese losses came from fighters, with the remainder from Chinese flak.) Ground attack, however, was dismal. A formation of twelve Shrikes attacked a Japanese landing party at Poashan on the Yangtze River, flying at an altitude of only fifty feet; intensive Japanese small-arms and machine-gun fire shot down nine of the twelve during the first pass. The remaining three wisely did not return. Elsewhere, the Chinese used the two-seat Shrike as a fighter, shooting down two Japanese bombers during one mission, but losing a Shrike in return. The Shrikes do not figure prominently in CAF accounts of the air war; presumably all twenty had been lost by the end of 1937. Early in 1938, thirty Vultee V-11 attack bombers (the export version of the Army's A-19) arrived in China, but were quickly destroyed on their airfields by Japanese raids. As a rule, what ground-attack sorties were carried out tended to be by individual fighter pilots flying the Hawk II and III aircraft on strafing attacks against Japanese troop columns, river patrols and barge traffic, and mechanized forces. In the autumn of 1938, during the battle for Wuhan, a critical communications, supply, and airfield complex, both the CAF and the JAAF emphasized air-ground operations; while the Japanese won the battle they lacked the air resources to exploit it more fully and extend their operations even further westward into China, because they still had not knocked the CAF out of the war. In part, this was due to the infusion of Soviet airmen and aircraft. The Russian advisors and combat aircrews flying for China did not hesitate to use their bombers in a direct support role, as they had in Spain. For example, in September 1938, Soviet-flown bombers supported an attack by Chinese troops at Loshan, carrying a mix of fragmentation and high-explosive bombs (fully 70 percent were fragmentation). By good fortune, they arrived over the battlefield as the Japanese attempted a counterattack to break up the Chinese assault. The SB's, escorted by fighters, blasted enemy positions and troop concentrations, causing a Japanese withdrawal. Meanwhile, as one group of fighters flew cover for the bombers, a second group strafed a column of Japanese troops attempting to reinforce Loshan from the east.[71]

Japan clearly could not win in China, if for no other reason than it was so big. But such did not temper Japanese militarism, and Japan went from error to error, finally culminating in the attack upon Pearl Harbor. But before that ultimate folly, Japanese troops attacked Soviet forces in 1938 and 1939, triggering two savage and intense "incidents."[72] The first, the Changkufeng/Khasan incident, involved a struggle between a single Japanese infantry division against larger Soviet forces supported by aircraft for control of a strategic hill in a disputed border area between the Soviet Union and what is now North Korea. Aviation made little difference at Changkufeng—indeed, the Japanese did not even mount an air effort—though the Soviets did, on occasion, undertake ground attacks with formations of upward of 100 airplanes. Eventually, Soviet ground forces prevailed, and, following a truce, the battered Japanese forces withdrew. Even this lesson was ignored, and the next year, a larger and far more violent incident grew out of Japanese provocations. In that conflict, the Nomonhan/Khalkhin-Gol incident, air operations were far more extensive and influential.[73]

The Nomonhan Incident

The Nomonhan fighting lasted from May through September of 1939, and when it was over the Japanese 6th Army had been shattered, taking total casualties of 19,714 out of 59,925 men deployed in battle. One division, the 23d, had 11,958 casualties out of a total force of 15,140. In part, Japan had been betrayed by its own stereotype of the Soviet soldier as an unimaginative and poorly led successor to the troops trounced in the Russo-Japanese War, but the Soviet style of warfare—particularly deep multiechelon defense, and combined-arms offensives dominated by massive applications of artillery and aerial bombardment followed immediately by infantry-armor assault—proved overwhelming for opposing Japanese forces. In part, this was because Soviet forces in Mongolia had learned well the lessons of Changkufeng and had changed their combat style from the linear defense systems that had proven vulnerable the previous year. But an additional factor not to be ignored was the tactical generalship of the Soviet force commander, General (and subsequently Marshal of the Soviet Union) Georgi Zhukov.[74]

At first, the JAAF was able to inflict serious losses on Soviet air forces operating over the front; for example, Japanese troops advancing toward the Halha river on July 3

> saw Japanese fighters protect infantrymen from about twenty Soviet aircraft which were trying to bomb and strafe the ground troops at low level. When one Soviet aircraft exploded and crashed, the rest fled across the Halha. Encouraged by this successful conclusion to the first clash of arms they had ever witnessed, the troops picked up their pace and covered the ground along the ridgelines more rapidly despite scattered sniper fire.[75]

Although one has to be careful accepting outright victory/loss figures from Japanese and Soviet accounts, it is worth noting that the Japanese claimed to have shot down 481 Soviet aircraft for a loss of 41 of their own during July, a 12:1 kill ratio. But this dipped even by Japanese sources to slightly less than 6:1 during August. Soviet figures differ greatly, admitting a 4:1 loss ratio to the Japanese for May, a 3:1 victory ratio in June, a 4:1 victory ratio in July, and a 10:1 victory ratio in August. The truth, of course, necessarily lies between the two sets of figures; presuming an equal tendency to exaggerate, one may conclude that the true July figures were probably about 3.25:1 in favor of the Japanese, while August was about 1.57:1 in favor of the Russians. Thus, while the Japanese shot down more aircraft at Nomonhan than did the Soviets, over time they began to lose the air superiority war, and with it the opportunity to ensure that their own bombardment forces could participate in the air-land battle. Further, Japanese forces had to draw upon units and equipment in China, and upon new production straight from Japan, substantially lessening their chance for winning the China war. With a smaller pool of available airmen, losses had a greater impact than with Soviet air forces, while the Soviets had the opportunity of rotating in large numbers of fresh aircrew built around a core of veterans of the Spanish war and utilizing the latest production models of the I-16 and I-15 family, including ten I-16's capable of firing 82mm RS 82 solid-fuel air-to-air/air-to-ground unguided rockets, and the new—and ultimately disappointing—I-153 biplane. These aircraft had little difficulty gaining superiority over the older Kawasaki Ki-10 fighter biplane, and their numbers advantage helped overcome the performance advantage of the later Nakajima Ki-27 fighter monoplane.[76]

Zhukov's aggressive command style, coupled with the infusion of air and armor personnel with combat experience in Spain (and serving under Spanish war veterans such as air chief Schmutchkievich and armor expert Dimitri Pavlov), quickly resulted in Japan's offensive being blunted and reversed. While Japan had done well in the initial air war, Japanese ground units were roughly handled by their Soviet opposites, and as fighting extended into July and August, the pace of war intensified, particularly as the Russians resorted to combined armor and air attacks such as one on July 3, when a Japanese regiment was attacked by tanks and seventeen I-16 fighters. By the end of the month, Japanese division commanders were openly admitting the superiority of Russian low-level ground-attack tactics to those of the JAAF. Typically the Soviets would use combined artillery and air assault to fix the enemy and hold him until Soviet infantry-armor forces could fight and finish him.[77]

When the massive Russian-Mongolian offensive of August 20, 1939 began, Zhukov started artillery preparation at 5:45 in the morning, with the guns concentrating on Japanese antiaircraft and machine gun positions. At 6:30 (with, as Zhukov recalled, "the roar of our aircraft . . . becoming ever more deafening"), 150 bombers and 100 fighters attacked Japanese positions between the Halha and Holsten rivers, occasionally bombing on smoke markers fired by Soviet artillery. The bombardment and ground-attack operations paused for an intensive artillery and mortar assault at 8:15; at 8:30 the air assault resumed and continued when, at 8:45, red signal rockets announced the advance of Soviet ground forces. Zhukov recollected:

> The strike of our air force and artillery was so powerful and successful that the enemy was morally and physically suppressed and during the first hour and a half could not even open artillery fire. The observation posts, communications lines and fire positions of Japanese artillery were destroyed.[78]

Zhukov's memoir likewise quotes from the diary of a certain Japanese officer Fakuta, who perished in the attack:

> The fighter and bomber planes of the enemy, some fifty of them, appeared in the air in groups. At 6:30 a.m. the enemy artillery opened massive fire. It is horrible. The observation posts are doing everything possible to spot the enemy artillery,

but without any success, because enemy bombers and fighters bomb and shell our troops. The enemy is triumphant all along the front. . . .

[On August 21] A multitude of Soviet-Mongolian aircraft are bombing our positions, their artillery is worrying us all the time. After the bombing raids and artillery fire the enemy infantry charges to attack. The number of the killed is increasing. During the night the enemy aviation bombed our rear positions.[79]

The ferocity of the Soviet assault stunned the Japanese, who wondered where their own aviation and armor forces were; so unexpected was the attack that Japanese reconnaissance pilots had difficulty persuading their headquarters that a Soviet infantry-armor assault was underway, recollecting Hans Kundt's unwillingness to believe his own air recon during the Chaco war. After an abortive attempt to neutralize Soviet air power by striking at Soviet bases, the JAAF played an increasingly ineffectual role in the battle. Zhukov's forces rapidly turned the southern Japanese flank, destroying the Japanese Army's 23d Division in the process, fought a more protracted northern battle before rolling up the right flank, and might well have destroyed the entire 6th Army, except that Soviet forces were under orders not to extend their assault beyond the recognized border between Mongolia and Manchukuo. As at Changkufeng the previous year, a cease-fire on September 16 brought Nomonhan to an end. It had been a brutally instructive lesson for the Japanese government which sought, from that point onward, to do everything possible to ensure peace with the Soviets—a fact that the Sorge spy ring in Tokyo picked up subsequently and transmitted to Stalin, easing his concern over fighting a two-front war once Germany invaded Russia in 1941. For the Soviets, Nomonhan confirmed in a way that Spain had not—and could not—the basic soundness of evolving Soviet air-land battle strategy. Zhukov's next command following Nomonhan was command of the Kiev Military District; following an influential war game in December 1941 (where Zhukov played the role of an attacking German force, uncannily highlighting some of the very problems Soviet defenders would encounter six months later), the Politburo appointed him as Chief of the General Staff.[80] He would subsequently have numerous opportunities to duplicate his pattern of attack—and his success at Nomonhan—in a game of far higher stakes.

PART FOUR

The Second World War

The literature of the Second World War is voluminous, and the works that deal with the air war between 1939 and 1945 make up a large proportion of this material. Generally speaking, however, researchers have devoted far more attention to the air superiority war and to strategic bomber operations than they have to other topics, such as research and development, special operations, and battlefield air support.

Though virtually all the combatant nations used aircraft for ground support missions, the four that did the most to develop and exploit this field were Nazi Germany, Great Britain, the United States, and the Soviet Union. Thus, this study examines cases from the experiences of these nations in various campaigns and theaters, emphasizing the blitzkrieg of 1939–40;[1] the Anglo-American experience in the Western Desert, Southwest Pacific, Italy, and Western Europe;[2] and the struggle on the Eastern Front, particularly during and after the "mature" phase (Kursk and subsequently) of Soviet air-land operations.[3]

One caveat must be offered concerning the studies in this part: the relative lack of discussion of the war in the Pacific, specifically the development of Navy-Marine close air support, and Gen. George Kenney's Fifth Air Force operations. Since most operations in this theater concerned island assaults (essentially "snipping up" the Japanese empire), the dominant superiority and concentration of force that the United States and its allies were able to bring to bear in these attacks, the nature of the support provided, and the battlefield environment all differed significantly from other theaters, such as the Western Desert, Europe, or the Russian Front. For example, General Kenney's Fifth Air Force made extensive use of "hit and run" low-level attacks by heavily armed A-20's and B-25's in the Southwest Pacific; such aircraft assaulted Japanese airfields, dropping parachute-retarded fragmentation bombs (dubbed parafrags) and strafing with batteries of machine guns. The peculiar circumstances of Southwest Pacific air warfare enabled them to get away with relatively light losses, something that would not have been possible in more densely defended European or Mediterranean skies. Where Pacific developments foreshadowed some of the issues and procedures that appeared in the major force-on-force engagements in these other theaters (such as the experience of the Solomons campaign), those developments and experiences are covered. When later campaigns such as the Philippines and Okinawa occurred, however, they came after the maturation of wartime air support as demonstrated in France and in Russia. This is not to denigrate the work of those who labored to create effective air-land operations in the Pacific, and some of the developments, such as amphibious command ships, did have important implications for post-1945 conflicts, notably in Korea and, subsequently, Vietnam. As a general case, however, battlefield air support in the Pacific war is set aside in favor of a more detailed accounting of the North African, European, and Russian theaters.

11

The Blitzkrieg

The Invasion of Poland

On September 1, 1939, Hitler's forces attacked Poland, triggering the outbreak of a general European war that had smoldered since the unsatisfactory peace of Versailles twenty years before. As mentioned in Part Three, the Luftwaffe's basic prewar orientation was strategic rather than tactical, but in Poland (as in Spain) it was called upon to perform largely in the tactical role. Very quickly, despite the stunning series of victories that marked the Polish campaign, both the Luftwaffe and the German Army recognized that significant problems existed that affected the outcome of air-land operations. For example, no common radio frequencies were assigned to air and land units—a generic problem, it would seem, to virtually all nations that have a history of air-land operations. Liaison problems between air and land commanders generated confusion about the location of friendly units (as did the rapid pace of German ground advance, which tended to go beyond pre-briefed map demarcations), resulting in friendly casualties from misplaced air strikes. Not surprisingly—and again a lesson that would be discovered by the Allies themselves in subsequent fighting—battlefield air support proved most useful when applied to fixed fortifications and emplacements, and in attacks upon the enemy's rear areas (true battlefield air interdiction). When friendly forces were advancing at speed, air support became more haphazard, less coherent, less effective, and more likely to generate friendly casualties. F. W. von Mellenthin, a General Staff officer (and subsequently a Generalmajor) had a notable experience during the Polish campaign that reflected such problems:

Very early in the campaign I learned how "jumpy" even a well-trained unit can be under war conditions. A low-flying aircraft circled over corps battle headquarters and everyone let fly with whatever he could grab. An air-liaison officer ran about trying to stop the fusillade and shouting to the excited soldiery that this was a German command plane—one of the good old Fieseler Storche ["Storks," a light short-takeoff and -landing aircraft]. Soon afterwards the aircraft landed, and out stepped the Luftwaffe general responsible for our close air support. He failed to appreciate the joke.[4]

On the whole, of course, German air support of the army during the Polish campaign was spectacularly successful, which reflected in great measure upon the organizational genius of Generalmajor Wolfram Freiherr von Richthofen. Von Richthofen, while Fliegerführer (Air leader) of the so-called Nahkampfverband z.b.V. ("close battle association for special duties" discussed in Part Three, subsequently VIII Fliegerkorps), had drawn on his experiences in Spain and, recognizing the special challenge of ensuring effective cooperation between air and armored forces, had arranged to assign air liaison officers to specially equipped command cars—usually modified eight-wheel armored cars—operating at the apex of armored thrusts. Germany's attack on Poland involved a pincer assault by northern forces (*Heeresgruppe Nord* comprised of the 3d Corps, and the 3d and 4th Armies), and three southern armies (*Heeresgruppe Sud* consisting of the 8th, 10th, and 14th Armies). Von Richthofen's forces, really a tactical air force consisting of fighter, dive bomber, attack aircraft, and recce aircraft, supported the drive into Poland of the 10th Army, which thrust north of the Carpathian Mountains toward Lublin, then hooked north behind Lodz toward the confluence of the Bzura and Vistula rivers.

Much as Heinz Guderian would do later in the French campaign, the German 10th Army, led by General Walther von Reichenau, punched through Polish defenses with a coordinated air-land assault, and then roamed virtually at will in Polish rear areas, rolling up Polish forces toward the Bzura. The pincer of *Heeresgruppen Nord und Sud* eventually trapped the Polish army between the Vistula and Bzura rivers around Kutno, the ring tightened, and on September 19, 100,000 troops (representing the disorganized remains of nineteen divisions) surrendered to the German 8th Army. During these operations, Stuka dive bombers concentrated their attacks against Polish fortifications and rear areas, while Henschel Hs 123 attack biplanes furnished battlefield support, often operating from forward air strips—which, in reality, were usually any flat, clear stretch of ground "verified" as being suitable for aircraft operations by driving an automobile across it at speeds up to 30 mph! The Hs 123's proved surprisingly effective, and though already out of production, remained in service as a result of the Polish (and subsequent French) experience with *schlacht* units until the late war period. They dropped small bombs and incendiaries (dubbed *flambos*); in one case, Polish forces occupying a ten-square-mile area were "showered with 4000 incendiary bombs in five minutes, and [were] completely neutralized."[5] Pilots discovered that varying the Henschel's engine setting while flying just over ground level resulted in a shattering noise that induced panic in horse-equipped Polish formations. Demoralized, some Polish troops abandoned their positions and weapons, and fled the battlefield—a situation to be repeated with even greater effect during the assault on France the following spring. In summarizing the lessons of the German assault on Poland, an Army War College study team concluded that "the German air operations in Poland included every form of independent operation that the most enthusiastic advocate of air power has visioned, but they showed up outstandingly well in cooperative missions with ground forces."[6]

Germany's invasion of Poland did not go unchallenged, and however ineffective Polish resistance may have been, it was, nonetheless, sharp, courageous, and determined. The Polish air service remained largely intact and was not, in fact, destroyed on the ground during the Luftwaffe's initial air strikes. Seriously outclassed by the Messerschmitt Bf 109, Polish PZL P-llc fighters proved incapable of defending Polish air space and preventing attacks upon Polish ground-attack, bomber, and reconnaissance aircraft. Losses from air-to-air combat were high, and intense flak (antiaircraft fire) claimed many of those aircraft that did manage to survive or evade interception and pressed their attacks against German ground forces; it was in the air-to-air and air-to-ground war that the Polish air service was eventually overcome and destroyed.

The majority of German antiaircraft forces were an "organic" part of the Luftwaffe, but were subordinated operationally and controlled by the Army unit to which they were assigned. Additionally, however, the Army had its own antiaircraft forces (the *Heeresflak*). The basic antiaircraft unit was the battalion; battalions were either heavy (equipped with heavy artillery such as the 88mm cannon), mixed (with heavy and light—usually 20mm—cannon), or light (with a mix of 37mm and 20mm cannon). For night operations, each battalion included appropriate searchlight batteries. A Panzer division typically would be accompanied by a Luftwaffe mixed antiaircraft battalion consisting of three heavy batteries of up to six 88mm cannon apiece with two 20mm cannon for site defense, as well as two light batteries each consisting of twelve 20mm cannon and four searchlights. Heeresflak forces were assigned either to infantry or to artillery branches; infantry flak forces utilized 20mm and 37mm weapons and light machine guns on self-propelled mounts. Artillery flak was mechanized, and usually consisted of three heavy batteries (four 88mm cannon each) and two light batteries (either twelve 20mm or nine 37mm cannon each). Infantry forces received training in using personal weapons such as rifles and carbines against low-flying planes, in addition to machine guns and specialized antiaircraft cannon; unless under severe attack, columns were instructed to remain on the move. German flak cannon were unusually well suited for dual antiair and antiarmor or anti-infantry roles. The 88mm cannon—introduced in the Spanish war and subsequently the finest all-around multipurpose cannon of the Second World War—was a fearsome antitank weapon, but primarily intended for medium- to high-altitude air defense. It could fire twelve to fifteen 20-lb high-explosive projectiles (either time or percussion fused) to an effective ceiling of nearly 35,000 feet. When firing armor-piercing shells, its extremely flat trajectory, and a muzzle velocity of 2755 feet per second, made it eminently suited for antitank and antistrongpoint purposes. The 20mm cannon, particularly in its four-barrel Flakvierling 38 model, was devastating against low-flying aircraft. The Flakvierling 38 could fire 800 high-explosive and armor-piercing rounds per minute to an altitude of over 7200 feet, and like the 88, was dangerous to infantry, strongpoints and fortifications, and vehicles. German forces had a wide range of other antiaircraft weapons, but the most effective remained the 88 and the rapid-firing 20mm for defense against low-flyers.[7]

The Polish air service attempted to emulate the German success in ground attack with attacks of its own, but to little avail. Indecision in the Army high command resulted in the Polish air service making uncoordinated and sporadic assaults upon German forces though, in truth, it is difficult to see how any Polish air attack, no matter how well executed, could have made a difference. On September 2, eighteen PZL P-23 Karas bombers set out to attack German forces, losing seven aircraft in the process. The next day, twenty-eight Karases had better success in strikes around the Radomsko region, but claims that the attacks succeeded in "putting out of action some 30 percent of the enemy's vehicles" seem exaggerated.[8] That same day, September 3, PZL P-11 single-seat fighters joined in attacks on German armored columns, "with small results and heavy losses."[9] On September 4, following an inexplicable Polish High Command decision not to employ them in combat previous to this time, Poland's medium bomber brigade, composed of PZL P-37 Los twin-engine bombers (a modern and indeed excellent design) began an intensive series of attacks in conjunction with the remaining Karas force—still over 100 machines—against German columns on the central and northern fronts. In the first day of operations, Poland lost eight of twenty-eight Loses dispatched, and, unfortunately, this established a precedent for subsequent operations.[10] These attacks by Poland's Los bomber and Karas attack aircraft allegedly succeeded in causing delay and loss to 10th Army's armored forces; based on intelligence reports, a study team at the U.S. Army War College concluded:

> About the only effective employment credited to Polish bombardment formations were attacks delivered against German mechanized units, from altitudes around 600 metres. Some of these attacks held up German advances for from 12 to 24 hours, but when repeated against the same objectives, usually encountered effective defense measures.[11]

These attacks may have been responsible for the "plane shy" attitude von Mellenthin subsequently witnessed while accompanying the invasion with the 3d Corps. In part, the success that Polish attack and bomber forces did achieve—namely disruption and delay rather than mass destruction—likely stemmed from a basic problem encountered in the Polish campaign: army infantry and armored forces advanced too rapidly for the bulk of their flak protection—the Luftwaffe's flak formations—to keep up with them. Thus, the army had to rely on its own organic Heeresflak, which was, of course, still quite devastating. Thus, follow-on attacks likely encountered stiffened antiaircraft resistance due to flak units catching up with delayed columns, as well as from heightened German awareness of Polish intentions that resulted in denser combat air patrols by Messerschmitt Bf 109's and Bf 110's over the battlefront.[12]

By mid-September, the Polish air service had gone down fighting, shot out of the sky by the Luftwaffe's fighters and by flak. A summary in an official history of the Polish air service prepared by a Polish airmen's organization offers a glimpse of conditions as seen from the Polish perspective:

> The Polish bomber squadrons had a heart-breaking task. They were too weak to bomb enemy airfields or more distant bases and lines of communications. They had to concentrate on close support of the land forces. Even so, the

multiplicity of tasks was out of all proportion to the available aircraft. Losses were very heavy: one squadron lost nine out of 17 aircraft in an attack on an armoured group on 2nd September. Most of the aircraft which remained were so damaged by flak as to be unserviceable for further operations that day.

By the seventh day of fighting the Bomber Brigade had lost 47 per cent of its crews and aircraft and had no replacements available. Thereafter the situation deteriorated very rapidly. The Army made ever more numerous and more insistent demands for close support against the attacks of German tank formations. The number of serviceable bombers dwindled with headlong speed. . . . Within little more than a week organised operations were out of the question. . . .

The Bomber Brigade had lost 90 per cent of its aircrews and aircraft by 14th September. The Fighter Brigade was in the same plight. . . .

A few sorties were made on damaged aircraft, but almost all were shot down. On 17th September the few remaining operational aircraft flew across the Rumanian frontier and landed at Cernauti. On that day, therefore, fighting in the air over Poland ended.[13]

The Battle for France

The quick war in Poland was but a precursor of how the Blitzkrieg would function nine months later during the German assault on France. The attack on France was but the biggest challenge of what overall was called *Fall Gelb* (Case Yellow), a bold plan for the invasion of Holland, Belgium, and France. In this operation, three German army groups would play widely differing roles. One, Army Group A (composed of three armies and one Panzergruppe), would punch through French positions and (hopefully) cross the Meuse at Dinant. Army Group B (consisting of three Panzer divisions and airborne forces) would invade Holland and Belgium, putting on such a display of force as to convince the Allies that this was the main thrust of the German assault and thus draw attention away from the cocked fist—Army Group A. Army Group C was the southeasternmost German force, fronting the much-vaunted Maginot Line defense system. To support this assault, von Richthofen's newly renamed VIII Fliegerkorps was assigned to Army Group B for the early stage of the campaign against Belgium, until Belgian airfields had been destroyed. Then it would be shifted to support Army Group A's drive across the Meuse. Overall, the Luftwaffe forces arrayed against the three nations included approximately 1300 bombers (of a total Luftwaffe strength of 1758) and 380 dive bombers (out of a total of 417 in Luftwaffe service), as well as approximately 1210 Bf 109 and Bf 110 fighters (out of a total Luftwaffe fighter strength of 1736). Though once again the Allies would only see a dismayingly effective and smoothly functioning *"attaque générale aéro-terrestre,"* it is the difficulties that are perhaps of greatest interest—the continued general inability of German air and armored forces to communicate on a single-voice network, or even, for that matter, for German bombers and fighters to communicate with each other on a common-voice network. Given the poor state of the French military, however, these differences proved merely historically curious, not operationally critical.[14] Heinz Guderian, in charge of a Panzer corps of three divisions that would have actual responsibility for forcing the Meuse crossing, recognized the necessity of good air-land cooperation in spite of these deficiencies, and arranged to be in touch with appropriate air commanders via the established network of air liaison officers. He recalled:

> In order to establish a sound basis for co-operation as quickly as possible, I had invited the airmen to my planning exercises and I also took part in an air exercise that General [Bruno] Loerzer [commander of II Fliegerkorps] organised. The principal matter discussed was the Meuse crossing. *After detailed study we agreed that the air force could best be employed in giving the ground forces continuous support during the crossing; that meant no concentrated attack by bombers and dive bombers, but rather, from the very beginning of the crossing and throughout the whole operation, perpetual attacks and threats of attack against the enemy batteries in open emplacements; this should force the enemy gunners to take cover both from the bombs that were dropped and from the bombs that they expected to be dropped. The time schedules for these attacks, together with the targets were marked on a map* [emphasis added].[15]

During the actual attack upon France, the armored forces duplicated their success in Poland. Overhead, German fighters flew protective combat air patrol flights, shielding the movements of ground forces—particularly the thrust through the Ardennes, a feat held impossible by the French High Command—from the prying eyes of French and British reconnaissance aircraft, or shooting down Allied bombers and attack aircraft attempting to interfere with the German advance. If the aircraft somehow eluded the German fighter forces, the intense antiaircraft fire (as will be seen) made Allied ground attacks prohibitively costly. Once again the very fluidity of the ground situation created problems for air-land coordination, particularly in an environment characterized by weak or nonexistent direct communication links between air and land forces. At one point (following the crossing of the Meuse), Stukas attacked German troop concentrations at Cheméry, inflicting, as

Guderian recalled, "heavy casualties."[16] It was in the attack across the Meuse that German combined air-land assault tactics paid handsome dividends; here, on May 13, three days into the campaign, France suffered a mortal wound that would prove fatal the following month.

The crossing at Sedan—and another crossing farther north by a smaller force led by Erwin Rommel—were not masterpieces of smoothly working, combined-arms strategies accomplished in the face of stiff and dedicated resistance. Rather, there were elements of luck and chance that worked in the Germans' favor, together with a generally abysmal French counterstrategy. Constantly searching for the "methodical" battle and seeking to find a situation in which—as French prewar doctrine held—defensive firepower would predominate, the French were kept off-balance by the rapidly unfolding and constantly fluid nature of the German assault. The Meuse crossing, achieved on the 13th of May, was consolidated on the 14th and 15th. On the 16th the breakout commenced, and from this point on, France was on the skids. Having said this, however, it is interesting to note that the success of the German thrust seems to have surprised the German high command almost as much as it did the Allies; indeed, Guderian's thrust resulted in his being berated and very nearly relieved of command! Concern over "risking" the advance resulted in an overly cautious policy that ultimately delayed the advance of Guderian's Panzers, buying France a little time. But had the Allied defense been stronger and better organized—impossibilities in the France of 1940—possibly this hiccup in the German attack could have become the basis for a Marne-like counterattack that would have—as in 1914—brought the rapid blitzkrieg to a halt.[17]

Ironically, the first crossing of the Meuse, north of Sedan at Dinant, was made by Erwin Rommel's 7th Panzer Division, without any air support whatsoever, in the face of demoralized French resistance; Rommel displayed fully the bold and audacious quality he had first revealed at Caporetto in the First World War, and which he would manifest in the Western Desert. Guderian's thrust had dominated air support planning, and he was looking forward to continuous air attacks to substitute for a substantial lack of artillery—his artillery column was still enroute through the Ardennes. He received a shock, therefore, when his superior, General von Kleist, arranged with the Luftwaffe's Hugo Sperrle (commander of Luftflotte 3) to undertake a single massive air attack to be coordinated with what little artillery preparation Guderian could generate. Guderian protested to von Kleist that only continuous air attack could be expected to neutralize French defensive fire, but von

Kleist was unmoved. Sperrle's orders went to II Fliegerkorps air chief Bruno Loerzer for execution. Guderian, fuming, returned to his headquarters, nearly becoming lost over French territory due to an inexperienced pilot flying near dusk over unfamiliar terrain. The attack—framed by operational orders identical to those issued during training exercises at Koblenz—occurred the next day, May 13, 1940.[18]

Guderian's three divisions, the 1st, 2d, and 10th Panzer, were poised for assault on either side of the looping Meuse from Donchery in the north, past Sedan, and down to Wadelincourt in the south. The next morning, at 0700, Loerzer's II Fliegerkorps and von Richthofen's tactically oriented VIII Fliegerkorps began pounding French positions, starting with a series of level bombing attacks by Dornier Do 17 bombers, and followed by demoralizing dive-bomber attacks by Stukas. The incoming bombers were covered by a roving combat air patrol of Messerschmitt Bf 109 and Bf 110 fighters, though, in any case, French attempts to break up the attacks were not enthusiastic. The strike aircraft could attack free from worry of hitting friendly positions, as the river served as a natural demarcation line. At 1530, Guderian arrived at a forward observation post; precisely at 1600 the artillery preparation began, together with intensified air assault. Guderian noted with pleased surprise that the air attacks were precisely what he had arranged with Loerzer—continuous operations saturating French defenses and constantly keeping the French defenders under the threat of air assault. Meanwhile, his flak artillery and armor engaged French fortifications across the river with devastating point-blank cannon and automatic weapons fire. Despite well-situated defensive positions, a superiority in artillery, and a clear view of the attacking German forces, the French defenses crumbled, largely due to the morale impact of the howling dive bombers. The Henschels during the Polish campaign had flown with over-revving engines to demoralize defenders; the Ju 87's at Sedan flew with sirens to accomplish the same end. Though, ironically, the actual damage caused by the dive bombers was less than could be expected, the morale impact was all out of proportion to their attack. One German Panzer noncom wrote that the Stukas launched

> a regular rain of bombs . . . the explosion is overwhelming, the noise deafening. . . . A huge blow of annihilation strikes the enemy, and still more squadrons arrive, rise to a great height, and then come down on the same target. We stand and watch what is happening as if hypnotised, down below all hell is let loose! At the same time we are full of confidence.[19]

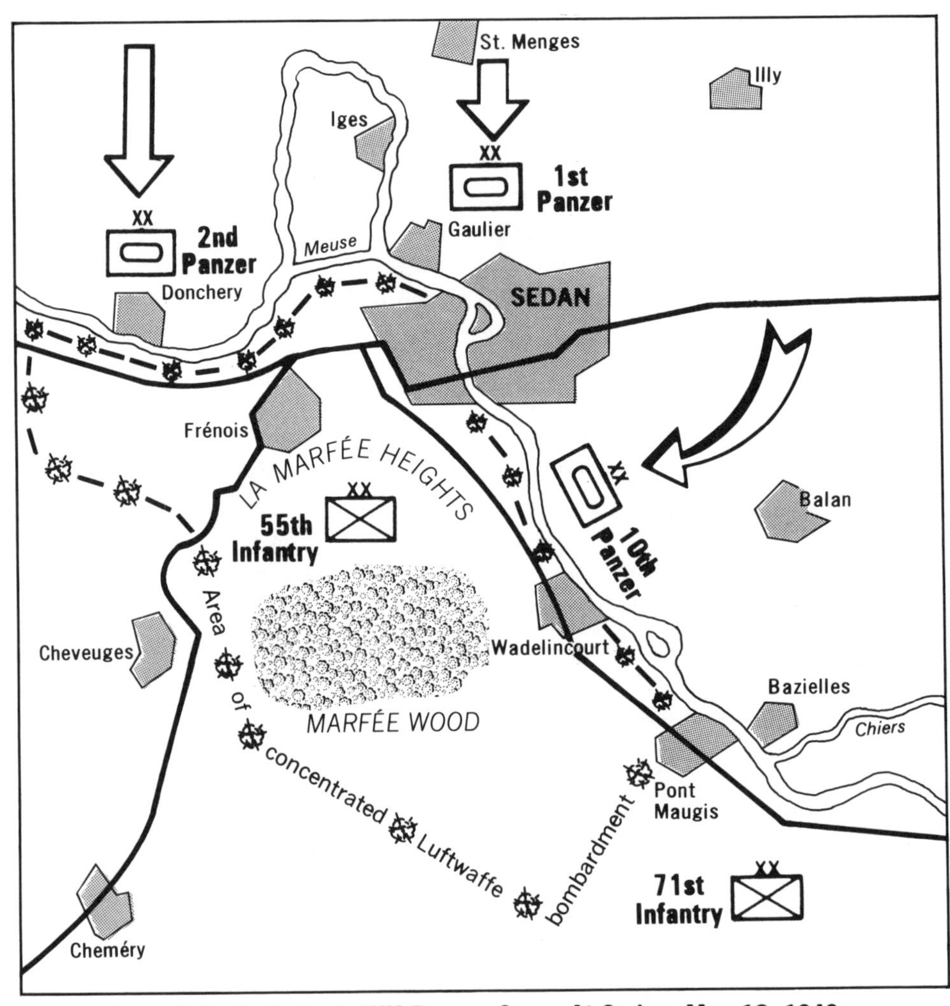

Assault By Guderian's XIX *Panzer* Corps At Sedan, May 13, 1940

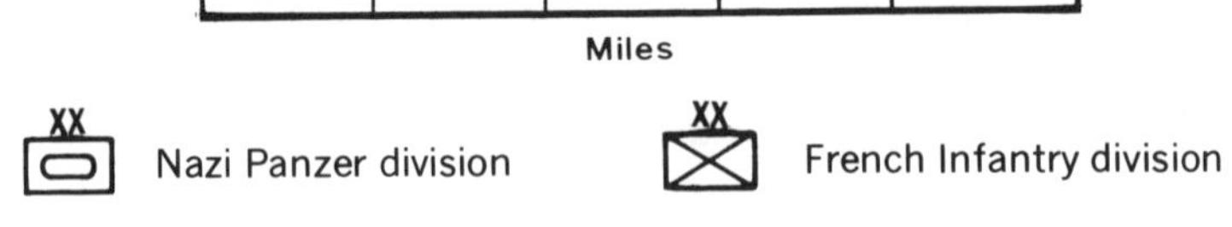

Writing after the French collapse on the effect of the dive bomber upon French morale, the historian Marc Bloch (subsequently tortured and shot by the Gestapo for his resistance activities) stated:

> Nobody who has ever heard the whistling scream made by dive bombers before releasing their load is ever likely to forget the experience. It is not only that the strident din made by the machines terrifies the victim by awakening in his mind associated images of death and destruction. In itself, and by reason of what I may call its strictly acoustic qualities, it can so work upon the nerves that they become wrought to a pitch of intolerable tension whence it is a very short step to panic. . . . No matter how thickly bombs may be sown, they never, in fact, register hits on more than a relatively small number of men. But the effect of bombing on the nerves is far-reaching, and can break the potential of resistance over a large area.[20]

As the Germans launched their rubber dinghies, the dive bombers shifted their targeting farther behind the actual river fortifications, attacking, for example, artillery positions near the Bois de la Marfée. Though the initial assault wave took heavy casualties, the morale of the French defenders had been shattered. French artillery failed to intervene in any meaningful way against the attackers, and French infantry—with a few gallant, isolated examples—surrendered or retired. By evening, assault forces of the 1st, 2d, and 10th Panzer were established on the opposite bank of the Meuse. By midnight, Guderian's engineers had spanned the river with a bridge and the first tanks were on their way across, together with accompanying flak and artillery.[21]

Still, at this point, the thrust could have been blunted and turned aside by prompt and decisive action. Air chief Loerzer had received the orders from Sperrle changing the form of preparatory air attack and had wisely decided to ignore them, less, it seems, from concerns of changing an attack even as it was about to unfold than from a generalized belief that Guderian was right: air support should be sustained and applied in a continuous fashion. Despite Loerzer's and von Richthofen's best efforts, French defenders had taken a heavy toll of the first assault wave across the Meuse. Now, with the Luftwaffe called away with more urgent tasks—the final subjugation of Holland—the opportunity existed for a concerted French attack, supported by Anglo-French air power, to still salvage something from the fighting on the Meuse, to still bring an end to the headlong German attack.

The Allied Riposte

Allied air resources had been devastated by the abrupt German air assault on May 10. In addition to the Armée de l'Air's fighter, bomber, and assault formations, the Royal Air Force had two RAF components present in France: the Royal Air Force Component of the British Expeditionary Force (essentially a group of observation and reconnaissance squadrons with four fighter squadrons for air control), and the Advanced Air Striking Force (AASF) consisting of ten squadrons of Fairey Battle and Bristol Blenheim light bombers charged specifically with attacking advancing German columns. The AASF had two "organic" fighter squadrons to escort its bombers and protect its own bases in the event of German attack. While the quality of RAF aircrew was excellent, and fully the equal of the best the Luftwaffe had to offer, the same could not be said of its aircraft. The Fairey Battle was a first-generation monoplane bomber that was entirely too slow and awkward for the conditions of 1940. In recognition of their vulnerability, Battle crews were directed to undertake low-altitude sorties to prevent fighters from attacking them from below; unfortunately, this did nothing to help them confront the multiple light flak that accompanied German formations. The Blenheim was likewise obsolescent and vulnerable to both intercepting fighters and ground defenses. The RAF's Hawker Hurricane fighters were marginally inferior to the Bf 109 and were, in any case, outnumbered.[22]

From the outset the Germans took a heavy toll of RAF and French attackers. On May 10, for example, thirty-two Battles attacked German columns in Luxembourg through "a storm of machine gun and small arms fire," losing thirteen outright and resulting in damage to all the rest. The twin-engine Blenheims fared little better, five out of six being lost to Bf 110's during one raid. On May 11, eight Battles left on another low-level attack; only one returned. On May 12, during ill-fated attacks on bridges in the Veldwezelt-Maastricht-Tongres area, five out of five Battles were shot down, the crew of one receiving a tongue-lashing from a German officer on the futility of making attacks against heavily defended targets. In the first three days of the war for France, the Battles never took less than 40 percent losses; by the end of May 12, the number of available Battle and Blenheim bombers had dropped from 135 to 72, the vast majority of these lost during low-level attacks against German columns. The French fared no better; ten Lioré et Olivier LeO 45 (an elegant and modern twin-engine bomber) attacked the Maas (Meuse) bridges at the Albert Canal and surrounding roads, losing one outright and with the other nine so riddled from light flak that only one was airworthy the next day. The fast little Breguet 693 attack bombers did even worse; five of six were blasted out of the sky like so many ducks during a low attack on Tongres, and the one that survived returned only to be junked. (One surviving Breguet gunner recollected initially thinking of the low-level attack as "a grand game!" but as they passed over the German column—"a lovely target!"—the flight was engulfed in "a hell-like outburst of fire and steel and flames, which grew bigger. I saw clearly the bursts of small caliber shells climbing towards us by the thousands" [exclamations in original text].) Disconcerted by low-level flak, French and British bomber crews shifted to medium altitudes, taking fewer losses but suffering a marked degradation in accuracy. Five Breguets returned to Tongres at medium altitude, losing two of their number to flak, the rest being damaged. Twelve LeO 45's attacked Tongres in the early evening, avoiding losses by bombing from altitude, but not hitting anything either—and they still returned badly shot up. Overall, French sources concede that the attacks near the Albert Canal were *"très coûteuse"* (very costly).[23]

Despite the change in Allied bombing tactics from low-level nap-of-the-earth flying to medium-altitude approaches followed by shallow diving attacks, things did not improve when the Allies tried to intervene at Sedan. On the 14th, the day after the crossing, the French and British threw themselves at the German bridgeheads, but were decimated by fighters and flak. Forty of seventy-one British Battles and Blenheims sent over Sedan that day were shot down. The Battles began their attacks at 0430 and 0630, mercifully evading fighters at that early hour and catching even the flak unawares; it was the only good news of the day. Attacks by French aircraft began at 0900 with a raid by nine Breguet 693's, followed by a raid at noon by old Amiot 143 bombers—a twin-engine product of the early 1930s already considered antique. In mid-afternoon, between 1500 and 1700, the Battles and Blenheims were committed in force, taking between 66 and 100 percent losses depending on the unit employed. Twenty-eight out of thirty-seven aircraft dispatched were lost in these two hours. In the early evening, the Blenheims attacked again, sustaining losses that were moderate—25 percent—in comparison with the carnage of attacks made earlier when the Bf 109 and Bf 110 combat air patrol was at full strength.[24]

Interestingly enough, even these largely ineffective attacks had a strong morale impact on German ground forces. They actually complained, in fact, that their fighter screen was too weak! Undoubtedly the bombing did contribute to a general uncertainty that bothered German ground forces. Guderian's memoirs contain references to several attacks that annoyed him, including one, on May 12, in which Allied bombers successfully bombed an engineer supply column, triggering many secondary explosions and resulting in a boar's head tumbling from a wall in his headquarters and missing him "by a hair's breadth."[25] Generaloberst Gerd von Rundstedt, commander of Army Group A, was caught with Guderian during the French midday attack on the Sedan crossing on May 14, unflappably asking Guderian dryly "Is it always like this here?"[26] Ironically, it was often German aircraft that posed the greatest danger to the ground forces, particularly once the Meuse was crossed and aircrews could no longer rely on it as the delineation between friendly and enemy forces. As previously mentioned, a Stuka attack near Cheméry caused heavy losses when the dive bombers mistook advancing Panzers for an expected French armored counterattack. Perhaps because of this episode, when a group of German airplanes attacked the 2d Panzer Brigade at Querrieu, near Amiens, the unit's flak opened fire without hesitation. "It was perhaps an unfriendly action on our part," Guderian recalled,

> but our flak opened fire and brought down one of the careless machines. The crew of two floated down by parachute and were unpleasantly surprised to find me waiting for them on the ground. When the more disagreeable part of our conversation was over, I fortified the two young men with a glass of champagne.[27]

It would have taken a well-thought-out combined-arms defensive to overcome the German assault into France. But doctrinal deficiences, weak morale, misuse of equipment, poor strategy, and poor tactical response negated any chance for success. In particular, French fighter and bomber sortie rates were extremely low—0.9 sorties per aircraft per day for French fighters versus upward of four sorties per aircraft per day for German fighters, for example. *Ten* squadrons of Breguet attack bombers flew only 484 sorties (with forty-seven losses) during the war; likewise, *eight* squadrons of Martin 167F attack bombers flew only 385 sorties (for fifteen losses) before the collapse. When French Morane-Saulnier M.S. 406 fighters were assigned to attack German tanks in late May, they lacked armor-piercing ammunition for their 20mm cannon, which could have enabled them to penetrate the vulnerable upper armor of the German Panzers used in the French campaign. Instead of gunning tanks, they were, in turn, decimated by German light flak.[28] As historian Faris Kirkland has shown, the failure of the French air force in 1940 was not, as is popularly believed, a matter of inferior technology (in fighter vs. fighter combat, in fact, the French air force seems to have consistently outperformed its Luftwaffe opponents); individual French aircraft were, in fact, every bit as good as their opposites. The failures lay in the kind of air war the French ran (with, for example, its "zone" employment of air power, which prevented the kind of mass and concentration of French air resources that might have made a difference), logistical support, and the weaknesses enumerated above. Arguments that French airmen could have salvaged this situation and "won the war" simply ignore existing realities. Undoubtedly, available Anglo-French air power could have been utilized with greater effectiveness in the battle for France. But to say this is not to imply that it could have, on its own, salvaged the situation. The collapse of France in 1940 highlighted the fact that nations, once launched on a course of military inferiority, cannot expect to make up for years—two decades in the case of France—of military neglect in the short span of months, particularly during the crucible of combat.[29]

The Lessons of Blitzkrieg

The performance of German air-land warfare during the blitzkrieg war of 1939 and 1940 had been most impressive, and inspired Great Britain, the United States, and the Soviet Union to spend a great deal of time thinking about the nature and future of air-land campaigns.[30] What was noteworthy about the German campaign was its synergistic blending of *firepower* on the battlefield—from infantry, armor, artillery, and air—coupled with *concentration, surprise, speed,* and *continuity.* The Wehrmacht's planners anticipated that unimpeded Allied air attack could be deadly, and so they emphasized acquiring air superiority by both deep operations (such as fighter sweeps and deep strikes against Allied airfields) and by operations over the battle area. The insightful employment of massive amounts of flak further negated Allied air effectiveness, and, of course, the flak itself proved useful in the actual conduct of the land battle. When operating against static defensive positions—as at Sedan—the close support forces functioned well; they did less well when operating in support of rapidly advancing land forces, as evidenced by confusion on the location of friendly and enemy forces, and occasional friendly bombings.

Already the stuff of myth because of their effectiveness in the blitzkrieg, Stuka operations had nevertheless revealed the fatal helplessness of the dive bomber when caught unprotected by high-performance fighters; in one case, a patrol of five French fighters had destroyed an entire formation of a dozen Ju 87's without loss to themselves. This vulnerability manifested itself in even greater Ju 87 losses during the Battle of Britain, and signalled the total unsuitability of the Stuka for combat in conditions other than the Eastern Front (where the air situation was so fluid that one could still get away with this kind of operation) or where the Luftwaffe could ensure air superiority. There is great irony here, for as the Germans were already beginning to discover the limitations of the special-purpose dive bomber, the Allies, impressed by the blitzkrieg, were about to waste vast amounts of time, money, and energy discussing it and even purchasing some for use in land warfare. Subsequently, Allied dive bombers operated in limited numbers in Sicily, New Guinea, and Burma, where their performance was generally less than spectacular.

As a result of the Polish and French campaigns, the Luftwaffe placed air-ground control teams in corps and division headquarters and with advancing infantry and Panzer units, equipping the latter with their own armored vehicles; improved intelligence information dissemination; emphasized coordination among the air units themselves—particularly the various Fliegerkorps—to ensure that close support and battlefield interdiction operations functioned smoothly; and, finally, emphasized protective fighter coverage of ground-attack and dive-bomber missions. The air-ground control teams consisted of an air signal liaison team stationed with the Ia (equivalent to a G-3—Plans and Operations) of German divisions and corps. Specialized fighter and ground-attack control detachments accompanied smaller forces into combat, operating typically at the regimental and Panzerkampfgruppe level. These teams controlled air strikes, responding to the control of the air signal liaison team which, in turn, acted as an important communications hub between division and corps requests, and the appropriate Fliegerkorps. The control detachments operating with the combat troops relayed information from the higher command of the Luftwaffe, advised and guided strike flights via radio and pyrotechnic signals, and generally kept the ground forces informed as to the "air's" intentions. Drawing on von Richthofen's experience in the Spanish war, these teams were occasionally accompanied by the Fliegerkorps commander himself—particularly during a major offensive or defensive battle—to ensure that

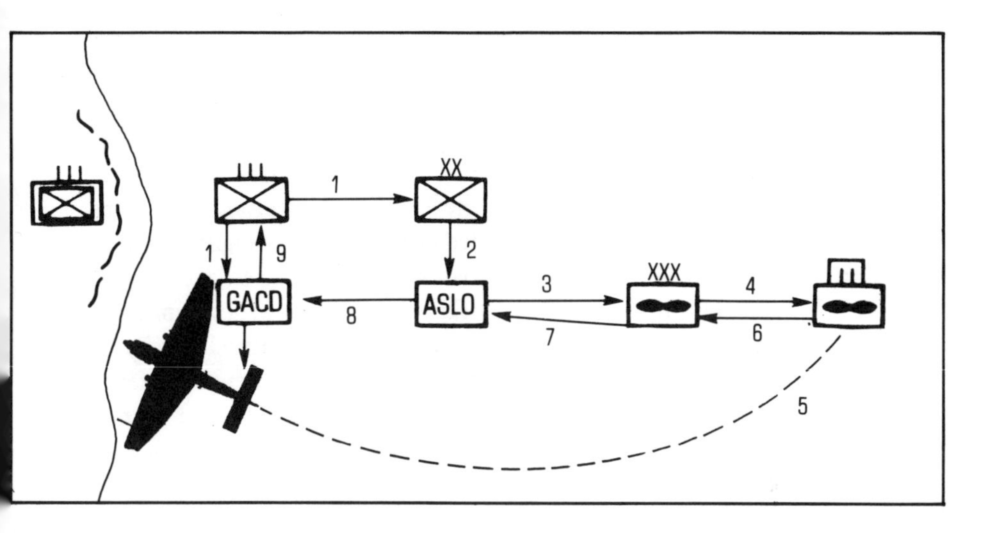

German Battlefield Air Support System

1. German infantry regiment confronting enemy infantry regiment in entrenched defensive position requests air support from its division headquarters, informing its attached Ground Attack Control Detachment (GACD) for coordination purposes.
2. Division HQ approves request and passes it to Air Signal Liaison Officer (ASLO) who heads an Air Signal Liaison Team.
3. ASLO relays requests to Fliegerkorps headquarters.
4. Fliegerkorps headquarters allocates air assets by contacting supporting air task force and ordering an air strike.
5. Air task force launches strike.
6. Air task force confirms launching strike to Fliegerkorps.
7. Fliegerkorps informs ASLO of strike, giving expected time, composition, and other relevant details.
8. ASLO relays strike information to GACD.
9. GACD relays strike information to infantry regiment.
10. GACD contacts strike flight leader and relays pertinent information to expedite success of strike.

Note: If supporting armored or mechanized forces, the GACD would be vehicle-mounted.

air-ground liaison was accomplished smoothly and effectively.[31] German air-ground operations were not rendered completely satisfactory as a result of the lessons learned in the blitzkrieg; many more changes would be made over the length of the war, one of the most important being the creation of a special *General der Schlachtflieger* in October 1943 to recognize the uniqueness of the attack aviation forces, separate them out from the fighter forces (headed by a "General of Fighters"), and more clearly delineate their position vis-a-vis the bomber forces (headed by a "General of Bombers").[32] But if all these changes and developments necessarily remained to be done, they nevertheless confirmed the significance of air-ground support operations. It was a lesson stated briefly, bitterly, and in sadness by France's Pierre Cot: "The Battle of France demonstrated the importance of air power in modern warfare; it proved that an army can do nothing without the support of an adequate air force."[33]

12

The Genesis of Anglo-American Air Support: The British Experience in the Western Desert

From Peace to War

Despite prewar doctrinal notions, both Great Britain and the United States developed their respective air support systems based on combat experience. Further, the American system, as employed in the Normandy breakout, was a system that owed its origins to British trial and error in the Western Desert. The swift German victory on the European continent stunned military professionals. In Great Britain, it resulted in a pronounced bureaucratic "turf war" between the British Army—which desired diversion of as many air assets as possible in support of army forces and, ultimately, creation of its own organic ground support air forces—and the Royal Air Force—which remained wedded to the interrelated notions of strategic air warfare involving bombardment of an enemy's homeland and air defense of Great Britain using high-performance interceptor fighters such as the Hurricane and the Supermarine Spitfire. Across the Atlantic, the American army echoed its British counterpart to a remarkable—and understandable—degree. (There, of course, the Army Air Forces were still a part of the Army per se, though the connection between the air and land forces was growing steadily weaker.) The realities of the European war forced both the RAF and the USAAF to come to grips with the problem of air support of land armies. In 1941, this manifested itself in a series of Army-AAF war games held in the United States, coinciding with actual British combat experience in the Middle East.[34]

The war games in the United States revealed limitations as well as potentialities of various combat forces. Not surprisingly, cavalry were shown to have little more than ceremonial value in modern war. Air operations during the Carolina and Louisiana games, however, also focused attention—particularly in light of the European experience—on future air-ground coordination and air support of the land battle. During the games, fully 60 percent of AAF sorties went toward interdiction missions, 22 percent to strike at armored and mechanized forces in rear areas, and 18 percent for "miscellaneous" missions including direct battlefield support; air units refused to support forces when the concentration of the "enemy" appeared too small to warrant profitable attack, or when the "enemy" was within the range of friendly artillery.

Control proved awkward; requests went from field units to division or corps level, were passed to an air liaison officer who approved or disapproved each request, and then passed it to an "air support command" that again reviewed it for approval or disapproval. When flights were en route to targets, the only air-ground communications link was via the air liaison officer at division or corps level; thus the actual front-line forces could not (as in the German system) communicate directly to the strike flight. Despite its deficiencies and a general uneasiness within the Army's ground forces, this system was elevated from operational concept to tactical reality via the issuance of Field Manual 31-35; its key element was the establishment of "air support commands" to work with ground forces. These commands actually had only observation aircraft on strength, and had to rely on the theater air commander to supply combat aircraft when needed; in 1942, fighters, dive bombers, and medium and light bombers were added as "organic" air support command elements. Air and land commanders both were uneasy with the arrangement. Air commanders saw it as drawing off strength that could best be used in the strategic air war against the enemy's heartland and in interdiction missions. Ground commanders saw the air support allotted to them as being too fragmentary, sporadic, and sparse to be much good. These concerns ironically mimicked arguments then raging through the British defense establishment as well, particularly since the American system roughly followed the inspiration of a British RAF-Army system which was, even at that time, undergoing profound revision as a result of lessons learned in France and the Western Desert.[35]

Britain's antipathy for direct involvement of aircraft on the battlefield—aside from its qualified experience and use of aircraft for such purposes in the "air control" operations of the 1920s and 1930s—was quickly swept aside by actual events. The perceived lack of effort of the RAF to assist the ground forces in France and in Greece—an unfair, but understandable, charge from soldiers exposed to bombing and strafing from an apparant all-triumphant Luftwaffe—was one reason. (After the Greek collapse, British soldiers referred to the RAF as the "Royal Absent Force.") But the other was the "overtaken by events" syndrome: the conduct of actual war operations resulted in the RAF, whether it liked it or not, being increasingly drawn into the land battle.

For example, a brief revolt by pro-Nazi elements in Iraq led by Raschid Ali resulted in a month-long siege of an RAF flying training school at Habbaniya, Iraq, in May 1941, an action bearing some comparison with the subsequent debacle at Dien Bien Phu and the besieging of the Marines at Khe Sanh. Using a motley assortment of modified trainers and a few genuine combat aircraft, the RAF contingent bolstered the meager ground defenses at Habbaniya by conducting 1600 sorties, dropping 100 tons of bombs (including 5500 20-lb antipersonnel bombs), and firing approximately 250,000 rounds of ammunition. Air attack proved critical in eventually routing the rebel forces and preventing both the overrunning of the camp and the likely collapse of the British position in that part of the Middle East. These strikes were conducted in concert with the RAF's armored car force on the ground and available infantry, and were characterized by operations well within 100 yards of friendly ground forces. In one case, Iraqi gunners moved a 40mm antiaircraft gun into a position behind a target marker's hut on a rifle range bordering the Habbaniya encampment. From there it could fire both at incoming aircraft and the ground defenders themselves. The task of taking it out was assigned to an elderly Fairey Gordon biplane. "I shall never forget," an eyewitness recollected,

> seeing the Gordon coming down virtually standing on its nose, flattening out, and then streaking across the airfield at about 6 feet, hidden completely from the pom-pom by the marker's hut. Then, lifting up slightly to clear the building, it laid its egg on the other side. The pilot swore he could see the look of horror on the gun crew's faces. I doubt he scored a direct hit, but that particular pom-pom site fell into immediate disfavour with the Iraqis.[36]

The brief Habbaniya campaign had demonstrated a degree of working together between air and land forces engaged in protracted battle previously unknown in the British experience. But it had been a static siege situation; now, British forces had a greater challenge—wielding an air-land weapon to confront the German and Italian forces challenging them in the Western Desert.

Western Desert Beginnings

The achievement of a workable air-ground support system is generally (and fairly) credited to Air Vice Marshal Arthur "Mary" Coningham, whose nickname was a corruption of "Maori," in recognition of his New Zealand origins. In fact, both the RAF and British Army in the Western Desert had been interested in such matters prior to Coningham's arrival, but it took his forceful touch to work out an understanding. An important but little-known precedent existed: even before the Habbaniya experience, a small force of South African aircraft had worked well in supporting British forces during the East African campaign which resulted in the expulsion of Italy from that part of the world. The South Africans had established a small "Close Support Flight" consisting of Hawker Hartbee and Gloster Gladiator biplanes. The theater air commander (an RAF Air Commodore, equivalent to a British Army Brigadier) chose to co-locate himself (with his own communications network) with the ground forces commander (a lieutenant general). Thus, any requests from the Army were immediately acted upon by the RAF commander. This system worked with great effectiveness during East African operations, and though it seems not to have had a major impact upon subsequent air-ground doctrine formulation, it is nevertheless a noteworthy forerunner to the kind of air-land operations conducted during the victorious campaign in Europe from 1943 through 1945.[37]

When Coningham arrived in the Western Desert, Erwin Rommel had already embarked on his rampaging campaign that would eventually last until the Nazi's final collapse in Tunisia in mid-1943. Rommel's activities further fueled the debate between the RAF and the British Army on the proper role of air forces over the battlefield. In November 1940, following the collapse of France, the RAF had created a special "Army Co-operation Command" intended to develop procedures and undertake training of RAF personnel for joint air-land operations. In fact, ACC never amounted to much, and the Army, indeed, saw in its withering a basic RAF hostility toward the direct air support notion. ACC eventually was disbanded at the beginning of June 1943, but by that time it had been replaced by a genuine working organization: the Tactical Air Force. Debates over the "proper" functions of air power consumed much time and generated much passion during higher headquarters meetings back in Britain, and eventually were elevated not merely to the service chief level, but to the level of the Prime Minister for decision. Charged with the responsibility of confronting the Luftwaffe and with supporting the Army during its grapplings with Rommel, Air Chief Marshal Arthur W. Tedder, in charge of RAF Middle East Command (himself intimately involved in the debate), was heartened by a missive from Prime Minister Winston Churchill in September 1941. Churchill's directive attempted to stake out a middle ground between the Army view (as enunciated by Chief of the Imperial General Staff Sir Alan Brooke) that the Army needed to control air assets over the battlefield and the RAF should provide a continuous air umbrella over Army forces engaged in battle, and the view of the RAF (as enunciated by Chief of Air Staff Sir Charles Portal) that the proper function of an air force was waging a strategic air offensive against the enemy that would ensure the safety of the Army on the battlefield by destroying the enemy air force's ability to wage war. (This same debate was echoed with a bit more tolerance and understanding at the theater level, between Tedder and the then commander-in-chief of Middle East, General Sir Archibald Wavell.) Churchill strongly stated:

> Never more must the ground troops expect, as a matter of course, to be protected against the air by aircraft. If this can be done, it must only be as a happy make-weight and a piece of good luck. Above all, the idea of keeping standing patrols of aircraft over moving columns should be abandoned. . . . Upon the military Commander-in-Chief in the Middle East announcing that a battle is in prospect, the Air Officer Commanding in Chief will give him all possible aid irrespective of other targets, however attractive. . . . The Army Commander-in-Chief will specify to the Air Officer Commanding in Chief the targets and tasks which he requires to be performed both in the preparatory attack on the rearward installations of the enemy and for air action durirg the progress of the battle.[38]

Churchill's memo came at a critical point, for the record of joint air-land operations between the RAF and the British Army in the Western Desert up to this time had been characterized by frustration and disappointment that boiled over into interservice bickering. In May and June of 1941, the Army, under Wavell, had attempted two operations to relieve the siege of Tobruk: Brevity and Battleaxe. Both were dismal failures, and inadequate air support had exacerbated the already tense relations between the Army and the RAF. Several key disagreements existed. The Army wanted aircraft to bomb German tanks, despite previous indications—including experience with British tanks being bombed by Axis aircraft—that tanks were most difficult to hit and destroy with aerial bombs. The RAF—if compelled to operate over the battlefield—felt much better about using bombs against more vulnerable motor transport and troops, particularly by showering enemy forces with 20-lb small antipersonnel bombs à la the Western Front and Palestine in the Great War. But the RAF was uneasy about bombing at all, since procedures governing establishment of a "bomb line," beyond which the RAF could bomb without fear of hitting friendly forces, were lacking.

Likewise, despite the previous experience of the RAF and British Army working together in the First World War and in the interwar years, much work remained to be done on developing common RAF-Army methods of designating targets, and marking friendly forces. Communication and coordination between the air and land commanders was a real problem; in Battleaxe, the air and land commanders had been separated by no less than *eighty miles* between their two headquarters! The critical problem was ensuring adequate air and ground communication, and it went beyond merely establishing an air-land radio net. The Army itself had inadequate radio communications with its own forces, and often could not be certain of the precise location of units so as to order up direct air support from the RAF with any degree of confidence that the strikes would not result in friendly casualties.

As a result of these experiences and the general controversy on air support, the Army and RAF established a joint working group in Cairo that did much to illuminate the various problems and difficulties. Thanks to strong direction from the two key commanders—the RAF's Tedder and the Army's General Sir Claude Auchinleck (who had replaced Wavell)—RAF and Army communication staffs hammered out a joint communication system to cover air-land operations. It went into effect at the end of September 1941 and consisted of mobile radio-equipped control parties called Forward Air Support Links (FASL). Located at Brigade and Division level, they maintained radio communication with reconnaissance and strike aircraft, and with their own Corps headquarters. At Corps level was a co-located Air Support Control Headquarters which would consult with Corps as requests for support came up, and if Corps determined that a strike was required, would then contact RAF Air Headquarters, Western Desert. If RAF Air Headquar-

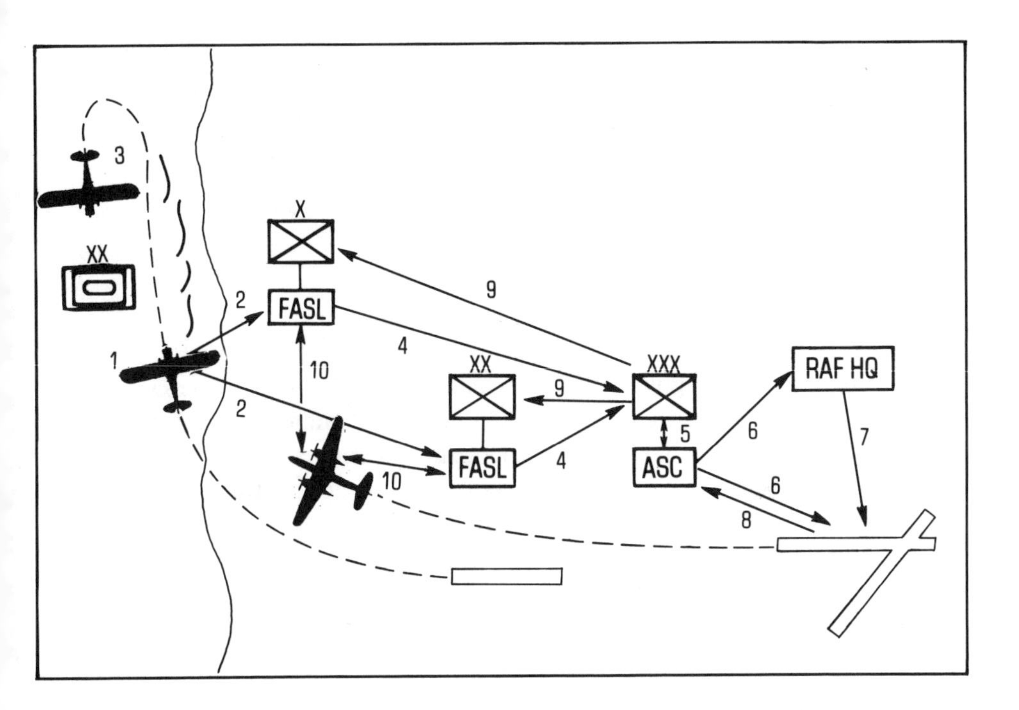

British Direct Air Support System (September 1941)

1. Recce aircraft from advanced airstrip spots potential threat.
2. Recce aircraft informs Brigade (only one shown for clarity) and Division Forward Air Support Links (FASL).
3. Recce aircraft remains on station (if possible) to furnish updates.
4. Brigade and Division FASLs inform Corps Headquarters.
5. Corps HQ coordinates with co-located Air Support Control Headquarters (ASC HQ) and determines course of action.
6. ASC HQ requests support from RAF Group HQ, simultaneously alerting rear airfield as to a possible mission requirement.
7. RAF Group HQ approves strike and sends order to airfield.
8. Airfield confirms strike with ASC HQ, which informs Corps HQ.
9. Corps informs Brigade and Division of mission.
10. Strike aircraft coordinate with Division and Brigade FASLs to receive updated information and targeting instructions.

ters concurred that a strike was required—and because of the mutual trust and confidence between Tedder and "the Auck," as well as the Churchill directive, this was an all-but-foregone conclusion—Air Headquarters would order a strike. The strike group would proceed to the battle area, coordinate their attack and receive final tactical updating from the Forward Air Support Links, and then attack the target.[39]

The Crusader Experience

Coningham, a strong-willed New Zealander who had taken over command of the Western Desert Air Force in July, complemented Tedder perfectly. To Coningham fell the opportunity—and responsibility—for trying out the new system in November 1941, during the Crusader offensive. Crusader aimed at destroying Rommel's forces, relieving Tobruk and opening Tripolitana to invasion. After a good start, British forces bogged down around Sidi Rezegh; though the British did manage to lift the siege, Rommel's sharp and decisive counterattack forced the Eighth Army back toward the Egyptian frontier, and paved the way for yet another Rommel offensive several weeks later. On the ground, then, Crusader was yet another disappointment. But in the air it was something else again. Though the new ground support system had more than its share of difficulties to be worked out, in general it functioned well. Coningham, assisted by providential weather, quickly established air superiority over the battlefront. Unusually heavy rains grounded German airplanes and bogged down Panzers and trucks alike, making perfect targets for attacking Bristol Beaufighter, Bristol Blenheim, and Martin Maryland (the British version of the USAAF's A-22) light bombers and attack aircraft. Coningham also had a new aircraft on strength: a bomb-carrying version of the Hawker Hurricane nicknamed the Hurribomber. Hurribombers harassed German columns, dropping two 250-lb bombs and strafing with their cannon, together with Curtiss Tomahawks, the export model of the American P-40.[40]

The era of the Allied fighter-bomber had arrived, and Crusader is memorable for that introduction. The British fighter-bomber was a natural outgrowth of the Battle of Britain. At the end of that conflict, the inability of the Luftwaffe's bomber forces to survive in England's skies resulted in the German air service introducing the *Jagdbomber* (literally, "fighter-bomber"): the bomb-loaded Bf 109 (and soon the Focke-Wulf FW 190) used on harassing "tip and run" raids across the Channel.

RAF Fighter Command, looking for a mission to fulfill following its brilliant performance in the air defense of Great Britain in 1940, saw the bomb-loaded Hurricane (and subsequently the Spitfire) as an ideal means of forcing German fighters in France to rise up off their airfields and do battle: it was thus an outgrowth of the air superiority war that triggered this fighter-bomber development, together with the practical inspiration of the German experience. Hurricanes proved surprisingly effective for missions ranging from antibomber nighttime intruding over the Continent to daytime low-level attacks (dubbed "Rhubarbs") with bombs (and subsequently rockets) and strafing against a variety of targets. These strikes established practices and precedents for the mid-war attacks across the Channel by twin-engine Westland Whirlwind fighter-bombers and, in 1944–45, by Hawker Typhoons and American Republic P-47 Thunderbolts operating initially across the Channel from bases in England, and then, following the Normandy invasion, from advanced airfields on the Continent.[41]

The development of the fighter-bomber addressed two bankruptcies of prewar air doctrine thinking: first, that fighters would be *defensive* aircraft operating against bombers in an attempt to prevent them from fulfilling the dictum of "always getting through," and, second, that battlefield air applications would be the province of the specialized attack aircraft. For European and Middle East air warfare, the ground-attack role belonged increasingly to the fighter-bomber, beginning with Coningham's Hurribombers in Crusader. As the Royal Flying Corps in Flanders had distinguished between "trench strafing" (close air support) and "ground strafing" (battlefield air interdiction) in the First World War, the RAF now distinguished between "direct support" (blending what is now thought of as deep interdiction and battlefield air interdiction) and "close support" (which were attacks in support of the forward line of troops, identical in concept to modern close air support). Crusader, in late 1941, offered examples of both.

Crusader, which ended in late January 1942, offered a cautionary note for future air-ground operations. First, despite the new control system, communication problems and confusion between the location of friendly and enemy forces abounded. In part, this was due to the intensely fluid nature of the battle, which, in some respects, resembled more an engagement between two fleets at sea than the grappling of land armies. But it was also due to lack of established recognition and identification procedures, the fact that the Germans were operating large numbers of captured British trucks, and inexperience with the air-ground system. The *average* time from a request for air support to the actual attack on enemy forces in response to the call was between *2½ and 3 hours.*[42] More distressing was a perceived lack of will in the ground forces to come to grips with the enemy and defeat him—and here the shoe was, for once, on the other foot; after France and Greece, it had been the RAF that had been dubbed the "Royal Absent Force," and worse. Now, to Coningham and Tedder, and many aircrews down the line, it was the Army that now absented itself from the battlefield. The official RAF history frankly states that

> the Royal Air Force had clearly achieved a very fair degree of superiority and the Luftwaffe had in no material sense affected the outcome of the battle. Yet the Army had still retreated. What the soldiers had previously imputed to the Air Force's wilful refusal to give proper protection, the airmen now imputed to the Army's faint-heartedness and incompetence.[43]

Clearly, no matter how successful it might be, an Air Force could not win a land battle when the land forces were not able, for whatever reason, to shoulder their burdens and come to grips with the enemy. Crusader was fought with a workable but untried air-ground system. It had functioned as well as such a new system could reasonably be expected to. But much better would have to be expected of air-ground cooperation in the future if Rommel were to be defeated.

From Crusader to El Alamein

The year 1942 has been called "The Year that Doomed the Axis." It was also the year that America became involved in the air-ground war, and the year that Coningham and Tedder had an opportunity to demonstrate decisively what the new air-ground system could accomplish. Its impact upon future American operations would not come until the middle of the following year, and then, with growing skill and understanding, the Allies would move on to Italy and, in 1944, to France. During 1942, fighting in the desert vied with renewed Axis threats against Malta, the staggering Japanese success in the Pacific and Southeast Asia, and the needs of the Soviet Union for scarce resources and attention. In the midst of this, in May, Rommel launched an offensive at Gazala that would carry the Afrika Korps to El Alamein, perilously near to Alexandria. Badly led—eventually Auchinleck himself would be forced to take direct command of the army and craft the victory of first El Alamein—the British Army fell back before Rommel's onslaught. Without strong land forces, the RAF could not be expected on its own to wrench victory from defeat, but it made a noticeable impact on the land battle, and its impact was of critical importance. Rommel was always conscious of air attack, and made frequent mention of it in private papers and in letters to his wife Lu. In the fighting that raged in this offensive, the RAF and Luftwaffe both undertook extensive operations against opposing ground forces, and Rommel's writings reflect those in the full, noting the decreasing strength of the German air force as attrition took its toll and the continuing presence of the RAF which appeared daily to grow more in strength and confidence. During bitter fighting at Bir Hacheim, the RAF threw its full weight into the battle, but was unable to prevent Rommel from defeating a gallant French garrison force.[44]

Coningham unveiled yet another wrinkle in desert air support in this battle: the Hurricane IID, a special antitank version carrying two Vickers 40mm cannon. The new Hurricane dated to an Air Staff decision in May 1941 to equip a modified Hurricane with a new 40mm antitank cannon developed by Vickers for the ground forces. It carried two cannon, each with fifteen rounds of armor-piercing ammunition per gun, and had two small .303 machine guns carrying tracer bullets to use for sighting in the larger weapons. This aircraft anticipated later German and Russian cannon-armed antitank aircraft. The Hurricane IID was very effective in its anti-Panzer role; on one mission, for example, four IID's destroyed five tanks, five trucks, and an antitank gun near Mersa Matruh, attacking from altitudes so low that one Hurricane received severe damage from colliding with a tank it had just destroyed. Despite this good start, however, the IID did not prove as robust as its Russian cousin, the Shturmovik, and though the IID subsequently did a great deal of good work, its reputation never approached that of the more famous Soviet or German "tank busters."[45]

The near-total collapse of British forces in the wake of the fall of Bir Hacheim resulted in "the Auck" taking over direct command of the Eighth Army from its commander, General Sir Neil Ritchie, and in the last week of June, the fighting in the desert neared its crisis point. Auchinleck ordered a retirement to pre-established defensive positions (the so-called El Alamein line) and the Eighth Army dug in, awaiting Rommel's assault, which generated bitter fighting during almost all of July. Strained by Bir Hacheim, pressed to the limit by the intense operations involved in the seizure of Tobruk, the Luftwaffe could not maintain the pace of operations necessary to support Rommel's continuing thrust toward Alexandria; in contrast, the RAF flew nearly

15,400 sorties (not including antishipping strikes) in July, nearly 5500 in the first week of July alone.[46] Rommel's writings became even more pessimistic; during the fighting around El Alamein, he wrote of "continuous round-the-clock bombing by the RAF," and referred to having set up his command post on Hill 31 "under relay bomber and low-flyer attack."[47] Worse was to follow. Auchinleck had made a near-miraculous stand at El Alamein, and had managed to bring Rommel's offensive to a close, but he had fallen victim to a Churchillian purge and, though the true victor of Alamein, was replaced by Lieutenant General Bernard L. Montgomery. Tedder, an Auchinleck admirer, was sorry to see him go; though Auchinleck's taking direct command of Eighth Army had resulted in the army headquarters and the air headquarters being located apart, Auchinleck's personal strong interest in the air issue and his working relationship with Tedder (and through Tedder with Coningham) had ensured that operations went smoothly. For example, during first El Alamein, the delay between Eighth Army's request for air support and the launch of a strike flight diminished to thirty-five minutes—a far cry from Crusader's 2- to 3 hours. Subsequently, Montgomery generally accepted Coningham's notions of air war, though both Tedder and Coningham faulted him—and rightly so—for his timorous and plodding pursuit of Rommel after second El Alamein.

Rommel attempted to restore his offensive with the battle of Alam Halfa in August 1942; he ran afoul of heavy minefields and intense battlefield bombing attacks that killed key staff and disrupted his units. Supply shortages added to his woes, and he ultimately had to call off his offensive, writing petulently to his wife that "we had to break off the offensive for supply reasons and because of the superiority of the enemy air force—although victory was otherwise ours."[48] In his papers, he went even further, complaining that the offensive had failed because of strong British defenses, supply shortages, and because

> non-stop and very heavy air attacks by the RAF, whose command of the air had been virtually complete, had pinned my army to the ground and rendered any smooth deployment or any advance by time-schedule completely impossible.[49]

He went on to state:

> We had learnt one important lesson during this operation, a lesson which was to affect all subsequent planning and, in fact, our entire future conduct of the war. This was that the possibilities of ground action, operational and tactical, become very limited if one's adversary commands the air with a powerful air force and can fly mass raids by heavy bomber formations unconcerned for their own safety.[50]

Rommel noted in particular "the paralysing effect which air activity on such a scale had on motorized forces," and "above all, the serious damage which had been caused to our units by area bombing."[51] He concluded by bitterly stating:

> Anyone who has to fight, even with the most modern weapons, against an enemy in complete command of the air, fights like a savage against modern European troops, under the same handicaps and with the same chances of success. . . .
>
> The fact of British air superiority threw to the winds all the tactical rules which we had hitherto applied with such success. In every battle to come the strength of the Anglo-American air force was to be the deciding factor.[52]

In late October, 1942, Montgomery launched an offensive against Rommel at El Alamein, forever ending Rommel's hopes of reaching the Suez Canal and Alexandria. Rommel was in Austria on medical leave, but returned to the battle in time to participate in its final stages, imposing an order on the German retreat largely due to the famous strength and charisma of his own personality. During this offensive, the RAF's air power was less significant than the overwhelming superiority Montgomery had in material and tanks. In slightly over one week, the RAF flew 10,405 sorties, with American airmen flying nearly 1200 of their own; in contrast, the combined German-Italian total sorties for this time period was approximately 3120. The air offensive campaign at "second" El Alamein encompassed strategic bomber operations from Palestine against Benghazi, naval air operations (based on Ultra signals intercept intelligence information) to interdict Axis supply ships attempting to reach Rommel, and tactical aircraft from Coningham's Desert Air Force operating over the battlefield. Tedder had already welcomed some the of signs of Montgomery's personal leadership style—Montgomery, for example, had set up an operations room in Eighth Army headquarters patterned on the RAF model ("most refreshing" in Tedder's view), evidence of his committment to battle management—something that had been sorely lacking in his predecessors. Likewise, Tedder was pleased to find the RAF and the USAAF working together well at the operational level (though at the command level both Tedder and Portal complained about the "difficulties" of dealing with "the Americans"). Accordingly, Tedder—not without justification—expected air-land cooperation during the battle to function much more smoothly than previously. Generally, he was correct in his assumption; RAF and Army officers worked together with mutual confidence of a degree not previously experienced. Still, some difficulties remained: the traditionally minded staff of Montgomery's command center, for example, closed up shop midway through the battle, compelled by curiosity to go forward and see for themselves how the battle was progressing. But this was but a slight hiccup on an otherwise smoothly unfolding operation.[53]

The reality of Middle East operations reflected what was happening back in England in senior-level discussions between the Army, RAF, and Churchill. Approximately three weeks before, on October 5, Brooke (the CIGS), Portal (CAS), and Churchill did battle one more time on the air support question that had gnawed at RAF-Army relations since the Dunkirk withdrawal in 1940. It was Churchill who, in a memo issued two days later, had the last word: "Whenever our Army is established on land, and is conducting operations against the enemy, the system of organization and employment of the Royal Air Force should conform to that which proved so successful in the Western Desert."[54] (He was thinking, of course, of an upcoming invasion of Europe—specifically, a return to France.)

Following second El Alamein, Coningham drafted a pamphlet on air support for Montgomery, which Montgomery subsequently issued as his own; it was intended to give soldiers a realistic appreciation of the role of battlefield air support. Given the battles back in England, such a document might have proven most difficult to prepare; given the blooming RAF-Army relationship from Crusader onward, it was natural that it would come out of the Middle East. It emphasized that the RAF and Army must "live and work side by side" via adjacent headquarters, and included a discussion on command and control, types of support, the necessity of good intelligence and communications, the role of air support in "set-piece" and fluid battles, emergency procedures, the necessity of a bomb-line, and similar matters. But at its core was a critical paragraph that reflected more the RAF view than that of the General Staff, and which thus represented an acceptance by the British Army of a lesson learned over the painful months of desert warfare—a lesson that was at once a prime element of the RAF credo:

> It follows that control of the available air power must be centralized and command must be exercised through Air Force channels. Nothing could be more fatal to successful results than to dissipate the air resources into small packets placed under command of Army formation commanders with each packet working on its own plan. *The soldier must not expect or wish to exercise direct command over air-striking forces* [emphasis added].[55]

13

The Necessary Interlude: Doctrine and the American Experience in the Pacific, Tunisian, and Italian Campaigns

In April 1942, the U.S. War Department had issued Field Manual FM 31-35, entitled "Aviation in Support of Ground Forces." This manual attempted to create a workable ground-air support system, but, in truth, it merely generated the appearances of such a system. With its network of Air Support Parties, Air Support Control Centers, and communications between the ground forces and the air forces (consolidated within an Air Support Command), it looked like a tactical air control system. In fact, however, it was cumbersome, and flawed in both concept and execution. Conceptual weaknesses were its emphasis on Corps-level air support; the Corps commander and staff in effect had their own mini air force on call for their use. In execution, this resulted in a tendency to be overly concerned about one's own forces, and not as concerned as to what was happening elsewhere. Further, there was a built-in tendency to try to stem enemy air and ground attacks at the main line of resistance, rather than working farther back in enemy territory, disrupting communications, logistical lines, supply and strong points, airfields, and the like. It was, in short, the kind of system that the British Army wanted in 1939–41, until experience in the Western Desert taught otherwise. Ominously, it was quite close to the French system of "zone" air power, where ground commanders were able to parcel out—and hence dilute—the French air effort so that it could not be concentrated to best effect to meet the onrushing German forces. While FM 31-35 by no means implied a consensus of what needed to be done or what system

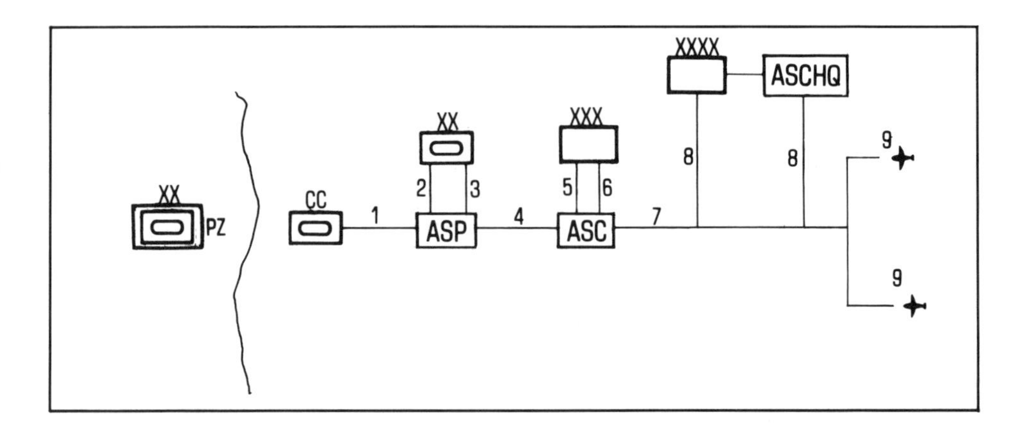

U.S. Air Support System, April 1942–February 1943

1. An armored Combat Command requests an air support mission from Air Support Party (ASP) attached to Division HQ.
2. ASP relays request to Division Commander for approval.
3. Division Commander approves request and notifies ASP.
4. ASP relays request to Air Support Control (ASC) at Corps HQ.
5. ASP relays request to Corps Commander for approval.
6. Corps Commander approves request and notifies ASC.
7. ASC orders mission from rear-area airfields.
8. Army and ASC HQs monitor network, intervening if necessary.
9. Mission is launched from rear-area airfields.

should exist for air-ground operations, it nevertheless was the system under which American air and ground forces went to war against Rommel in Tunisia in 1943.[56] Ironically, by that time, American air and ground forces had already demonstrated on the other side of the globe that they could wage effective joint service and coalition warfare while engaged against the Japanese—particularly tenacious and fanatical foes.

The Pacific Experience: Guadalcanal, New Guinea, and Bougainville

In the Solomon Islands and New Guinea campaigns, the procedures of FM 31-35 went out the window in favor of "adhocracy" solutions tailored to meet the special circumstances at the time.

For example, in Guadalcanal fighting, the AAF's 67th Fighter Squadron, a unit equipped with the Bell P-400 Airacobra (an export version of the P-39) was placed under the operational control of the 1st Marine Air Wing. The P-400, like the P-39, proved unsuccessful in air-to-air combat with the Japanese Mitsubishi A6M Type 0 fighter (the infamous Zero, or, as it was code-named by Allied technical intelligence, the Zeke). Thus, in an attempt to find a use for the plane, and as a consequence of the desperate conditions on the island, the 67th was put to work strafing and bombing on behalf of embattled ground forces, carrying a 500-lb bomb or depth charges for use as antipersonnel weapons, in addition to its guns (which included alternatively a 20mm or 37mm cannon firing through the propeller hub). The P-39/P-400 proved a spectacular ground-attack airplane; the AAF judged it "invaluable when employed in close coordination with the Marine troops," and it subsequently figured prominently in many of the bitter Guadalcanal battles. During fighting at Bloody Ridge, three P-400's devastated Japanese concentrations with intense bombing and strafing, though ground fire so damaged two of the aircraft that they had to be landed "deadstick" (i.e., sans power) back at Henderson Field. Distinguishing enemy from friendly forces proved difficult in the thick jungle conditions on the island, and following unsuccessful use of colored identification panels, the Marines assigned a radio-equipped "air forward observer" team to front-line forces to control and direct incoming air strikes, with much greater success. To get around the problem of joint USN-USMC-USAAF communications (since naval and AAF aircraft operated on different frequencies), a radio from a grounded P-400 was modified with two microphones and placed adjacent to radios used to control naval aircraft so that all strike flights could be controlled simultaneously. Drawing on their small-wars experience, the Marine air and ground commanders in Guadalcanal—most of whom were Nicaraguan veterans—used their combined air and land strength to confront and ultimately defeat their Japanese opponents. The control methods and joint-service working partnerships first explored at Guadalcanal continued, with some elaboration, throughout the island-hopping war of the Southwest Pacific.[57]

New Guinea fighting added new air support experience, and, in most respects, resembled that of Guadalcanal, though on a greater scale. Here the Americans fought in close partnership with the Australians, and here the AAF got its chance to use its own dive bomber—the Douglas A-24 Dauntless (an "Army-ized" version of the Navy's carrier-killing SBD) which arrived in the New Guinea theater in April 1942. The experience soured the AAF on the dive bomber once and for all, for the A-24 revealed all the weaknesses of the Ju 87 when operated in a land campaign in which supporting air superiority was lacking. Only one out of seven returned from a mission on July 29, 1942, despite P-39 escort; after this, the AAF quickly withdrew it from combat, sending the rest back to Australia's relative safety. Operating with protective escorts better able to handle the Japanese fighter threat (such as the Navy's Grumman F4F and Vought F4U fighters), the attractive and elegant Douglas Dauntless was a most formidable weapon; operating in the far less secure conditions of New Guinea, the Dauntless was at a distinct disadvantage. As in Guadalcanal, ground forces quickly sought ways to develop cooperative support arrangements with Allied air forces. Infantry and artillery officers, for example, were assigned as air liaison officers and gave strike crews pre-mission briefings. The Australians went further, and developed airborne strike coordinators and controllers, anticipating the post–World War II forward air controller system utilized in Korea and Southeast Asia. Using two-seat Commonwealth Whirraway tactical reconnaissance and liaison aircraft, Royal Australian Air Force pilots and observers led strike flights to ground targets. Subsequently, the RAAF introduced the Commonwealth Boomerang, a specialized army cooperation and ground support fighter which operated like a "fast FAC" over the edge of battle, leading strikes and marking targets for attacking aircraft. The thick jungle conditions of New Guinea made distinguishing friendly from enemy forces very difficult, and there were numerous cases of friendly bombings. For their part, Japanese forces in New Guinea feared artillery and mortar fire far more than air strikes. But over time, Allied air-ground effectiveness improved markedly.

During the landings on Bougainville in December 1943, close cooperation between air and ground forces resulted in some devastating air strikes. Admittedly, the success of these strikes was largely due to both air and ground forces being within the same service. At Hellzapoppin Ridge on Hill 1000, Marine Grumman TBF Avenger torpedo bombers flying as low as fifty feet dropped forty-eight 100-lb delayed-fused bombs in a criss-cross series of passes against Japanese defenders as close as seventy-five yards from attacking Marine infantry from the 3d Marine Division. The infantrymen judged the strikes as "the most effective factor in taking the ridge." Generally, the Marines used pre-mission briefings by infantry officers on terrain peculiarities and the tactical situation, flew ground liaison officers in attacking strike aircraft, and assigned air liaison officers with rifle companies being supported from the air. Communications and control were thus effective and appropriate; further, ground forces marked their positions with colored smoke and designated target areas with white smoke. While the density of jungle did hinder visual acquisition of targets, the extensive attempt at coordination and the closeness of the attack itself—something that would not have been possible had the aircraft been attacking more heavily defended positions, or under conditions where air superiority did not exist—ensured that every effort was made to identify hostile from friendly forces. In the ten close air support strikes conducted during the Bougainville landing, none of which was farther than 500 yards (and seven of which were less than 200 yards), Marine support aircraft decimated Japanese defenders. The price, unfortunately, was two Marines killed and six wounded by one plane whose crew misidentified its target and struck friendly forces, demonstrating that despite the greatest of precautions, battlefield air support accidents could still happen. The experience of Guadalcanal, refined amid the Bougainville and Northern Solomons campaigns, formed the basis for Marine air-ground doctrine and employment for the rest of the war, and, in turn, that experience carried over into Korea and onward into Vietnam and the present day.[58]

Kasserine and Its Aftermath

In January 1943, Gen. Dwight D. Eisenhower, Allied Commander-in-Chief in North Africa, wrote to Brig. Gen. Russell P. Hartle, expressing

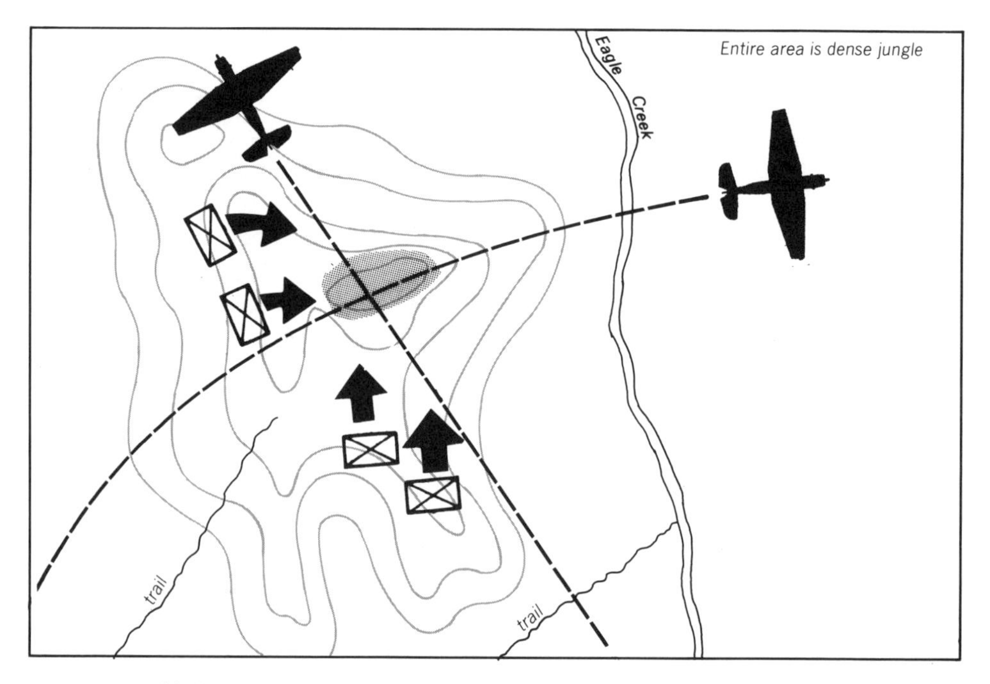

Air-Land Attack On Hellzapoppin Ridge, December 18, 1943

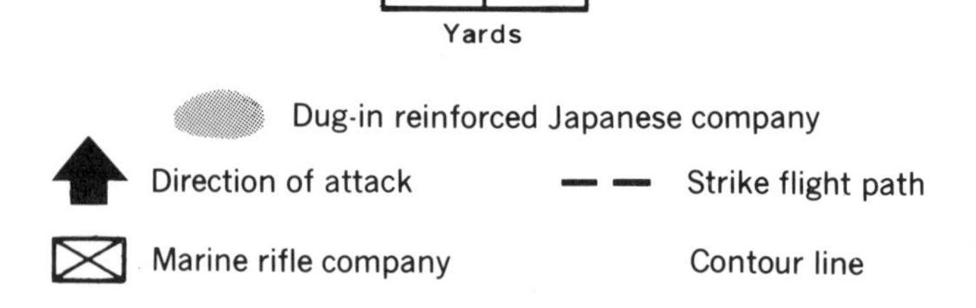

grave misgivings over the degree of training and discipline of the American troops arriving for combat in the Middle East. He emphasized in particular teaching troops to fire back at attacking aircraft with any available weapon, training ground and air forces to recognize each other and distinguish each other from German forces, and concluded that

> training in air support of ground troops must be conducted—and must be applicatory. We have a published doctrine [FM 31-35] that has not been proved faulty. The air and ground forces must be together in training, or they will not be able to do so in combat.[59]

Unfortunately, Eisenhower's fears would be realized in a few short weeks, during the battle of Kasserine and subsequent fighting to expel the Axis from North Africa. Kasserine exposed critical weaknesses in training, doctrine, and leadership and forced a change in the operations of American air units charged with responsibility for support of ground forces. FM 31-35 *did* "prove faulty." In retrospect, as Eisenhower himself recognized, the signs of the impending debacle had been visible in the weeks of fighting that preceded Rommel's bold thrust through Kasserine Pass.

Following the invasion of North Africa in November 1942, Allied forces had raced eastward from the landing sites in an attempt to seize Tunis. Unfortunately, for a variety of reasons, such did not occur, and as 1943 dawned, the British, Americans, and Free French forces operating to the west and east of German-held Tunisia recognized that the next few months would be marked by stubborn Nazi resistance. American forces, as yet unblooded in combat against the Wehrmacht, were infected with brash enthusiasm; such was particularly true of II Corps, commanded by Maj. Gen. Lloyd R. Fredendall, upon whom Rommel's Panzers would wreak decisive defeat, despite the availability of intelligence indications via Ultra that just such an attack might take place. Kasserine, retired General of the Army Omar N. Bradley subsequently wrote, "was probably the worst performance of U.S. Army troops in their whole proud history."[60]

Accompanying II Corps into battle was Col. Paul L. Williams's XII Air Support Command, built around squadrons of modern A-20 attack bombers, Supermarine Spitfire and Bell P-39 fighters, and North American B-25 medium bombers. Other medium bombers (such as the Martin B-26 Marauder), long-range Lockheed P-38 fighters, some P-40's, and even heavy Boeing B-17 Flying Fortresses were available from other commands, should an emergency arise. Operating in accordance with the dictates of FM 31-35, II Corps and XII ASC representatives soon got into a succession of spats over the quality of air support being delivered. Because the kind of coordination and control facilities present in the Coningham system or the later AAF Tactical Air Command system were lacking, information transfer between II Corps and the ASC was deficient. Accordingly, the ASC often did not find out about Army needs until after the fact, and the Army complained about its vulnerability to German air attack. During the retreat into Tunisia, the Luftwaffe had been increasingly active, a function of the shrinking of the Axis pocket so that its dwindling strength could actually assume greater significance. Tied to perceived Corps needs, the ASC was not free to undertake the kind of deep sweep operations that might have eliminated German air strikes by shutting them off at their source—over their own territory, at their airfields. Instead, American air power tended to be frittered away in "penny packets," with minimal result and nagging constant losses. For his part, Fredendall had so much faith in the air commander that he willingly told Williams to undertake on his own such missions as he saw fit; while showing a commendable faith in air power, Fredendall's detachment helped deepen the intelligence morass afflicting II Corps–XII ASC relations, so that, increasingly, ASC's missions bore little relationship to the real needs that II Corps actually had. Meanwhile, Army forces experienced enough German air attacks to make them nervous about any airplane above them, inculcating a "shoot first, ask questions later" attitude; for their part, airmen chafed under what they perceived as unnecessary restrictions tying them to land forces, as well as the recognition that they were apparently as likely to be fired upon by friendly forces as they were by the enemy.[61]

Thus was established a condition ready-made for disaster, and that disaster came in mid-February 1943 when Erwin Rommel launched a surprise attack westward toward Kasserine Pass. Failures within the ground forces contributed to his success as the offensive picked up momentum, crushing American resistance and resulting in large numbers of captured vehicles and prisoners. American air support was desultory at best, while the Luftwaffe threw itself into the fray with elan and vigor. The offensive opened on the 14th, and over the next nine days, the German advances forced American units back a total of twenty-one miles. Five forward airfields had to be abandoned, greatly complicating air support missions; the closest base was now Youks les Bains, across the Algerian border. Weather conditions from the 18th on precluded much flying by both the Allies and Luftwaffe, so that Kasserine's critical period was fought as virtually exclusively a land battle. Despite the weather, B-17's attempted to bomb the famous pass, but got lost and struck an Arab village more than *100 miles* from the battle area by mistake—a not uncommon navigational wandering in "strategic" air operations by virtually all the combatant nations in the early to mid-war period. Finally, on February 22, Rommel abandoned the offensive, less from Allied defensive measures than from internal dissension within the Nazi command itself. The Panzers returned to the defensive Mareth line, and the war in Tunisia went on to an eventual Allied victory in May, by which time Rommel had prudently left the scene.[62]

Kasserine resulted in major changes, not the least of which was the subsequent elevation of Lt. Gen. George S. Patton as commander of II Corps. But it resulted in the sacking of FM 31-35 as well. Midway during the battle, a previously planned Eisenhower reorganization of the Allied effort in the Middle East took place that affected both ground and air units. This opportune reorganization aimed at centralizing air and ground commands, and created an 18th Army Group under the command of Gen. Sir Harold Alexander, which incorporated both British and American units. On the air side, Tedder became AOC of Mediterranean Air Command, again a command incorporating both American and British subordinate organizations. Under Tedder, USAAF Lt. Gen. Carl W. Spaatz assumed command of the Northwest African Air Forces (NAAF), one of which was the newly formed Northwest African Tactical Air Force, headed by Air Marshal Sir Arthur Coningham. Coningham arrived at the front before the end of the Kasserine debacle, and immediately imposed his own command style upon the players, essentially scrapping FM 31-35 in favor of his own air-ground coordination and control system that had worked so well previously. In the future, air requests would have to go above Corps level to the highest Army level, and tactical air units—specifically the XII Air Support Command and RAF 242 Group—would be commanded by Coningham directly. The mission of the tactical air forces would be, first and foremost, the neutralization and destruction of enemy air forces; next would be the destruction of enemy columns by light bombers and roving fighter-bombers. Tanks, because of the difficulty in destroying them, would be left alone, to be dealt with by the ground forces. Coningham had no illusions that his ideas would be accepted with enthusiasm by ground force commanders; accounts written by such since the war are filled with invective against Coningham, and his alleged bias toward support of British, rather than American, units.[63] But he found willing listeners among American airmen who were frustrated in the extreme over their misuse since Operation Torch, the invasion of North Africa. Three of those influenced were particularly bright and aggressive fighter leaders: Brig. Gen. Elwood Quesada, in charge of XII Fighter Command, Col. Philip Cochran (the real-life inspiration for cartoonist Milton Caniff's fictional Flip Corkin), and Col. William W. Momyer. These three men would adopt Coningham's philosophy—when they had the opportunity—to the advantage of their own units later on: Quesada's IX Fighter Command (later the IX Tactical Air Command), Cochran's First Air Commando Group of China-Burma-India fame, and Momyer's command of Seventh Air Force in Vietnam and, subsequently, USAF Tactical Air Command itself.[64]

In the multitude of changes that resulted from America's Kasserine experience, many personnel in both the AAF and ground forces remained in the North African theater or returned to the United States to become influential spokesmen for various viewpoints. Basically, the airmen in Tunisia, aside from implicitly recognizing merit in Coningham's actions, saw within them the affirmation of beliefs long held within the Air Corps Tactical School before the Second World War: namely, the first and most important mission of an air force was to seek out the enemy air force and destroy it. Without air superiority, this viewpoint ran, all other missions could not be accomplished. (It was not merely an American notion; it was something, rather, that the air leaders of virtually all the combatant nations would have agreed with, despite postwar instant analysis by the victors that tended to denigrate the committment of the Axis air forces to that viewpoint.) Second, once having secured air superiority, the air force could next turn its attention to interdiction and deep strike. Only then should the air force be applied to direct battlefield support of an army. It was this viewpoint, for example, that Brig. Gen. Lawrence S. Kuter, who had been Eisenhower's air operations officer and subsequently deputy commander of the North African Tactical Air Force (NATAF), brought back to Washington when he returned to the States in May 1943 to become Hap Arnold's planning chief. In light of Kasserine, Eisenhower assembled a study group of air and ground officers to examine doctrinal implications of the Kasserine experience and prepare a new draft field manual on air power. The result of this was a document that had a staggering impact upon the future nature of air force–Army ground force relations: FM 100-20, "Command and Employment of Air Power."[65]

FM 100-20: The AAF's "Declaration of Independence"

In retrospect, there is a somewhat hysterical tone to FM 100-20, as if its authors were frustrated that their message through the years on the separateness of air power had been so missed that they had to ensure that even now, even after the Kasserine debacle and the shining beacon of Coningham's efforts, the message would not be ignored yet again. So, for example, FM 100-20's first section is typeset entirely in upper case, starting with the inflammatory words: "LAND POWER AND AIR POWER ARE CO-EQUAL AND INTERDEPENDENT FORCES; NEITHER IS AN AUXILIARY OF THE OTHER." It went on, equally boldly, to state that land forces were so vulnerable to air attack that the most important mission of a friendly air force was to neutralize the enemy air force. The missions of a tactical air force were rank-ordered in priorities; FM 100-20 stated that an air campaign must consist of three phases, corresponding to the three priorities:

> (1) *First priority.*—To gain the necessary degree of air superiority. This will be accomplished by attacks against aircraft in the air and on the ground, and against those enemy installations which he requires for the application of air power.
> (2) *Second priority.*—To prevent the movement of hostile troops and supplies into the theater of operations or within the theater.
> (3) *Third priority.*—To participate in a combined effort of the air and ground forces, in the battle area, to gain objectives on the immediate front of the ground forces.[66]

FM 100-20 hit with all the impact of one of Rommel's Panzers. Issued in July 1943, "this manual was known at the Pentagon—and viewed with dismay by the Ground Forces—as the Army Air Forces' 'Declaration of Independence.'"[67] In many ways, it was a declaration of independence for the air force, but in reality, it constituted more a recognition of a growing separateness that had evolved over the years. Although FM 100-20 officially only replaced the latest edition of FM 1-5 ("Employment of Aviation of the Army," issued in January 1943), in reality it furnished the final nail in the coffin of FM 31-35 as well. In so doing, it did not substitute a body of tactical recommendations or procedures to replace 31-35; as can be noted above, the "third priority" quoted previously is vague in the extreme. But it was this very vagueness that in many ways enabled FM 100-20 to enfold and—in the broadest possible sense—encompass the air support needs set forth in 31-35.

The American tactical air commands and air forces that prosecuted the war against the Nazis in 1944–45 developed differing support systems that were generically similar but which varied in structure, terminology, and procedures. Not until the issuance of a new FM 31-35 in August 1946 would a standardization of these systems be spelled out between the air and land forces. But the absence of that postwar manual did not mean that the needs of the land forces for effective air support were not being met during the 1944–45 time period. Rather, those needs (as will be seen) were being met spectacularly well, as even the ground forces were forced to recognize in their own analysis of late war operations. And they were being met within the framework of FM 100-20, which had originally engendered so much controversy and anxiety within the ground forces community.[68]

In retrospect, it is certainly easy to see why FM 100-20 was controversial. For example, it had been adopted by the War Department without the consultation or approval of Lt. Gen. Lesley J. McNair, the commander of Army Ground Forces and the organizational co-equal of AAF chief Hap Arnold. Rather, Eisenhower's group had produced a draft document that was routed to and approved by Gen. George C. Marshall, the Army chief of staff, bypassing the ground forces completely. (For his part, Marshall had been made aware of the Coningham-Tedder-Montgomery system of air support when the Assistant Secretary of War for Air, Robert Lovett, had brought Montgomery's pamphlet on air support to his attention; Marshall's favorable reception of it is evident in the subsequent adoption of FM 100-20.) If nothing else, this was an extraordinary happening and indicative itself of the serious impact that the Tunisian fighting—Kasserine in particular—had had upon the Army and "its" air force. FM 100-20's teachings were put into the curriculum of the Infantry School at Ft. Benning a month before it was formally issued, in July 1943. McNair regarded the issuance of FM 100-20 as an order to be carried out, but never ceased to view the doctrinal statements of the manual with the gravest reservations. He believed, as did most Army ground forces officers, that the issuance of FM 100-20 reflected a basically unhealthy notion: the innate unwillingness of the AAF to be diverted from its strategic air war objectives to come to the assistance of the ground forces. It is shockingly ironic, then, that this man—who had so many reservations about the application of battlefield air power—should have been killed by a "short" bomb strike by AAF strategic bombers operating in support of Army ground forces during Operation Cobra, the Saint-Lô breakout in 1944.[69]

The Aftermath of Tunisia and FM 100-20

There is greater irony in the Tunisian campaign itself, and perhaps the glimmering of a larger truth. Both sides complained bitterly about the effectiveness of the other side's air attacks, and consequently denigrated the performance of their own aviation forces. After having been exposed to air attack by a formation of twin-engine Junkers Ju 88 bombers, Omar Bradley developed a profound appreciation for air attack that colored his future thinking about the effectiveness of the AAF: he criticized it continuously as not able to keep enemy "air" off the back of his troops. Certainly, these feelings were even more pronounced in the troops who had been exposed to multiple attacks by German aircraft before, during, and after Kasserine. They led to a well-publicized intemperate exchange between the fiery George Patton and the equally short-fused Coningham that eventually required resolution from no less a personage than Eisenhower himself. Until the Axis collapse in Tunisia, and sporadically during Sicily and Italy, Allied forces came under German air attack—and when they did, to a greater or lesser extent, many of the arguments about "dedicated" air support of the Army were dusted off and used again. The irony, of course, is that the level of this German activity was virtually negligible in terms of what would have been needed to decisively affect the ground battle, and it never approached—nor could it have—the level of Allied air support activity employed during Normandy and the breakout across France, or, for that matter, during the rest of the war to the final collapse of the Third Reich in May 1945. Indeed, it did not even approach the level of Allied air activity within the Western Desert and Tunisian campaigns. The smoothly functioning, always-available, devastating Nazi air support of Kasserine was smoothly functioning, always-available, and devastating only in the minds and eyes of inexperienced troops unused to what *real* intensive air operations could do; in the eyes of the Luftwaffe and Wehrmacht high command, the shoe was on the other foot: they saw vast Allied air superiority, and their own disorganization, conflict over roles and missions, technical and logistical deficiencies—in short, the same problems that United States forces believed were unique to themselves.[70]

In 1944–45, the Allies so dominated the skies over Western Europe that the Luftwaffe, for all practical purposes, ceased to be a factor over the battlefield. Wary of German air attack, and then not experiencing it, Army ground commanders became so effusive in their praise of air support operations that they tended, if anything, to attribute almost a miraculous strength to it. Yet nothing had really changed between 1943 and 1944–45 except for one thing: the Luftwaffe was unable to appear over the battlefield except under the rarest of circumstances. When an isolated air attack *did* occur, it attracted disproportionate attention. For example, after Normandy, Quesada was once called on the carpet by Bradley (with whom he had a very good working relationship) for not being able to prevent a "severe" German air strike that had completely paralyzed an Army Corps from launching an attack. Shocked, Quesada checked out the situation; anger replaced anxiety, and he arranged to visit the affected unit with Bradley, keeping to himself the details of the "severe attack" so that Bradley could hear a first-hand account of the attack by the unit commander. Two German fighters had wandered across the front and strafed a regiment in bivouac. The two enterprising pilots burned a halftrack, but the regiment had experienced only a single casualty: a cook collected some shrapnel in his buttocks. This was the "severe" attack that had, in true mythological fashion, grown so out of proportion that Bradley had been informed it had imperiled the operations of an entire Corps. Infuriated, Bradley returned to his headquarters with Quesada, subsequently issuing a letter telling his ground commanders "in rather aggravated language that they must not expect to go through a war being immune from the German air force."[71]

The larger truth in all of this is that regardless of nationality, raw troops exposed to air attack, no matter how slight, tended to be seriously demoralized by it. One had the tendency to see the enemy as all-powerful, superbly equipped, superbly organized, and smoothly functioning, all, of course, to the implicit detrimental comparison with one's own air forces. Only when the expectation of air attack really had been removed—as with the Allied drive across Europe in 1944–45, did ground forces begin to believe that their own air forces were doing a good job; even then, any sporadic air attack awakened latent fears, causing ground commanders to charge their supporting air forces with abandonment.

Unfortunately, one cannot state that the events of the first few months of 1943 ended all problems with air support between the air and ground forces, and that the Allies, firm in unity and purpose, marched onward ever-victorious to Rome and Berlin. Difficulties remained with applying air over the battlefield. In many cases these difficulties were born of suspicion between air and land "communities" as a result of FM

100-20. Others were psychological, as with new troops exposed to sporadic enemy air attack, or with airmen who, having had friends shot at or lost to friendly ground fire, regarded all "ground pounders" as trigger-happy and unappreciative of the work that the airmen did. But there were very real problems that had to be addressed by the air and land forces, and the most important of these were in the control, communications, and intelligence arenas. Many of these were highlighted in the Sicilian campaign and the Italian campaign that followed, and as a result of their being ironed-out, American air and land forces approached Normandy with a much better appreciation of what air-land coordination could and could not accmplish. It was this interim campaign—the Sicilian and, particularly, the Italian experience—that demonstrated that FM 100-20 did not constitute a fatal dichotomy of purpose between air and land forces.

Sicily . . .

The Sicilian campaign (July 10–August 17, 1943) started inauspiciously, and though air-land cooperation improved over the length of this brief struggle, overall it was disappointing. Sicily (and subsequently Italy) marked the only attempt by the AAF to employ a dive bomber in the European theater: the North American A-36 Invader. The A-36 was a dive bomber derivative of the initial production P-51A Mustang fighter, which had been designed to meet a Royal Air Force specification, and only afterward adopted for service by the AAF. The A-36 differed from the P-51A in having retractable dive brakes to slow it during a diving attack. It had the distinction of being the highest performance dive bomber ever built; at low altitudes it could dogfight on equal terms with opposing fighters, something the Stuka could sorely have used during its own lifetime. During the Sicilian and Italian campaigns, the A-36 flew extensive support sorties, and though used as a dive bomber on occasion, most Invader pilots flew it as fighter pilots would subsequently operate the P-47 and P-51 over France and Germany: they made shallow diving attacks, and hedgehopping strafing runs at high speed. Indeed, for the most part, ground crews wired the dive brakes shut to minimize drag thus, in effect, restoring the airplanes to the P-51A standard.[72]

Unfortunately, whatever good the A-36's may have done in Sicily, they gained an unenviable reputation for bombing friendly forces, even when those ground forces were identified with luminescent identification panels. Bradley was strafed by A-36's at the command post of Maj. Gen. Terry Allen—his third strafing by friendly air that day. During fighting for Troina, A-36's dive-bombed the headquarters of the British XXX Corps, having mistaken it for a German-held position they were to attack in support of Allen's forces. A group of American infantry and artillerymen hard-pressed by German infantry and tanks on Monte Cipolla was attacked by seven A-36's that killed or wounded nineteen men—two bombs were dropped on the embattled command post itself—and destroyed the unit's last four howitzers, helping generate conditions forcing a retirement to friendly territory. In another case that eerily recalls Guderian's experiences in France, Bradley described how a flight of A-36's attacked a column of friendly tanks:

> The tankers lighted their yellow smoke bombs in a prearranged recognition signal. But the smoke only caused the dive bombers to press their attacks. Finally in self defense the tanks turned their guns on the aircraft. A ship was winged and as it rolled over, the pilot tumbled out in a chute.
>
> When he landed nearby to learn that he had been shot down by *American* tanks, he bellowed in dismay.
>
> "Why you silly sonuvabitch," the tank commander said, "didn't you see our yellow recognition signal?"
>
> "Oh ____" the pilot said, "is that what it was?"[73]

During the first forty-eight hours of the Sicilian invasion, no close air support sorties were flown in support of the Seventh Army's operations, and II Corps received none for the first seventy-two hours. There is a tendency in the official Army history of the Sicilian campaign to imply that this should not have been so, though the authors do discuss the many other missions that air had to perform in the critical opening days of the Sicilian offensive. The Sicilian invasion did not go unopposed by Germany's Luftwaffe and Fascist Italy's Regia Aeronautica; during the initial invasion, for example, a German air attack against the landing forces sunk an American destroyer and a minesweeper, a brutal reminder that the Axis air threat was a real one. Accordingly, the Allied air effort understandably was directed primarily against the Axis air forces, with a view to establishing Allied air supremacy as quickly as possible. This did not mean that the needs of ground forces were ignored; rather the Allied air forces struck first at the Axis air threat, next undertook interdiction of German forces attempting to reach the landing areas, and only then—two to three days into the invasion, depending on the sector of the front one examines—undertook direct support operations of ground forces. In short, then, the employment of air power in Sicily was "scheduled" in accordance with the dictates of FM 100-20—and this prioritizing of air effort would be utilized again, at Normandy, on a much vaster scale. The AAF itself was aware that air support in Sicily left much to be desired, and the official AAF history of the war admits the problems that Sicilian operations highlighted. Chief among these were communications, identification of friendly forces, and improved ground-air liaison. This latter point received some attention during the Sicilian campaign when the II Corps experimented with "fighter-control parties" equipped with Jeeps carrying VHF radios for directing fighter-bomber strikes against enemy positions. Their success resulted in adoption for the subsequent Italian and Normandy campaigns. Light spotter aircraft did excellent work observing for artillery; these too, would appear subsequently in Italy and France, and would take on an additional function as well—acting as airborne forward air controllers for close support and battlefield interdiction air strikes. In sum, then, Sicily was not Tunisia, and if air support had not been perfect, it nevertheless had come a long way from the Kasserine debacle.[74]

. . . and Italy

The history of the Italian campaign, which began in early September 1943, is one of frustration, disappointment, and dashed hopes. Under Generalfeldmarschall Albert Kesselring, the German forces in Italy made a remarkable stand against the Allies, fighting a defensive war up the length of the Italian peninsula that lasted to the eve of the collapse of Germany itself. Some of the major battles of that conflict—the landing at Salerno, the struggle around Cassino, the landing at Anzio, have become classics of military history in their own right. Throughout the campaign, air activity in support of ground operations constituted a hallmark of Italian combat. Much of this activity was in the form of interdiction strikes, in a vain effort to stem the flow of supplies from Germany and Austria via the Brenner Pass into Italy. But there was also a significant close support character to Italian fighting and even many of the so-called interdiction operations were tied directly to what was occurring on the battlefield.

The German air force rapidly declined as an effective force over the length of the Italian campaign, something that German commentators and reporters noted with dismay. In particular, the effectiveness of German ground-attack formations was nil. During the Nettuno landings, the I and II Gruppen of Schlachtgeschwader 4 went into combat escorted by two Geschwadern of fighters from Germany. The two FW 190 attack formations got away without loss so long as the fighters were present, but the fighters were sent back to Germany for air defense duties a few days later. Immediately even these powerful Jabos were unable to undertake their attack missions. Ground fire and flak proved too intense for air-to-ground operations, including strafing, even though the FW 190's were accelerating from 65-degree dives to air speeds of about 500 mph as they "bottomed out" of the dive. To curb losses, S.G. 4 next attempted to operate the FW 190's as hit-and-run Jabos, but without great success; such were loss rates that by May 1944, fully 50 percent of all graduating pilots from the Schlachtfliegerschule in Germany were going to S.G.4. Unable to operate as ground-attack fighters, the FW 190's were, instead, used (ironically) as pure fighters, flying against Allied intruders. In this role they performed well—but it represented a clear recognition that they had failed in their stated purpose: destruction of advancing Allied ground forces. Finally, in June 1944, S.G.4 left Italy for the more conducive environment of Northern Russia. It would appear that a unit of this sort could have been most valuable during the Allied invasion of Normandy and the critical fighting of June–August 1944 in the *bocage.* However, so intense was the air defense threat surrounding Allied forces in Normandy that the III Gruppe of S.G.4, which *was* fighting in Normandy, was withdrawn to join the I and II Gruppen in Russia after only two weeks of combat! In sum, after Italy, German Schlachtflieger essentially dropped out of the war in the West. At the very end of the war, when the German perimeter had been choked down between the Allies advancing from the East and West, a few *schlacht* aircraft (including the ubiquitous Stuka) flew against American and British forces, but they were fitfully employed and generally massacred by fighters and flak. After the war, Generalmajor Hubertus Hitschold, the Luftwaffe's last General der Schlachtflieger, reflected on the lessons of Italy for future ground attack operations, noting that "the prerequisite for successful and lasting operations of ground attack units is air superiority."[75]

Such problems did not afflict the Allies. The Italian campaign was characterized by a range of air support operations that were to become commonplace in subsequent fighting elsewhere. These included use of radio-equipped forward control posts, of aerial forward air controllers, and of medium bombers to deliver close support strikes. As these uses evolved, they demanded corresponding procedural developments in communications and liaison with ground forces, and these, too, were standardized for subsequent operations.

During the Italian campaign, the British instituted the so-called Rover system, which subsequently was adopted by American forces as well. Rover was a communications "tentacle" consisting of an RAF controller and Army air liaison officer equipped with VHF radio who went "roving" as needed from brigade to brigade at the front, operating a control post (either "visual"—direct observation—or "forward"—acting on information received from other observers) for controlling fighter-bomber strikes against enemy forces immediately in front of friendly forces. Three basic kinds of Rover control existed: Rover David and Rover Paddy, a control system for British strike aircraft, Rover Joe, an equivalent for American strike aircraft using American controllers and liaison personnel, and Rover Frank, which was not for air at all, but, rather, a control system for emergency counterartillery fire against German artillery.[76]

A second innovation introduced in Italian fighting was the use of a light aircraft acting as a forward air controller; this early FAC received the appellation Horsefly. Horsefly's origins were unusual. In May 1944, XII Tactical Air Command (redesignated from the XII Air Support Command the previous month) established a system, dubbed Pineapple, to seek out and destroy enemy motorized transport operating between the front line and the bomb line. Whereas previously a forward control station at VI Corps headquarters called in fighter-bombers of the 324th Fighter Group to attack transport caught below the bomb line, Pineapple allowed aircraft to be controlled from the main control centers at Anzio and Naples, and put on standing patrols for antitransport hunting, to be called in as needed by any aircraft—for example, a tactical reconnaissance aircraft such as a F-6 (a photo-recon version of the P-51 Mustang), or a roving fighter-bomber—that spotted vehicles below the bomb line. To a small degree Pineapple broke important ground for the AAF in using on-scene aircraft to call in air strikes. Horsefly, the next logical step, owed its existence to the suggestion of an artillery-spotting pilot flying a Piper L-5 "Grasshopper" in support of the 1st Armored Division as it slogged its way along north of Rome. One day, low on fuel, the L-5 driver landed next to a XII TAC controller, Capt. William H. Davidson. The L-5 pilot asked why no one had ever thought of using an L-5 to direct fighter-bombers onto a target when artillery was unavailable to mark the target with smoke shells. It is unfortunate that we do not know the name of this L-5 pilot, as he may well be considered the father of all subsequent AAF/USAF FACs. Davidson passed along the idea to Brig. Gen. Glenn O. Barcus, the commander of the 64th Fighter Wing (and subsequently XII TAC commander), who in turn recommended it for approval to the XII TAC commander, Brig. Gen. Gordon P. Saville, who quickly added his own endorsement.

By mid-June 1944, XII TAC had procured L-5's and equipped them with SCR-522 radios. As operated in XII TAC, the L-5's were flown by fighter-bomber pilots on assignment to work with the Corps. Two L-5's supported the activity of each Corps, and would vector strike flights onto targets on the orders of the Division or Corps controllers. After some strike pilots had difficulty spotting the Horsefly L-5's from other "conventional" artillery-spotting L-5's, all Horsefly spotters were given distinctive markings. Each carried an infantry observer to help distinguish friendly from enemy forces, and if operating with armored forces, would carry an observer expert in identifying friendly and enemy armor. Horsefly FAC's operated at 3000 to 4000 feet, and dropped small smoke bombs to mark targets for fighter bombers. They roved as far as fifteen to twenty miles inside hostile territory searching for targets, and one enthusiastic pilot mounted two bazookas under the wings of his L-5 in hopes of bagging a tank. XII TAC took the Horsefly FAC from Italy to France when it deployed to support the invasion of Southern France, and thereafter Horsefly operated with XII TAC until the collapse of Germany. Horsefly FACs may be considered the predecessors of the Mosquito FACs of Korea, who, in turn, anticipated the FACs of Southeast Asia.[77]

The Italian campaign also witnessed the use of heavy and medium bombers—normally used for deep interdiction and strategic bombing operations—as direct support aircraft. The Martin B-26 Marauders of the AAF's 42d Bomb Wing proved particularly valuable, and, indeed, RAF Air Marshal Sir John Slessor called them "probably the best day-bomber unit in the world."[78] The 42d's medium bombers had been vigorously employed in virtually all phases of the Italian campaign, from the invasion itself to the Anzio operation, the reduction of Cassino, and the grinding interdiction campaign. Eventually, it would take its B-26's to France for final operations against the Siegfried Line in 1945. During the bitter fighting at Anzio, the 42d was asked how close it would be willing to bomb to the front lines; the wing, confident of its bombing accuracy, breezily replied "With suitable aiming points provided and good weather, will accept any target the Army needs eliminated," recommending a bomb line no closer than 1500 feet in front of friendly forces. On the evening of February 19, 1944, the Army sent in grid coordinates on three areas centered along an east-west highway, each measuring 150 yards wide by two miles long; the closest was but 1200 feet from the front line. The 42d subsequently reported:

> The operations on 20th February were executed as planned. The respective target areas were smothered by excellent concentrations of the 20 lb. fragmentation bombs. Something new had been added; medium bombers had bombed effectively within 1200 feet of the Allied lines without disturbing those people on the "ring side." Of the 144 attacking bombers, 3 were shot down over the target and 64 were damaged by the intense heavy flak that raked the formation.[79]

As a result of the February 20 mission, local Army forces and the 42d Wing enthusiastically set up an exchange program to further air-ground cooperation, and eventually this resulted in crews detailed for familiarization duty with front-line troops, and ground troops detailed to the AAF, including flying combat missions with the Wing. The mutual respect and understanding that resulted from this exchange was tremendous; it resulted not merely in psychological and comprehension benefits, but in development of new attack techniques and weapons. For example, ground officers told 42d ordnance personnel that the fragmentation bombs had worked well on above-ground targets, but not on troops in foxholes and open-air shelters, and asked if the 42d could develop air-burst weapons. The 42d began a series of experiments on a Sardinian bombing range and eventually employed air-burst bombs up to 1000 pounds, using them for the first time during an April 1944 raid against a German division being held in reserve at Itri; more than 1000 casualties resulted from the bombing. The 42d subsequently noted:

> There is a great difference in the mental attitude of aerial crews when "close-in" Army targets are attacked as opposed to the normal strategical target. Crews will fly through intense flak to a "close-in" target and do an excellent piece of work. Ten days later the same crews will fly into just as intense flak to attack a bridge or supply dump and do only a fair job.
>
> In spite of the losses and injuries sustained to date in Air-Ground cooperation work, morale always reaches a new high during such periods of operation.[80]

It must be noted, of course, that not every bomber or fighter wing was as good at—or as dedicated to—air-ground cooperation as was the 42d. Problems in operating fighters and bombers in close proximity to friendly forces remained in Italy, but compared to those problems in the earlier Sicilian and Tunisian campaigns, they were small. Nevertheless, attacks on friendly forces still caused concern. Heavy bombers striking

Cassino also hit Venafro by accident, "causing many casualties." Some strafing P-40's inflicted over a hundred friendly casualties during the advance on Rome, in one particularly nasty incident. In an attempt to minimize such incidents in the future, air units placed greater emphasis on the bomb line and on a new concept, the "close cooperation line." The bomb line (sometimes called the bomb safety line) was commonly fixed between five and ten miles ahead of friendly forces, and never closer than a distance estimated as ten minutes away by infantry advance from the front. Recognizing this, German forces operated with relative impunity between the bombline and the front, cognizant that they only faced attack by artillery. In response to this, the Army and AAF arranged to conduct close air support flights within the bomb line; some of these were Pineapples going after transport, but most were Rovers. To prevent friendly casualties, a new line, the so-called close cooperation line, was set up immediately in front of friendly troops by the G-3 (Air) officer at Corps headquarters acting on information received from his divisions at the front. As the front fluctuated, appropriate map references were sent to the senior air controller at Corps who would be reponsible for ensuring that close support strikes did not hit within the close cooperation line, but remained within the space between the close cooperation line and the bomb line. Beyond the bomb line, of course, the AAF and Allied air elements did not have to concern themselves whatsoever with integrating their operations with those of the forces on the ground. The rapidly evolving situation on the ground naturally demanded that both air and ground forces be constantly updated on the location of the close cooperation line. The controller at Corps would pass its current location to the headquarters of support wings, which, in turn, would pass it on to the support groups; the G-3 (Air) had responsibility for passing its location along to Army headquarters. The location of this line changed as much as *ten times* daily.[81] Experience with this system likewise transferred naturally into Allied operations across Europe in 1944–45.

The most publicized air campaign in Italy during the Second World War was Operation Strangle, undertaken to interdict critical road and rail segments and thence deny the German military vitally needed supplies. Under the overall doctrinal guidelines of FM 100-20, Strangle constituted "Second Phase" (i.e., interdiction) operations, but, as the XII TAC appreciated, there was a fine line on occasion between what constituted interdiction operations and "Third Phase" battlefield air support missions. As the XII TAC operational summary noted:

> Bombing a bridge one mile beyond the Bomb Safety Line might be considered a Phase 3 operation, whereas bombing a bridge two miles beyond the bomb safety line might be considered a Phase 2 operation. An armed recce might, on the same mission, hit an objective deep in enemy territory and attack a convoy proceeding to supply the front lines.[82]

Thus, it is wise to consider some aspects of Strangle in terms of the "Phase 3" air campaign waged at the front.

Strangle has gone down in the air power folklore as a classic interdiction campaign, one that seriously hampered the German war effort. It gave its name, in fact, to an abortive attempt in 1951 to interdict Communist resupply routes in Korea; Korea's Strangle died in 1952 amid clear evidence that it had not worked. Strangle really did not succeed in isolating the battlefield in Italy, either. Strangle was born of the idealistic notion that air power alone could force a withdrawal of the German armies up the Italian peninsula. To do this, planners relied on concentrated air attacks against the German supply network to force shortages that would lead to a contraction of the German war machine. Strangle, as RAND analyst F. M. Sallagar has noted, "did not achieve supply denial, which had been its objective, but it contributed immeasurably to the defeat of the German armies by denying them the *tactical mobility* [emphasis added] which was essential to them." Thus there was a serendipitous nature in the Strangle operations: its architects established it to starve the German war machine in Italy and thereby force an evacuation of the peninsula; instead, it hampered tactical movement, forcing German commanders to move by night, a situation that General Frido von Senger und Etterlin compared to a chess player able to make only one move to an opponent's three.[83] Strangle can be seen, then, as fitting into the contemporary military air-land battle formula of "find, fix, fight, and finish." German forces had been "found" by reconnaissance and intelligence; air power "fixed" them so that they could be "fought and finished" by combined air-land combat action.

It is ironic that Strangle "failed" in the eyes of its creators because it could not achieve its stated purpose, yet accomplished significant tactical results in hindering German movement in the battle area, doubly so because some of the stated "tactical" and "Phase 3" operations were much less successful. For example, when AAF medium and heavy bombers attacked around Cassino in March 1944, they cratered the local terrain so badly that German defenders gained additional cover and broken terrain for defensive fire positions. At the time, the airmen themselves weren't convinced of the value of such strikes; the strategic bomber community saw the effort as drawing away resources that could be better utilized in the strategic air campaign against Germany, and the medium bomber community itself recognized that the bombing operations were more beneficial to the defenders than to the attackers.[84]

In contrast, German accounts bitterly note the overwhelming Allied air superiority in Italy which resulted in the Luftwaffe being unable to intervene and protect German ground forces, and the demoralizing and paralyzing effectiveness of the fighter-bomber. In early May, Strangle gave way to Diadem, a combined air and ground offensive. Diadem succeeded in draining German resources more effectively than Strangle alone had, because here *air interdiction was combined and synchronized with ground maneuver warfare, forcing the Nazis to expose themselves to air-land attack.* Under these multi-threat circumstances, BAI attacks extracted a heavy toll from the German defenders. The German 10th Army's *Kriegstagebuch* (War Diary) reflects this desperate situation, with entries in May 1944 enumerating "ceaseless air attacks," "heavy fighter-bomber support for enemy ground forces," "enemy air dominates the battlefield," "unremitting Allied fighter-bomber activity makes movement or troop deployment almost impossible," and "fighter-bombers maintain constant patrol over all roads. . . . daytime movement is paralyzed."[85]

The Italian campaign continued up to the war's end, run less boldly and imaginatively on the ground than it had been in the air. Overall, Sicily and Italy were a necessary transition step between the development of air support in the Western Desert and Tunisia and its ultimately successful application during the Normandy operations and the break-out across France. Methods of command and control, the doctrine of FM 100-20, and the air leaders who would run the tactical air war until the collapse of the Nazi machine were all tested and evaluated in the proving ground of Mediterranean warfare. In Sicily and Italy, the AAF first employed the newer technology aircraft that would devastate the Luftwaffe and German ground forces in 1944–45, notably the Republic P-47 Thunderbolt and the North American P-51 Mustang. AAF Lt. Gen. Ira Eaker perceived the critical importance of the Sicilian and Italian experience (and that of Tunisia before it) when he wrote in February 1945:

> The Mediterranean theatre has been the primary crucible for the development of tactical air-power and the evolution of joint command between Allies. Ever since Alamein the Mediterranean has been a laboratory of Tactical Air Forces, just as England has been the primary testing ground of Strategic Air Forces.[86]

The time had now come to take the experiences of the Mediterranean and apply them where they could do the most good: in the hot furnace of Western European combat, in the midst of the greatest combined-arms operation conceived and executed to that time—and probably for all time: the invasion of Europe.

14

A Deadly Efficiency: Anglo-American Air Support in Western Europe

Preparations for Overlord, and the Götterdammerung of the Luftwaffe

Allied planning for the invasion of Europe occupied over two years, and in that planning, the role of air power came under profound scrutiny. In August 1943, the Combined Chiefs of Staff had given their approval to the general tactical plan for the invasion, dubbed Overlord. In February 1944, Gen. Dwight D. Eisenhower assumed command of the European theater, and to him would fall the heavy responsibility of carrying off this "1066 in reverse." Eisenhower succinctly summarized both the goals and the outcome of the invasion:

> Our main strategy in the conduct of the ground campaign was to land amphibious and airborne forces on the Normandy coast between Le Havre and the Cotentin Peninsula and, with the successful establishment of a beachhead with adequate ports, to drive along the lines of the Loire and the Seine rivers into the heart of France, destroying the German strength and freeing France. We anticipated that the enemy would resist strongly on the line of the Seine and later on the Somme, but once our forces had broken through the relatively static lines of the bridgehead at Saint-Lô and inflicted on him the heavy casualties in the Falaise Pocket, his ability to resist in France was negligible. Thereafter our armies swept east and north in an unimpeded advance which brought them to the German frontier and the defenses of the Siegfried Line.[87]

From the outset of Overlord planning, Eisenhower and the rest of the combined forces planners recognized that air power would be critical. The last thing planners wished to face was hostile air power over the battlefront. During the Dieppe raid, the Luftwaffe and Royal Air Force had grappled in the air as Canadian soldiers fought and died in an ill-conceived "reconnaissance in force." The RAF took heavy losses, but managed to prevent the Luftwaffe from attacking the embattled Canadians. But the requirement to ensure air superiority over Dieppe resulted in the RAF having insufficient air assets to itself participate in the land battle. Obviously, such a situation could not exist during Overlord; the Luftwaffe would have to be destroyed, but not at the price of sacrificing vitally needed air support missions for air superiority ones.

Fortunately, circumstances in early 1944 indicated that the Luftwaffe was on the skids. By the fall of 1943, Republic P-47 Thunderbolts equipped with long-range external fuel tanks (called drop tanks, because they could be used and then jettisoned before entering combat) were inflicting heavy losses on German fighters over Occupied Europe and to the German periphery. Then, in December 1943, the North American P-51B Mustang entered service. Featuring superlative handling qualities and aerodynamic design, and powered by a Rolls-Royce Merlin engine, the P-51B (and its successors, the P-51C and P-51D) could escort bomber strikes to Berlin and back, thanks in part to a symmetrical wing section that was thick enough to house a large quantity of fuel and streamlined enough to minimize drag.

What really should be remembered is less the relative contributions of these two fine aircraft and, instead, an overall truth of the strategic bombing effort: whatever else it may have accomplished—or failed to accomplish, for that matter—its primary value to the Allied cause was as an aerial magnet that drew up the Luftwaffe to be destroyed by the American fighter force.[88] The omnipresent Thunderbolts and Mustangs (and less frequently P-38 Lightnings) gave the Luftwaffe no respite over Germany, complementing the shorter-legged Spitfires and Hawker Typhoons and Tempests of the Royal Air Force. Between January and June 1944—the five months before D-day—the Luftwaffe was effectively destroyed: 2262 German fighter pilots died during that time. In May alone, no less than *25 percent* of Germany's total fighter pilot force (which averaged 2283 at any one time during this period) perished. Big Week, an air operation more precisely called Argument, targeted the German aircraft industry for special treatment; while production continued (something that has been overemphasized and taken out of the context of the entire air war effort in postwar accounts), the fighter force took staggering losses. In March 1944, fully *56 percent* of the available German fighters were lost, dipping to 43 percent in April (as the bomber effort switched to Germany's petroleum production), and rising again to just over 50 percent in May, on the eve of Normandy. No wonder, then, that the Luftwaffe's cursory D-day contribution to defending Normandy from the Allied invasion amounted to less than a hundred sorties. Put quite simply, the Allies did not merely possess air *superiority,* but air *supremacy* as well.[89]

Overall, the Allied air campaign for the invasion of Europe followed the dictates of FM 100-20; the campaign consisted of attacks aimed at destroying the Luftwaffe, followed by attacks aimed at isolating the battlefield via interdiction, and, finally, once forces were engaged in close combat with the enemy and the requirements of Phase 1 and Phase 2 operations had been fulfilled, Phase 3 attacks—battlefield air interdiction and close air support strikes—would become increasingly predominant. The requirements to keep the landing sites secret—particularly the deception to encourage the Germans to devote their greatest attention in the region of the Pas de Calais—complicated the air campaign problem. They forced strike planners to schedule vastly more operations across the sweep of likely landing sites rather than just at the site of Overlord. For example, rocket-armed Royal Air Force Hawker Typhoon fighter-bombers of the Second Tactical Air Force (2 TAF) attacked two radar installations outside the planned assault area for every one they attacked within it.[90]

Entrusted with the defense of Nazi-occupied Europe from the Allies, Field Marshal Erwin Rommel recognized that he faced a most critical challenge. The Panzer and Stuka units that he might wish to have to defend the West were, instead, needed for the Eastern Front; and, of course, aircraft like the Stuka simply could not be expected to survive in the face of intensive Allied air and ground defenses. Once before, in 1940, it had been France that confronted the spectre of defeat; now the shoe was on the other foot. The "Desert Fox" emphasized meeting and defeating the invasion forces on the beach; already he realized that if the Allies got a toehold on the continent, it would be at best extremely difficult to remove them, and likely impossible to do so. Rommel discussed the upcoming invasion frequently with his naval aide, Vice Adm. Friedrich Ruge, and the Allied air threat figured prominently in his thoughts. On one occasion, as staff members scattered at the low-level approach of two British fighters while Rommel inspected a gun battery on the coast, Rommel defiantly remained standing before them, in plain view. The planes, probably on a tactical reconnaissance mission, roared overhead. One wonders whether Rommel was subconsciously attempting to offset, by this theatrical (if foolhardy) gesture, the crushing Allied air advantage that he knew was deployed against him and his forces. On April 27, forty days from the invasion, Ruge confided in his diary that he found the disparity between the Luftwaffe and the Allied air forces "humiliating"; by May 12, he was reporting "massive" air attacks, though troops often exaggerated the amount of actual damage. On May 30, with "numerous aircraft above us, none of them German," Ruge narrowly missed being bombed into the Seine by a raid that dropped the bridge at Gaillon. At 0135 on June 6, as Ruge and other senior staff officers regaled themselves with tales of the Kaiser's army and real and imagined conditions around the world, the German Seventh Army reported Allied parachutists landing on the Cotentin peninsula. Overlord was underway. Time had run out for Rommel, and the countdown to the ignominy of the bunker in Berlin had begun.[91]

The Allied Tactical Air Forces

As Overlord embarked upon its preparatory phase, tactical air power increasingly came into play. The tactical air forces supporting the planned invasion of Europe were under the overall command of Royal Air Force Air Chief Marshal Sir Trafford Leigh-Mallory. The Tunisian experience had convinced the RAF—and, of course, the AAF as well—that the tactical air force approach was the best for supporting land armies. Two great tactical air forces existed to support the activities of the ground forces in the invasion; these were the AAF's Ninth Air Force and the RAF's Second Tactical Air Force. In addition, of course, Eisenhower and his ground commanders could call upon strategic aviation as required, in the form of the AAF's Eighth Air Force and Great Britain's Bomber Command.

The Ninth Air Force at the time of Overlord included various commands, one of which was the IX Fighter Command. The IX Fighter Command in turn spawned two Tactical Air Commands, the IX TAC and the XIX TAC. IX TAC had three fighter wings, and the XIX TAC had two. Each of these fighter wings contained at least three—and usually four—fighter groups, a group typically consisting of three fighter squadrons. Of the two, IX TAC was the "heavy"; it could muster no less than eleven fighter groups, while the XIX TAC could muster seven. IX Fighter Command had been intended primarily as a training headquarters, and functioned as such from late 1943 into early 1944 under the command of Brig. Gen. Elwood Quesada; eventually Quesada assumed command of the IX TAC, and Brig. Gen. Otto P. "Opie" Weyland took over XIX TAC. Originally called Air Support Commands, both the IX and XIX ASC had been redesignated in April 1944 as Tactical Air Commands to better recognize their purpose and function as part of the Ninth Air Force itself. At this point, no in-theater formalized structure linking the Ninth and its subordinate commands directly to specific land forces units existed, though there was a generalized understanding that the IX TAC would support the activities of the First Army, and the XIX TAC would support those of Lt. Gen. George Patton's Third Army once the Third became operational in France nearly two months after D-day itself. Eventually, on August 1, 1944, when both Patton's Third Army and Bradley's 12th Army Group became operational, this arrangement was formalized. Bradley's first letter of instructions, as chief of 12th Army Group, was that the Ninth Air Force, as a supporting force, would support the 12th Army Group; Quesada's IX TAC would directly support the First Army (as it already had in Normandy fighting), and Weyland's XIX TAC would support Patton's Third Army. By war's end, the Ninth Air Force had a XXIX TAC as well, and a 9th Bomber Division with three bomb wings, a total of eleven bomb groups of medium and light bombers.[92]

The accompanying organization chart details the command relationship between the Ninth Air Force and the 12th Army Group, as well as that between Great Britain's Second Tactical Air Force and the 21st Army Group. Additionally, it includes the First Tactical Air Force, activated in October 1944 to support the activities of the 6th Army Group; unlike the all-American Ninth Air Force or the all-British Second Tactical Air Force, the First Tactical Air Force was an Allied tactical air force consisting of American, French, and British subordinate commands.[93]

Great Britain's Second Tactical Air Force (2 TAF) had grown out of initiatives in mid-1943 to structure a "Composite Group" to support the invasion of Europe, and had sprung from the ashes of the moribund and never-satisfactory Army Co-operation Command. In January 1944, Air Marshal Sir Arthur Coningham took command of 2 TAF and then, in March 1944, took on additional duties as commander of the Advanced Allied Expeditionary Air Force (AAEAF). Ironically, at this critical juncture, two serious command problems arose: the relationship between the RAF commanders themselves, particularly Coningham, Leigh-

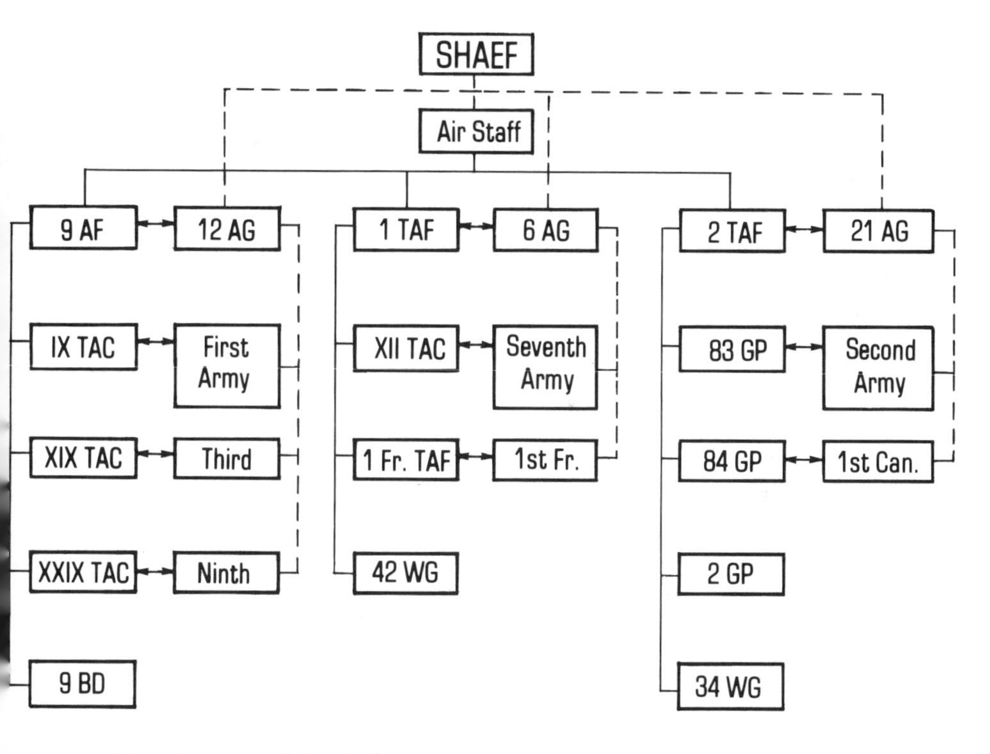

Abbreviations and Symbols

AF	Air Force
TAC	Tactical Air Command
TAF	Tactical Air Force
BD	Bombardment Division
AG	Army Group
Fr	French
Can	Canadian
GP	Group (RAF)
WG	Wing (RAF)
SHAEF	Supreme Headquarters Allied Expeditionary Force
↔	Liaison and Coordination
—	Air Chain of Command
---	Land Chain of Command

Mallory, and Arthur Tedder (Deputy Supreme Commander for Overlord) and then, and much more serious, a serious breakdown between the RAF commanders and 21st Army Group commander Field Marshal Sir Bernard L. Montgomery, who also wore an additional hat as commander of Allied ground forces during the invasion. It is ironic, in view of Montgomery's undoubted supportive view of air action in the Mediterranean, and his whole-hearted acceptance of Coningham's thoughts on air support—as evidenced by the famous pamphlet—that Montgomery and the RAF should now come to disagree over the relationship between the air and the land commander, but so they did. Despite Montgomery lipservice to the contrary, his actions in early 1944 clearly indicate that he considered his equals in the RAF as merely advisers. For their part, Coningham and Tedder nursed grudges going back to the advance after second El Alamein and Montgomery's slowness during the pursuit of Rommel's retreating forces. The critical question, in the airmen's views, was how rapidly Montgomery would advance to seize airfields that the tactical air forces could exploit so that Allied air power would not have to operate across the Channel, from bases in England. In fact, this issue turned out to be far less important than originally thought, for bases were hacked out of the Normandy terrain, often only a few thousand yards from opposing German forces. In further irony, Montgomery's planned advance from the beachhead (which the airmen considered too slow) turned out to be instead overoptimistic; the actual advance was even slower. It was in this case, of course, that Allied air power in Normandy proved all important. As John Terraine noted:

> History insists that the last word, in regard to the Battle of Normandy, must be that the quarrels did not, finally, matter: Allied air power was so overwhelming that the defeat of Allied intentions on the ground never threatened disaster, only delay, and that only in the early stages, well compensated later. But let us be quite clear about it: what made the ultimate victory possible was crushing air power.[94]

Britain's 2 TAF consisted of four RAF Groups: No. 2 Group, No. 83 Group, No. 84 Group, and No. 85 Group. Of these four, only the first three—2, 83, and 84—were really available for the air-land battle in Normandy; 85 Group was under the temporary operational control of No. 11 Group, attached to an RAF home command known as "Air Defence of Great Britain." 2 Group consisted of four wings of Boston,

Mitchell, and Mosquito light and medium bombers; 83 Group, exclusive of a reconnaissance wing and some light aircraft used for artillery spotting, consisted one Mustang wing, four Spitfire wings, and four Typhoon wings; 84 Group, again exclusive of recce and spotting aircraft, consisted of one Mustang wing, five Spitfire wings, and three Typhoon wings. As the campaign progressed, and in a fashion analogous to that of the AAF Ninth Air Force, 2 TAF's subordinate units directly supported subordinate units of the 21st Army Group. Thus, the British Second Army could rely upon 83 Group, and 84 Group supported the First Canadian Army. Another important relationship, however, was that formed between the Ninth Air Force's IX TAC and the 2 TAF's 83 Group. IX TAC's Elwood Quesada and 83 Group's commander, Air Vice Marshal Harry Broadhurst, worked well together. For example, after troops were ashore at Normandy, control of tactical aircraft passed from shipboard control centers to two land-based control centers: a IX TAC control center in the American sector of the beachhead, and an 83 Group control center located in the British sector. Coningham later praised the "excellent teamwork" between the two control centers. This teamwork would be refined even further in the weeks ahead.[95]

Altogether, the tactical air forces had 2434 fighters and fighter-bombers, together with approximately 700 light and medium bombers available for the Normandy campaign. This force first struck against the Germans during the preparatory campaign prior to D-day. At D minus 60 days, Allied air forces began their interdiction attacks against rail centers; these attacks increased in ferocity and tempo up to the eve of the invasion itself, and were accompanied by strategic bomber raids against the same targets. The bridge campaign, which aimed at isolating the battlefield by cutting Seine bridges below Paris and Loire bridges below Orleans, began on D minus 46. In this campaign, the fighter-bombers proved more efficient than medium or heavy bombers, largely because their agility enabled them to make pinpoint attacks in a way that the larger bombers, committed to horizontal bombing runs, could not. Further, the fighter-bombers had the speed, firepower, and maneuverability to evade or dominate the Luftwaffe; though ground fire and (rarely) fighters claimed some attacking fighter-bombers, the loss rate was not what might have been expected had conventional attack or dive bombers been assigned to such work. By D minus 21, German airfields within a radius of 130 miles of the battle area were being attacked, and these, too, continued to the assault on the beachhead.[96]

The Invasion and the Structure of Air Support

During the June 6 D-day assault itself, a total of *171 squadrons* of British and AAF fighters undertook a variety of tasks in support of the invasion. Fifteen squadrons provided shipping cover, fifty-four provided beach cover, thirty-three undertook bomber escort and offensive fighter sweeps, thirty-three struck at targets inland from the landing area, and thirty-six provided direct air support to invading forces. The Luftwaffe's appearance was so minuscule that Allied counterair measures against the few German aircraft that did appear are not worth mentioning. Of far greater importance was the role of aircraft in supporting the land battle. It is worth noting in this regard that as troops came ashore at Normandy, they made an unpleasant discovery all too familiar to the Marine Corps and Army operating in the Pacific campaign: despite the intensive air and naval bombardment of coastal defenses, those defenses were, by and large, intact when the invasion force "hit the beach." This was particularly true at Omaha beach, where American forces suffered serious casualties and critical delays, being hung up on the beach despite a massive series of attacks by B-17's and B-24's of the Eighth Air Force and medium bombers as well against German defenses in the early hours of June 6 prior to the troops coming ashore. In fairness, it must be added that the air commanders themselves had, in fact, predicted that the air bombardment and naval bombardment would not achieve the desired degree of destruction of German defensive positions. The general perception among the Army that air would cleanse the beaches before their approach, however, was shattered and replaced by a lack of confidence that only the subsequent success of fighter-bombers operating against the battlefield would cure. By and large, throughout the post-Normandy campaign—and, for that matter, speaking broadly of the Second World War as a whole—*the fighter-bomber proved overwhelmingly more valuable in supporting and attacking ground forces in the battle area than did the heavy or even the medium bomber.*[97]

Drawing upon experience ranging from the Western Desert and Tunisia through the Sicilian and Italian campaigns, Allied tactical air control in Normandy and during the subsequent European campaign was generally excellent. Greatly assisting it was the wartime evolution of radar. The Allied air forces had radar available to them from the very first day of Normandy operations, and it was soon incorporated into tactical air control as well as for early warning and air defense purposes. American radar control may be taken as typical for the Allied effort. In a generic sense, all the Tactical Air Commands followed similar organizational and procedural lines as regards radar utilization or, for that matter, the functioning of support system requests. Specific differences did exist, and standard terminology itself was lacking between the various TAC's, but on the whole, broad generalized similarities were more characteristic of American TAC operations than were the differences. (The standardization of air control procedures would come about after the war, with the preparation of a new FM 31-35 in August 1946.)

Radar had first been used for tactical air support control during the Sicilian and Italian campaigns, and now, in Normandy and the subsequent breakout, it reached new levels of refinement. Each TAC had a radar control group built around a Tactical Control Center (also called a Fighter Control Center), a microwave early warning radar (dubbed a MEW), three Forward Director Posts, three or four SCR-584 Close Control Units (the SCR-584 being a particularly fine precision radar used for positioning data and antiaircraft gun laying), and, finally four Direction Finding stations, dubbed Fixer stations. The MEW, considered the heart of the system, would be located within ten to thirty miles of the front. Originally developed for air defense purposes, this radar network now took on added importance for the control of tactical air strikes. For example, when an Air-Ground Coordination Party sent in a request for immediate air support, that request went directly to a Combined Operations Center functioning between the TAC and the Army. There, the Army G-2 and G-3 and the TAC A-2 and A-3 evaluated the request. Assuming it was considered legitimate, the Army G-3 and Air A-3 would each approve it, and the Air A-3 would relay it to the Tactical Control Center with a recommended course of action. Typically, the TCC would relay the request to airborne fighter-bombers, and a geographically appropriate Forward Director Post would furnish precise radar guidance and navigation information from the MEW and SCR-584 radars to the strike flight, vectoring them to the target area. Once in the target area, of course, the strike flight leader would communicate with the Air-Ground Coordination Party that had sent in the request for final details. For their part, the Air-Ground Coordination Party would arrange for artillery to mark the target with colored smoke and also, if possible, to undertake suppressive artillery fire against known enemy antiaircraft defenses. Radar was also used for so-called blind bombing in conditions of reduced visibility; SCR-584 control eventually enabled blind bombing strikes with accuracies on the order of 400 yards from the predetermined aiming point, notably during the

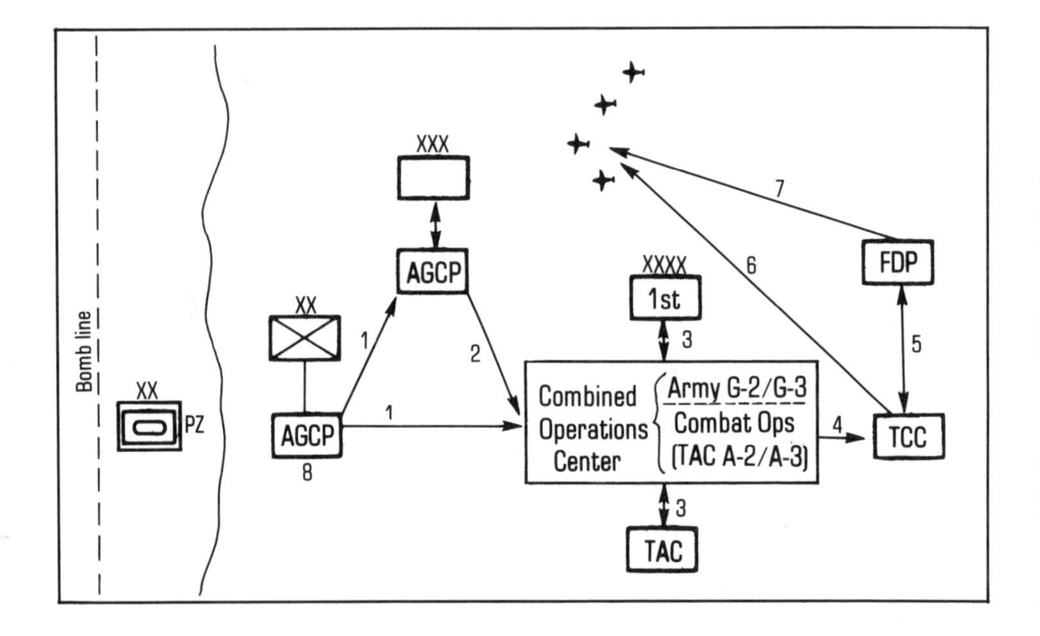

IX Tactical Air Command Immediate Support System, Summer–Fall 1944

1. Division Air-Ground Coordination Party (AGCP), staffed by Tactical Air Party Officer (TAPO) and Division G-3 (Air), send direct support request to Army G-3 at Combined Operations Center (COC), also informing Corps G-3 (Air) so Corps AGCP can monitor or intervene as necessary.
2. Corps AGCP monitors communications net.
3. COC, consisting of Army G-2 and G-3 together with Tactical Air Command's A-2 and A-3 (termed Combat Operations), consults with Army HQ and TAC HQ on request; then G-3 and A-3 each approve it.
4. A-3 at Combat Ops relays support request and recommended course of action to Tactical Control Center (TCC), also termed Fighter Control Center.
5. Forward Director Post (FDP), in constant communication with TCC, provides continuous updates on location of friendly and enemy air units using microwave early warning (MEW) radar tracking.
6. TCC relays strike request to airborne "on-call" fighter-bombers.
7. FDP, using SCR-584 radar, furnishes precise guidance and navigation information to en-route strike flight.
8. Division AGCP prepares for incoming strike flight by artillery fire to mark targets with colored smoke and to suppress enemy air defenses; AGCP will maintain communication with strike flight during attack.

Battle of the Bulge in winter 1944–45. Blind bombing on the battlefield, however, was far more typical of the Korean and Vietnam wars than it was of the Second World War. The accompanying diagram illustrates the control system used by IX TAC during the 1944 campaign.[98]

Air Pro and Contra Armor

Normandy's most noteworthy tactical air support development, however, was the close partnership between air and armored forces, typified by the "armored column cover" missions perfected by the IX TAC under Quesada. During the Italian campaign, the British had begun operating so-called contact cars that operated as mobile air-ground control posts with armored forces. Now, at Normandy, 83 Group under Broadhurst placed contact cars with leading British armored forces so that tactical air units would know precisely and at all times the location of friendly and enemy forces. The contact cars functioned in close cooperation with tactical reconnaissance aircraft, and reduced the time necessary to set up immediate support strikes. The contact car scheme proved its value particularly during the German retreat out of the Falaise "pocket." Quesada's development of a similar system for the American forces in Normandy stemmed from his interest in the Army's mission and his relationship with Omar Bradley, then commander of the First Army; "Bradley liked me," Quesada recalled years later, "and I liked him a hell of a lot."[99] Bradley admired Quesada's willingness to regard air support "as a vast new frontier waiting to be explored."[100] Because of this, these two strong-willed commanders got along exceptionally well, and felt confident enough to express frank opinions about how each other was doing his job. Shortly before the Saint-Lô breakout, Quesada became convinced that Bradley was reluctant to concentrate his armored forces because of the magnitude of German defensive forces along the front. So Quesada made a deal: if Bradley would concentrate his armor, IX TAC would furnish an aviator and an aircraft radio for the lead tank so that it could communicate with fighter-bombers that Quesada would ensure would be operating over the column from dawn until dark. Bradley immediately agreed, and a pair of M4 Sherman tanks duly arrived at IX TAC headquarters in Normandy (only a hedgerow away from Bradley's own command post) for trial modification. The modification worked, and became a standard element of First Army—and subsequently 12th Army Group as a whole—operations.[101]

By the end of July 1944, the Quesada armored column cover operations were receiving enthusiastic support from armor and air forces personnel alike. The 2d Armored Division, for example, had three air support parties, one with the division commander, and one apiece with each of its two Combat Commands. Combat Command A found the system particularly useful; their air liaison officer (from the armored forces) rode in a Sherman whose crew were entirely AAF except for the tank commander. The tank commander could communicate with his fellow tankers via a SCR-528 radio, while the air liaison officer had a SCR-522 to communicate with the column cover flight. Column cover consisted of four P-47's relieved by another flight every thirty minutes. CCA's liaison officer reported:

> The planes worked quite close to us, generally with excellent results. . . . Our best air [reconnaissance] information came from the column cover. On occasions G-2 asked me for specific information, and I asked the planes to get it. In most cases the pilots furnished information to me without request, especially that of enemy motor movements. Before leaving, the flight leader would report to me on likely prospective targets, and I would pass the information on to the incoming flight commander.
>
> On one occasion we made an unexpected move for which no air cover had been provided. Information was received of a group of hostile tanks in some woods three or four miles away. I called direct to a plane operating in the zone of another corps and asked him to relay a request to fighter control center for some fighters. Within 15 minutes about 12 planes reported in to me. I located my tank for the plane commander by telling him of the yellow panel [used for identification of friendly forces, and located on the back deck of the tank], then vectored him on to the woods where the enemy was reported. When he seemed to be over the target, I told him to circle and check the woods under him. He located the tanks, and they were attacked successfully.[102]

In a study done immediately after the war for the United States Strategic Bombing Survey, the Air Effects Committee of the 12th Army Group (a committee composed entirely of ground officers, and thus free of the kinds of built-in bias that might have afflicted a committee composed of AAF personnel) assessed the role of tactical air power in the European campaign. They examined a number of issues, generating a report (which Bradley signed) that is remarkable for the degree to which it endorses the air support system the AAF employed in assistance to the ground forces; indeed, one would not imagine, if reading solely this document, that but two years before the AAF and Army Ground Forces had been at virtual swords' points over the entire air support issue. Of armored column cover operations, the report stated:

Armored column cover . . . was of particular value in protecting the unit from enemy air attack and in running interference for the spearhead of the column by destroying or neutralizing ground opposition that might slow it down or stop it. . . .

The decision of the Ninth Air Force to give high priority to armored column cover in a fast-moving or fluid situation from the break-out in Normandy to the final drive across Central Europe made a successful contribution to the success of the ground units in breaking through and encircling the various elements of the German armies. . . . [After addressing immediate support needs] the flight leader patrolled ahead of the armored column, as deep as thirty miles along its axis of advance, in an intensive search for enemy vehicles, troops or artillery. This effort permitted our armor far greater freedom of action than would have been otherwise possible.[103]

Normandy operations, typified by Quesada's armored column cover and Broadhurst's contact cars, thus fulfilled a notion first attempted a quarter-century earlier, amid the mud of Flanders: the notion of the airplane as a *partner* of the tank, as a "counter antitank" weapon. In that war, then-Colonel J. F. C. Fuller, Great Britain's greatest armor advocate, had recognized that cooperation between air and armor forces was "of incalculable importance."[104] (Subsequently, in lectures in the early 1930s, he spoke perceptively of the "close relationship between mechanization on the ground and motorization in the air.")[105] It is a nice coincidence that Leigh-Mallory, the commander of Allied tactical air forces in Normandy, had commanded a squadron of tank-cooperation aircraft in the Great War; one wonders if this controversial, gifted airman (who died in a flying accident in November 1944) reflected back in his own mind, as the Normandy campaign unfolded, to those early days of open-cockpit biplanes and awkward, ungainly tanks, and the progression of both air and land warfare technology since that time.

There was, of course, another side to the airplane as well, and if the Allied Typhoons and P-47's were friends of British and American armored forces, they proved implacable enemies of German armored, mechanized, and infantry forces. This was an aspect of warfare—the airplane as *enemy* of the tank—that even the formidable Fuller had failed to prophesy, but of which the Second World War had already given much evidence. Opposing offensive mobile armor, as in North Africa, the fighter-bomber proved of limited use. Now opposing armor typically lying in defensive ambush, or retreating in constrained, restricted columns, the rocket- or bomb-loaded fighter proved devastating.

The Ninth Air Force and the Second Tactical Air Force had vast quantities of fighter-bombers. IX TAC, for example, had twenty-four squadrons of Republic P-47 Thunderbolts, while 2 TAF had eighteen squadrons of Hawker Typhoons. Both were beefy, powerful, macho aircraft and capable of absorbing considerable battle damage and still returning to base. Of the two, the P-47 was the more survivable, in part because it had a radial piston engine. The Typhoon had a liquid-cooled engine and "chin" radiator installation that was most vulnerable to ground fire. Affectionately known as the Jug, the P-47, on occasion, returned to base not merely with gaping holes from enemy defenses, but with whole cylinders blown off its engine. Pilot memoirs reveal that while the P-47 was regarded with affection and, indeed, fierce loyalty, the Tiffie (as the Typhoon was dubbed) was regarded more with an uncomfortable respect and awe bordering on fear.

Both had, for their time, prodigious weapons-carrying capabilities. Both could lug up to a 2000-lb bomb load, carrying one 1000-lb bomb under each wing. Typically, however, both operated with smaller loads. A P-47 would carry an external belly fuel tank and one 500-lb bomb under each wing; many were also configured so that the plane could carry air-to-ground rockets, typically ten 5-in HVAR (high-velocity aircraft rocket) rockets. P-47's on an armed reconnaissance mission would usually operate three flights, two armed with a mix of bombs and rockets, and the cover flight carrying only rockets. Over 80 percent of the bombs dropped by P-47's during the European campaign were 500-lb weapons; less than 10 percent were 1000-lb bombs, and the difference was made up by smaller 260-lb fragmentation bombs and napalm. While acknowledging the spectacular effects and destructiveness of rockets, the AAF considered bombs more effective for "road work" due to accuracy problems in firing the solid-fuel weapons. The British, on the other hand, preferred rockets, the Typhoon carrying eight having 60-lb armor-piercing warheads. (Possibly these differences stemmed from launching methods; the P-47's used "zero length" launchers while the Typhoons used launch rails. It could be expected that the rails would impart greater accuracy, stabilizing the rocket immediately after ignition until it had picked up sufficient speed for its tail fins to stabilize it. There is, however, an interesting report from Montgomery's 21st Army Group that calls into question the alleged success that British air-to-ground rockets enjoyed against tanks and motorized transport, as will be discussed subsequently.)[106]

In addition to their bomb and rocket payloads, the P-47 and the Typhoon each had powerful gun armaments. The Typhoon had four 20mm Hispano cannon. The P-47 carried eight .50 cal. machine guns with 400 rounds per gun, and it proved "particularly successful" against transport. It occasionallly even caused casualties to tanks and tank crews: its .50 cal. armor-piercing bullets often penetrated the underside of vehicles after ricocheting off the road, or penetrated the exhaust system of the tanks, ricocheting around the interior of the armored hull, killing or wounding the crew and sometimes igniting the fuel supply or detonating ammunition storage. At first, this latter point is surprising, given the typically heavy armor of German tanks. But Maj. Gen. J. Lawton "Lightning Joe" Collins, commander of First Army's VII Corps, was impressed enough by the success that P-47's had strafing tanks to mention it to Quesada.[107]

There were, of course, other fighter-bombers that operated in Normandy and across Europe, notably the Lockheed P-38 Lightning, North American P-51 Mustang, and Supermarine Spitfire. With the exception of the Lightning (which had a concentrated armament installation that made it a formidable strafer), all of these proved disappointing. All were quite vulnerable to ground fire in light of their liquid-cooled engine systems, and thus they were used far less for ground attack and much more for air superiority operations.

Allied Air over the Battlefield

Virtually immediately the tactical fighter bombers of the IX TAC and 2 TAF made their presence felt on the land battle. For the first four days of the invasion, the fighter-bombers flew from their bases in Southern England, but the first rough airstrips were available for use on the Continent on June 10. Eventually they numbered thirty-one in the British zone and fifty in the American. Two problems quickly manifested themselves in these early "at the front" air support operations. The peculiar thick dust of Normandy played havoc with the inline engines of the Spitfire and Typhoon, until mechanics fitted special air filters to the aircraft and engineers watered down the runway surface. Second, these forward strips were close to enemy positions, and thus came under frequent shelling. In one case, Typhoons operating from a forward strip attacked German tanks and fortifications a mere 1000 yards away from the runway, an operation calling to mind more the experience of the Marines and Army at Guadalcanal or at Peleliu than the European campaign.

The ordeal of the German Panzer-Lehr division offers a good case study in what awaited German ground forces in Normandy. Ordered north to confront the invasion, the division got underway in the late afternoon of June 6, and came under its first air attack at 0530 on June 7, near Falaise (which would soon establish its own place in air-ground history). Blasted bridges and bombed road intersections hindered movement, particularly of support vehicles. So intense were the attacks along the Vire-Beny Bocage road that division members referred to it as a *Jabo Rennstrecke:* a fighter-bomber race-course. Air attack destroyed more than 200 vehicles in this one day's activities alone. Despite the rainy weather—weather which had, in fact, endangered the landing on the beachhead itself—fighter-bombers continued to strike at the division, to the dismay of division personnel, who had believed the increasingly bad weather would offer them some respite. This was but the beginning of an ordeal that would last throughout the French campaign; Panzer-Lehr was in for some rough times in the near future.[108] This division was by no means alone in its trials. The 2d SS Panzer division Das Reich made its way from Toulouse to Normandy, encountering serious delays enroute and, in typical SS fashion, responding by murdering and otherwise brutalizing the civilian population of France. They got their taste of real war once they had crossed the Loire; as Max Hastings relates,

> questing fighter bombers fell on them ceaselessly. The convoys of the *Das Reich* were compelled to abandon daylight movement after Saumur and Tours and crawl northwards through the blackout. . . . [During a change of command] an Allied fighter bomber section smashed into the column, firing rockets and cannon. Within minutes . . . sixteen trucks and half-tracks were in flames. . . . Again and again, as they inched forward through the closely set Norman countryside, the tankmen were compelled to leap from their vehicles and seek cover beneath the hulls as fighter bombers attacked. Their only respite came at night.[109]

The night did offer some protection to the besieged Germans, but even the night did not grant total immunity. The 2 TAF used twin-engine De Havilland Mosquitos as night battlefield interdiction aircraft, sometimes having the "Mossie's" bomb and strafe under the light of flares dropped from North American Mitchell medium bombers used as flare ships. Later in the European campaign, when the German night air attack menace had largely disappeared, the AAF used Northrop P-61 Black Widow night fighters in a similar role. Overall, however, the inability to successfully prosecute night attacks to the same degree that they did attacks in daytime was a source of frustration to air and ground commanders alike. Bradley's air effects committee noted that there was "never enough" night activity to meet the Army's needs.[110]

Intelligence information from Ultra set up a particularly effective air strike on June 10. German message traffic had given away the location of the headquarters of Panzergruppe West on June 9, and the next evening a mixed force of forty rocket-armed Typhoons and sixty-one Mitchells from 2 TAF struck at the headquarters, located in the Chateau of La Caine, killing the unit's chief of staff, many of its personnel, and destroying fully 75 percent of its communications equipment as well as numerous vehicles. At a most critical point in the Normandy battle, then, the Panzer group, which served as a vital nexus between operating armored forces, was knocked out of the command, control, and communications loop; indeed, it had to return to Paris to be reconstituted before being able to resume its duties a month later.[111]

Generalfeldmarschall Rommel's reaction to being pinned to the ground by Allied tactical air was a repetition of the feelings he had expressed during the dark days of 1942 when scourged by the Desert Air Force. Already by June 9, Ruge was writing that "the air superiority of the enemy is having the effect the Field Marshal had expected and predicted: our movements are extremely slow."[112] In a letter to his wife the next day, Rommel wrote, "The enemy's air superiority has a very grave effect on our movements. There's simply no answer to it."[113] In walks with Ruge, Rommel continued to complain about the invasion situation, "especially the lack of air support."[114] Ruge concluded that "utilization of the Anglo-American air force is the modern type of warfare, turning the flank not from the side but from above."[115] As the campaign went on the situation turned increasingly bleak. By July 6, during a dinner party, a "colonel of a propaganda battalion" remarked that "where is the Luftwaffe?" ranked among the most common questions from soldiers.[116] In staff discussions about the future—as if one really existed for the Third Reich—Rommel and Ruge concurred that "the tactical Luftwaffe has to be an organic part of the army, otherwise one cannot operate," which showed how little the two men understood the evolution of Allied air power over the previous three years of the war. It was *precisely because* Allied air power was *not* subordinate to the armies that it was free to use mass and concentration to achieve its most productive ends—and thereby help the Allied armies the most.[117]

It is bitterly amusing to note how closely Rommel's complaints at this time mirror those of the British and American army leaders of 1941 and 1943, respectively. Rommel grew increasingly testy about air matters; during breakfast on July 16, Rommel was "incensed" at the presumptuousness of a Luftwaffe staff officer who intemperately accused the

German army of not taking fullest advantage of German air force attacks throughout the war.[118] The next day, as Rommel drove en route to his headquarters after a quick trip to an SS armored unit, two 83 Group Spitfires strafed him, killing his driver, wounding a passenger, and causing their car to plunge off the road, out of control. Rommel, thrown out, narrowly escaped death from a fractured skull. With that, Rommel's war effectively ended. He returned to Germany for treatment and recuperation, dying by his own hand that October when implicated—rightly or wrongly—in the officers' plot to assassinate Hitler, a plot that tragically went awry. Allied tactical air had removed from command the German commander best suited by experience and leadership to oppose the ground forces building up to the breakout from the Normandy lodgement area.[119]

Cobra: The Heavy Bomber in Air Support

Disconcertingly, the Allies found that once ashore at Normandy, their anticipated advance into France received a serious setback from the terrain. Farmers' fields were bordered by thick hedgerows, a *bocage* that proved a natural boon to German defenders, for it gave them cover and also forced the Allies to follow predictable paths of advance as they attempted to work their way around it. Eventually, a locally inspired solution was found to one of the most difficult problems of hedgerow fighting: preventing tanks from riding up over the hedge and exposing their vulnerable undersides to antitank fire. The solution was disarming; a sergeant fitted "tusks" to the prow of a tank, demonstrating that they pinned the tank to the hedge and held it in place as its engine punched it through the hedge in a shower of dirt. Now, this "absurdly simple" device (in Bradley's words) had given the Army the means to free up its armored forces for a fast-moving mobile breakout across France.[120]

Any breakout from the lodgement area would require the insightful and creative use of air power, including the application of strategic aircraft such as the American B-17 and B-24 and the British Halifax and Lancaster. Thus, another important aspect of the Normandy fighting was the use of heavy bombers in a tactical troop-support role. The results, to say the least, were mixed. Altogether there were six major raids by heavy bombers in support of breakout operations in Normandy. The first of these was by 457 Halifax and Lancaster bombers from RAF

Bomber Command on July 7, supporting Montgomery's assault on Caen; the second was an even larger raid by 1676 heavy bombers and 343 light and medium bombers on July 18. On July 25, a third attack by American bombers of the 8th and 9th Air Forces struck at Saint-Lô, preparatory to the First Army's breakout. A fourth attack on July 30 supported the Second British Army south of Caumont. A fifth Anglo-American raid on August 7–8 supported the attack of the First Canadian Army toward Falaise from Caen, and the sixth raid, again supporting the attack on Falaise, followed on August 14.[121]

Overall, the Allied high command considered these raids successful, and accounts by German forces caught in them confirm that they had a devastating (if short-lived) impact upon morale. Generalfeldmarschall Hans von Kluge (who had succeeded the wounded Rommel and who himself would shortly commit suicide over his alleged involvement in the plot against the Führer) complained that bomb-carpets buried equipment, bogged down armored units, and shattered the morale of troops. But, it might be added, the terrain disruption (as Cassino had shown earlier) worked both ways: it hindered the attacker as much as the defender, and, in fact, bought the defender time to regain his composure (to a degree) and dig in for the follow-on attack. If such air attacks were to be useful, they had to be followed immediately by a follow-on ground assault. When such occurred, ground troops did acknowledge that defenders were dazed and prone to surrender.[122]

These attacks were not without graver problems as well; the first two strikes on Caen resulted in numerous "collateral" casualties to French civilians. Far more serious were attacks on friendly troops from misplaced bomb strikes. The Normandy campaign was like other campaigns in that the air and land forces had a period of time in which each had to get used to the other. Bradley remarked after the war that "we went into France almost totally untrained in air-ground cooperation," though it is difficult to accept at face value the accuracy of that statement, inasmuch as the air and ground forces worked together with an unprecedented harmony that had been quite absent in Tunisia and often strained in Sicily (though less so in Italy).[123] Nevertheless, in the very early stages of Normandy some "disconnects" did occur between the air and land communities. Friendly troops experienced attacks from Allied fighter-bombers. To minimize such dangers, air and ground commanders arranged for friendly forces to pull back in anticipation of an air strike against German positions. But if communication failed and the strike did not come off, troops found themselves fighting twice for the same piece of real estate as German forces moved back into the gap. Commanders learned to follow-up air strikes with artillery barrages so that friendly infantry and armor forces would have the chance to close with the demoralized enemy before he recovered and redeployed. Eventually, six weeks after the Normandy landing, air and land forces were so confident of working together that fighter-bombers routinely operated as close as 300 yards to American forces. Such was not true, unfortunately, of strategic bomber operations in support of ground forces, as the strikes of late July and August clearly indicated.

The most publicized example of the difficulties of operating heavy and medium bombers in support of ground forces came during the preparatory bombardment for Operation Cobra, the breakthrough attack at Saint-Lô that led to the breakout across France, though the RAF's Bomber Command raid near Falaise on August 14 (three weeks after Cobra) killed a greater number of Canadian and Polish troops and, indeed, endangered the life of Coningham himself, who was up front in a command car observing the strike. The Cobra strikes killed slightly over 100 GIs (the actual figures vary from 101 to 118 depending on the sources) and wounded approximately 500 others. There is no question that the strikes were badly executed, and that serious command errors established conditions whereby the short-bombings occurred. The first came on July 24, a cloudy day, when Cobra had been initially set for launch. A postponement order reached the 8th Air Force commander, Lt. Gen. James H. "Jimmy" Doolittle, too late: the 8th's bombers were already airborne for France. Most wisely refrained from bombing due to weather and returned to base. Some found conditions acceptable and did drop. The friendly casualties occurred in three instances. A bombardier accidently toggled his bomb load on an Allied airstrip when another plane in the formation was destroyed by flak (German flak claimed three of the heavy bombers that day), damaging planes and equipment; a lead bombardier experienced "difficulty with the bomb release mechanism" and part of his load dropped, causing eleven other bombers to drop thinking they were over the target; and, finally, a formation of five medium bombers from the 9th Air Force dropped seven miles north of the target, amid the 30th Infantry Division. It was this latter strike that inflicted the heaviest casualties—25 killed and 131 wounded—on the first day that Cobra was attempted. The next day, in better weather, Cobra was undertaken, again with three unfortunate friendly bombings, by B-24's. In the first case, a lead bombardier failed to synchronize his bombsight properly, so that when he dropped—and eleven other

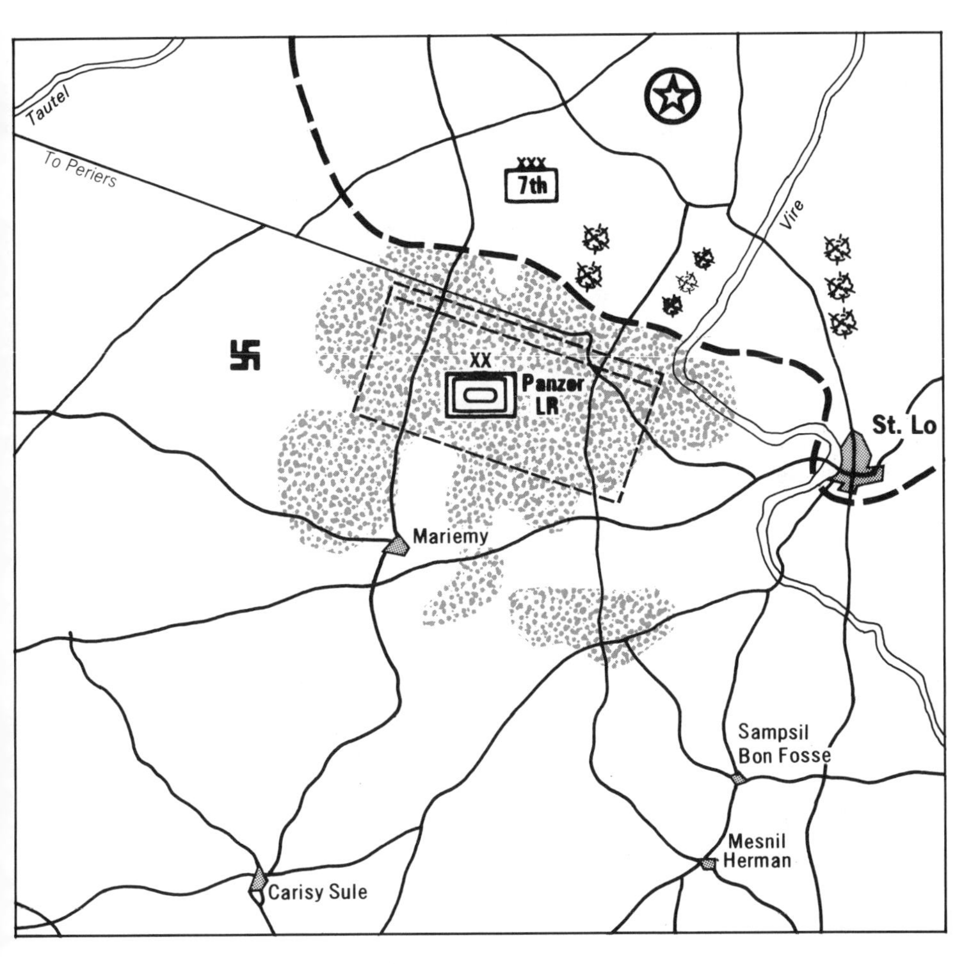

Operation *Cobra:* The Bombing Pattern

Areas of friendly bombings

Bombing pattern over German territory

Demarcation between American and German positions

Target zone, including narrow fighter-bomber zone

bombers dropped on his signal—a total of 470 100-lb high-explosive bombs fell behind the lines. The second occurred when a lead bombardier failed to properly identify the target and took the easy way out—bombing on the flashes of preceding bombs. A total of 352 260-lb fragmentation bombs fell in friendly lines. In the third case, a command pilot overrode his bombardier and dropped on previous bomb flashes; previous bombs had been off target but within a safe "withdrawal" zone. The pilot's bombs fell within friendly territory.[124]

All of the above errors were incidental to the real causes of the tragic bombings: confusion over whether the air attack would be flown perpendicular or parallel to the front lines, and the restricted size of the bomb zone. The Army wanted a parallel attack so that short bombs would not land in friendly territory (though, in fact, even such an approach would not guarantee an absence of friendly casualties). The AAF, concerned about the run-in to the target and enemy antiaircraft fire, wanted to fly a perpendicular approach. Further, the AAF bomber people recognized that the "heavies" were not as precise in such attacks as the fighter-bombers. They asked Bradley to keep friendly troops at least 3000 yards from the bomb line; Bradley compromised on a minimal distance of 1200 yards, with a preceding fighter bomber attack to cover the next 250 yards so that, in fact, the heavy and medium bombers would strike no closer than 1450 yards (a distance a heavy bomber would cover in approximately fifteen seconds). Getting a distinct aiming point and ensuring a split-second precise drop were thus critical. Despite Bradley's subsequent claims that the AAF was enthusiastic over the strikes, evidence indicates that the strategic bomber people were, in reality, anything *but* enthusiastic. As a general philosophical point, the strategic bomber people—British as well as American—believed that anything diverting them from their strategic air campaign against the Nazi heartland represented a weakening of effort. In addition, however, the AAF had strong feelings—communicated directly to Eisenhower—that the Cobra bombings were questionable if for no other reason than they would involve the dropping of a large quantity of bombs in the shortest possible span of time in a restricted bombing zone. But the AAF had been overruled and the operation went forward. When they did execute a perpendicular run, Bradley alleged that it flew in the face of what had previously been decided. After the short bombings of July 24, Bradley had ordered an immediate investigation of why the strike group had flown a perpendicular course. The response was that such a course had been previously agreed upon, and the AAF had, in fact, informed the ground forces of its intention to use a perpendicular run-in to the target. Bradley eventually charged that the "Air Force brass simply lied," though in earlier writings he had been far more temperate and one wonders whether this bold statement merely reflected the hardening of age.[125]

In any case, Bradley reluctantly concurred with AAF plans to make another attack on July 25 (though he has stated he did so because he was over an "impossible barrel"), and it was during this series of strikes that the most sensational casualty of Cobra occurred: Lt. Gen. Leslie J. McNair, former commander of Army Ground Forces and currently the "commander" of the fictional "1st Army Group," killed in his foxhole by a direct bomb hit as he waited to observe the follow-up ground attack. McNair's death and the other friendly casualties infuriated the ground forces, perhaps in part because they remembered his vociferous criticism of the organization of air support for army needs in 1942–43. But it seems not to have had a deleterious effect at higher command levels. Though Bradley has stated that Eisenhower informed him that he no longer believed that strategic bombers could be used to support ground forces, such is not evident from Eisenhower's written comments, and, in fact, American "heavies" were used subsequently in troop support missions, notably in the German winter offensive. In fact, Eisenhower's comments after Cobra's bombing were far *less* critical than might have been expected:

> The closeness of air support given in this operation, thanks to our recent experiences, was such as we should never have dared to attempt a year before. We had indeed made enormous strides forward in this respect, and from the two Caen operations [the strikes of July 8 and 18] we had learnt the need for a quicker ground follow-up on the conclusion of the bombing, for the avoidance of cratering and for attacks upon a wider range of targets to the rear and on the flanks of the main bombardment area. Our technique, however, was still not yet perfected, and some of our bombs fell short, causing casualties to our own men. *Unfortunately, perfection in the employment of comparatively new tactics, such as this close-support carpet bombing, is attainable only through the process of trial and error, and these regrettable losses were part of the inevitable price of experience* [emphasis added].[126]

In retrospect, Cobra was a masterful operation, but the preparatory bombing was tinged with faulty planning, sloppy execution, and bad luck. The precise train of culpability for the confusion surrounding the short bombings will probably never be determined with exactitude. What *is* certain is that the blame for the strikes is not entirely due to the

AAF. John J. Sullivan's incisive examination of the Cobra operation rightly concluded that there was no duplicity on the part of the AAF (much less "lies"), and that, in fact, the AAF had been most reluctant to undertake the operation at all. The ground commanders did not take adequate precautions to protect their troops, and thus, Sullivan concluded, Bradley and his fellow ground commanders bore "full responsibility" for the bombing casualties to exposed troops. But the airmen must share some blame, from Tedder and Leigh-Mallory, who did not supervise the operation as thoroughly as they should have, to the aircrews themselves who botched their runs.[127]

In sum, then, there is plenty of blame to go around for the Cobra bombings. But one must temper criticism of the strikes with an appreciation for what losses were occurring on the ground during the bitter hedgerow fighting, and the effect of the bombing on the German forces confronting the Allies. The small casualties incurred by friendly bombing, and the undoubted success of the bombing in shattering German resistance (even Bradley was forced to admit that Cobra "had struck a more deadly blow than any of us dared imagine") illustrate how petty the uproar surrounding the bombings really was, and how unfortunate as well: in the postwar folklore of air-land operations, the short bombing is all too often the only aspect of Cobra that seems to be mentioned. Because of this, it is refreshing to read Eisenhower's reasonable, mature, and admirable judgment expressed above. Eisenhower, one feels, never lost sight of the most important result of the Cobra bombing: it devastated German forces and paved the way for the breakthrough that would trigger the breakout and roll the Wehrmacht to the Third Reich itself.[128]

The main weight of the Cobra bombings fell opposite Maj. Gen. J. Lawton Collins's VII Corps, on Generalleutnant Fritz Bayerlein's already battered Panzer-Lehr division. The initial confusion of the July 24 strikes had misled the German defenders into thinking that they had, in fact, withstood and repulsed an American attack. They were not ready, then, for the whirlwind that descended upon them on the 25th. The bombing, as Collins subsequently recollected, "raised havoc on the enemy side."[129] Though VII Corps, hurting from the accumulated short bombings of two days, did not make great progress in its own ground attack on the 25th, Collins shrewdly realized that the German command and control structure had been badly disrupted from the air attack, and accordingly planned a full-scale assault for the next morning. There began the genuine breakthrough, as Combat Command A of the 2d Armored Division, ably supported by Quesada's IX TAC and building on the accomplishment of the 30th Infantry Division (which had taken the brunt of the short-bombings) cut through enemy defenses. Breakthrough now became breakout; the stage was set for the drive across Northern Europe.[130]

Bayerlein has left a remarkable account of the effects of the Cobra bombing and ground assault on his already war-weary command. In response to postwar interrogation he wrote:

> We had the main losses by pattern bombing, less by artillery, still less by tanks and smaller arms.
>
> The actual losses of dead and wounded were approximately:
>
> | | |
> |---|---|
> | by bombing | 50% |
> | by artillery | 30% |
> | by other weapons | 20% |
>
> The digging in of the infantry was useless and did not protect against bombing. . . . Dugouts and foxholes were smashed, the men buried, and we were unable to save them. The same happened to guns and tanks. . . . It seems to me, that a number of men who survived the pattern bombing . . . surrendered soon to the attacking infantry or escaped to the rear.
>
> The first line has [*sic*] been annihilated by the bombing. . . . The three-hour bombardment on 25.7—after the smaller one a day before—had exterminating morale effect on the troops physically and morally weakened by continual hard fighting for 45 days. The long duration of the bombing, without any possibility for opposition, created depressions and a feeling of helplessness, weakness and inferiority. Therefore the morale attitude of a great number of men grew so bad that they, feeling the uselessness of fighting, surrendered, deserted to the enemy or escaped to the rear, as far as they survived the bombing. Only particularly strong nerved and brave men could endure this strain.
>
> The shock effect was nearly as strong as the physical effect (dead and wounded). During the bombardment . . . some of the men got crazy and were unable to carry out anything. I have been personally on 24.7 and 25.7 in the center of the bombardment and could experience the tremendous effect. For me, who during this war was in every theater committed at the points of the main efforts, this was the worst I ever saw.
>
> The well-dug-in infantry was smashed by the heavy bombs in their foxholes and dugouts or killed and buried by blast. The positions of infantry and artillery were blown up. The whole bombed area was transformed into fields covered with craters, in which no human being was alive. Tanks and guns were destroyed and overturned and could not be recovered, because all roads and passages were blocked. . . .
>
> Very soon after the beginning of the bombardment every kind of telephone communication was eliminated. As nearly all C.P.'s [Command Posts] were situated in the bombed area, radio was almost impossible. The communication

was limited to [motorcycle] messengers, but this was also rather difficult because many roads were interrupted and driving during the bombardment was very dangerous and required a lot of time.[131]

By any standard, the Cobra bombing had an extraordinary effect on the German defenders, and as the official Army history of the Normandy campaign acknowledges, the Cobra bombing constituted the "best example in the European theater of 'carpet bombing.'"[132] This, of course, does not mean that the subsequent campaign on land was a pushover, for throughout the war, German combat formations showed a consistent and amazing resiliency, reforming, recuperating, and continuing to fight. So it was with the decimated Panzer-Lehr division, and so, too, would it be with many other battered Nazi units. Nevertheless, the Cobra operation as a whole put the Germany Army in France on the skids; ironically, it would be a Nazi command decision which would set the stage for its destruction in Northern France.

TacAir Omnipotent: Mortain and the Falaise-Argentan Pocket

Mortain and Falaise, like Wadi el Far'a, Guadalajara, and more recently Mitla Pass, have come to symbolize a particular form of warfare: the destruction of closely packed columns of troops and vehicles by constant and merciless fighter-bomber strikes in concert with action on the ground. Any chance of withdrawing with troops, equipment, and vehicles in good order was lost to the Wehrmacht due to the violence of the breakout from the beachhead at Normandy, and Hitler's order to von Kluge (a commander in whom Hitler was placing less and less faith, in light of the recent assassination attempt on his life and Kluge's alleged complicity) to stand firm in Normandy. As a result of Hitler's directive, the Wehrmacht launched a general offensive against Mortain, the weakest spot in the Allied line, on August 7. It failed amid stubborn resistance on the ground and intensive fighter-bomber attacks. The next step was the battering of German forces caught in the Falaise-Argentan pocket, fighting characterized by combined infantry-armor-artillery-air attacks directed against those units desperately attempting to escape eastward. As a result, though some German forces did escape through the ever-narrowing "gap" (thanks to yet another tardy follow-up by Field Marshal Montgomery), they did so without equipment and in a state of disarray and almost complete demoralization. By the end of

August, Allied forces had liberated Paris, advanced to the Seine, won the Battle of France, and set the stage for the Battle of Germany that would follow. Ahead lay some particularly bitter fighting—notably Montgomery's botched airborne invasion of Holland, and the ferocity of the German counterthrust in the Ardennes—but as of the end of August 1944, only the most ardent Nazi would have continued to believe in any expectation of an ultimate German victory.[133]

The attack on Mortain has gone down as an attack revealed by Ultra, so that the defenders—the American forces—were able to set up their defense in advance of the German thrust.[134] This might be termed the "myth of Mortain," for, in fact, Ultra *did not* offer a forewarning enabling the defenders to prepare for the attack. On August 2, Hitler had ordered von Kluge to prepare for an attack westward to the coastline, but this early indication of trouble ahead did not make its way from the Allied intelligence shop to Bradley's 12th Army Group. On the evening of August 6, orders went out for five Panzer divisions to attack through Mortain (which had already fallen to American troops) ninety minutes later—at 1830 hours. Ultra did not send out this message until midnight, but the German attack had itself been delayed in the field until just after midnight. Though the Allied signals arrived just immediately prior to the German attack, clearly they would not have offered the defenders any time whatsoever to make extensive plans or redeployments for the assault. Bradley, in his autobiography *A General's Life* is understandably testy about allegations that Mortain was predetermined by Ultra; his argument that he waged the battle without benefit of Ultra forewarning is borne out by the account of former Ultra intelligence analyst Ralph Bennett, who refreshed his own recollections by extensive research into the actual Ultra messages themselves. Bennett has stated that German update information during the Mortain fighting furnished "cheerful reading" to the analysts, but added little, if anything, to the information Bradley and Montgomery already had in the field from their own combat intelligence operations.[135]

When the Ultra message did come in, Bradley did order "all-out" air support the following morning, by which time the American 30th Infantry Division was locked in desperate and stubborn combat with the German tanks, but even here Ultra played a minor role, since the midnight attack would have triggered a day of Allied air support anyway, from battlefield requests. During this fighting, Broadhurst's rocket-firing Typhoons of 2 TAF had responsibility for defending the ground forces and attacking the German columns, while the AAF's 9th AF flew

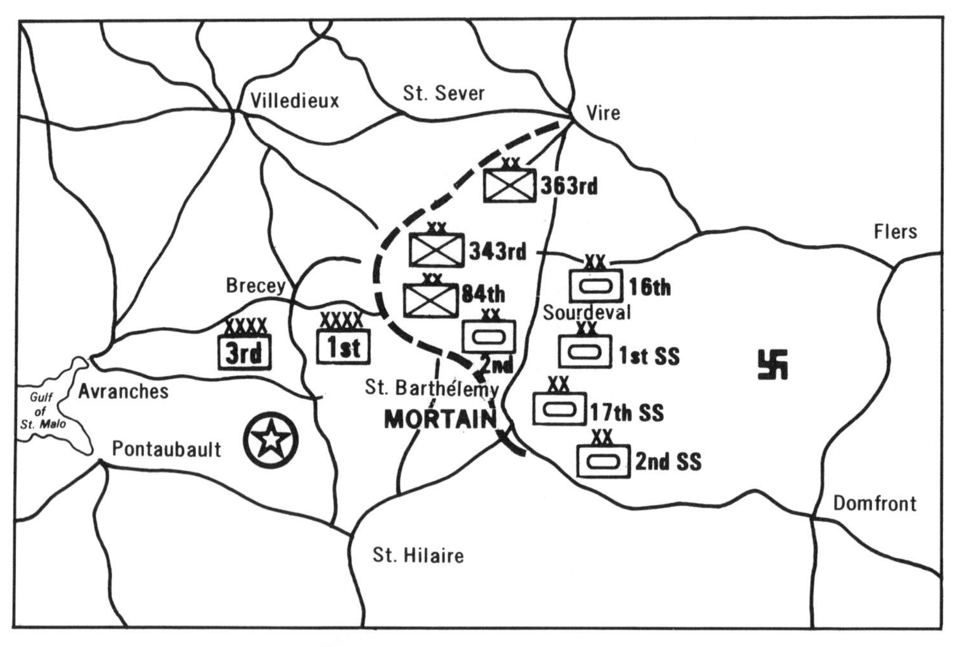

The Mortain Counterattack
(Situation as of 1200 hours, August 7)

— — Frontline between German and American forces

interdiction and air superiority sorties. For the Mortain operation, the Luftwaffe centralized its few fighter resources and attempted to intervene over the battlefield, but, in fact, the deep cover American air superiority sweeps gobbled them up as they took off, and "not one" (as Rommel's former chief of staff Generalleutnant Hans Speidel dismally recalled) appeared over the battlefield. The skies over Mortain belonged to the RAF; the weather was poor in the early morning, but as the day went on, the overcast lifted and patches of blue appeared. As the weather improved, Typhoons swarmed over the area, so many, in fact, that some got in each other's way, and several mid-air collisions apparently resulted. A morning recce flight located German tanks near St. Barthélemy, and thereafter, between the first engagements (just after noon) and late afternoon, Typhoons flew a total of 294 sorties over the battlefield. Typhoon pilot John Golley has left a graphic account of Typhoon operations at Mortain, particularly the battle between 245 Squadron (which was especially active) and the 1 SS Panzer division on the road near St. Barthélemy; their first attacks sprayed the tanks and transport with rocket and cannon fire, and a thin haze of smoke and dust spread slowly over the Norman countryside. The Typhoons broke off as they exhausted their ammunition and rockets, returning again and again to their strip to refuel and rearm. So intensive were the sortie rates that 245 Squadron, ever afterward, referred to August 7 as "The Day of the Typhoon."[136]

German commanders were shocked at the magnitude of the air attacks at Mortain, and which would be repeated before the month was out, at Falaise. On the ground, the 30th Infantry Division had stood firm, repulsing the German forces that did close to engage them. But it had been the air that had saved the day at Mortain, at least preventing a possible local German success that might have prolonged the campaign in France. As Eisenhower reported:

> The chief credit in smashing the enemy's spearhead, however, must go to the rocket-firing Typhoon planes of the Second Tactical Air Force. They dived upon the armored columns, and, with their rocket projectiles, on the first day of the battle destroyed 83, probably destroyed 29 and damaged 24 tanks in addition to quantities of "soft-skinned" M.T. [Motorized Transport]. The result of this strafing was that the enemy attack was effectively brought to a halt, and a threat was turned into a great victory.[137]

With the Nazi spearhead smashed, Mortain degenerated into a five-day slugfest. Foolishly, for a time the Germans continued to press toward Avranches, a move Bradley subsequently termed "suicidal," for

Collins's VII Corps was in position to attack the German flanks. Elements of the 2 SS Panzer division, operating south of the devastated 1 SS Panzer division, besieged Hill 317 (in whose shadow Mortain is nestled). The defenders, a lone battalion, stood firm. Supported by Allied air (including supply drops) and artillery, this battalion heroically held out until relieved by the 35th Division on August 12. Mortain came to an end; in the fighting after August 7, the 2 SS Panzer had joined the rapidly growing roster of German armored formations shattered by Allied combined air-artillery-armor assault. Generalmajor Rudolf-Christoph Freiherr von Gersdorff, the chief of staff of the German 7 Armee, subsequently concurred that the continuation of the counterattack toward Avranches was a "mistake."[138] Contributing to the German failure was the overemphasis of attacking north, between Mortain and Vire rather than farther south. In any case Mortain must be counted among the most important battles in the west, and recognized for what it was: a true example of air-land action. It set the stage for the next and even greater disaster to befall German arms in France: the battle of the Falaise-Argentan pocket, for after Mortain, the only course open to the Wehrmacht was headlong retreat toward the German frontier. In that retreat, Allied tactical air would offer no respite.

In retrospect, air was more critical—and under greater pressure—at Mortain than at the subsequent fighting in the Falaise-Argentan pocket, for Mortain was an Allied defensive battle whereas Falaise was an encirclement and an attempt to prevent the Germans from escaping out of the trap eastward. As the perimeter closed down, the pocket became a gap, and the Allies struggled to close it. The Falaise campaign may be said to have begun on August 7 (the same day as the German counterattack at Mortain) when Canadian troops launched a ground assault called Totalize toward Falaise. For the next two weeks, Allied troops—British, American, and Polish—harassed the German forces caught inside the pocket until finally, on August 21, the gap was closed. But by that time, what might have been a great encirclement echoing some of the pivotal battles on the Eastern Front was instead something less: a victory nevertheless, but a victory qualified by the number of German forces that had, in fact, been able to escape out through the gap. The fact that German forces did escape from Falaise outraged American commanders, from the even-tempered Eisenhower and Bradley to the mercurial Patton. They saw it as yet again an example of bad generalship by Montgomery, who pressured the pocket's western end, squeezing the Germans out eastward like a tube of toothpaste, rather than capping the open gap. Patton, ever aggressive, pleaded with Bradley for clearance to cut across the narrow gap, in front of retreating German forces, from Argentan north to Falaise. But Bradley wisely demurred, recognizing that the outnumbered Americans might be "trampled" by the German divisions racing for the gap. "I much preferred," Bradley recollected subsequently, "a solid shoulder at Argentan to the possibility of a broken neck at Falaise."[139]

Eventually, the Canadians pressed south from Falaise, the Americans north from Argentan, and both sought to narrow and close the gap by reaching the road network across and beyond the Dives River, at Trun, St. Lambert, Moissy, and Chambois. Beyond there the roads led toward Vimoutiers, funneling German forces into predictable killing grounds. Polish forces fought an especially prolonged and bitter struggle at Chambois that echoed Mortain's lone battalion. On August 19, they seized Chambois (soon dubbed "Shambles"), establishing defensive positions on Mont Ormel, to the northeast. Here was an ideal high vantage point to call in artillery and air strikes on the German forces streaming across the Dives and past their positions. As a consequence, extremely bitter fighting broke out between Polish and retreating German forces, but the Poles were able to retain control until the gap closed on August 21. The countryside around the Dives and Orne rivers was generally open, with sporadic patches of forested areas. The high ground across the Dives—specifically Mont Ormel—furnished an unparalleled vista of the entire gap area. In the third week of August 1944, this vista was marred by the near-constant bursting of bombs, rockets, and artillery, the ever-present drone of fighter-bombers and small artillery spotters (the latter especially feared and loathed by German forces), the corpses of thousands of German personnel and draft animals, and the burning and shattered remains of hundreds of vehicles and tanks. It was a scene of carnage without parallel on the Western Front.

In the days preceding the closing of the Falaise gap, the 2 TAF averaged 1200 sorties per day. The air war was particularly violent from August 15 through the 21st. Typhoons and Spitfires attacked the roads leading from the gap to the Seine, strafing columns of densely packed vehicles and men. Under repeated attack, some of the columns actually displayed white flags of surrender, but the RAF took "no notice" of this since Allied ground forces were not in the vicinity, and "to cease fire would merely have allowed the enemy to move unmolested to the Seine."[140] Typhoons typically would destroy the vehicles at the head of a road column, then leisurely shoot up the rest of the vehicles with their

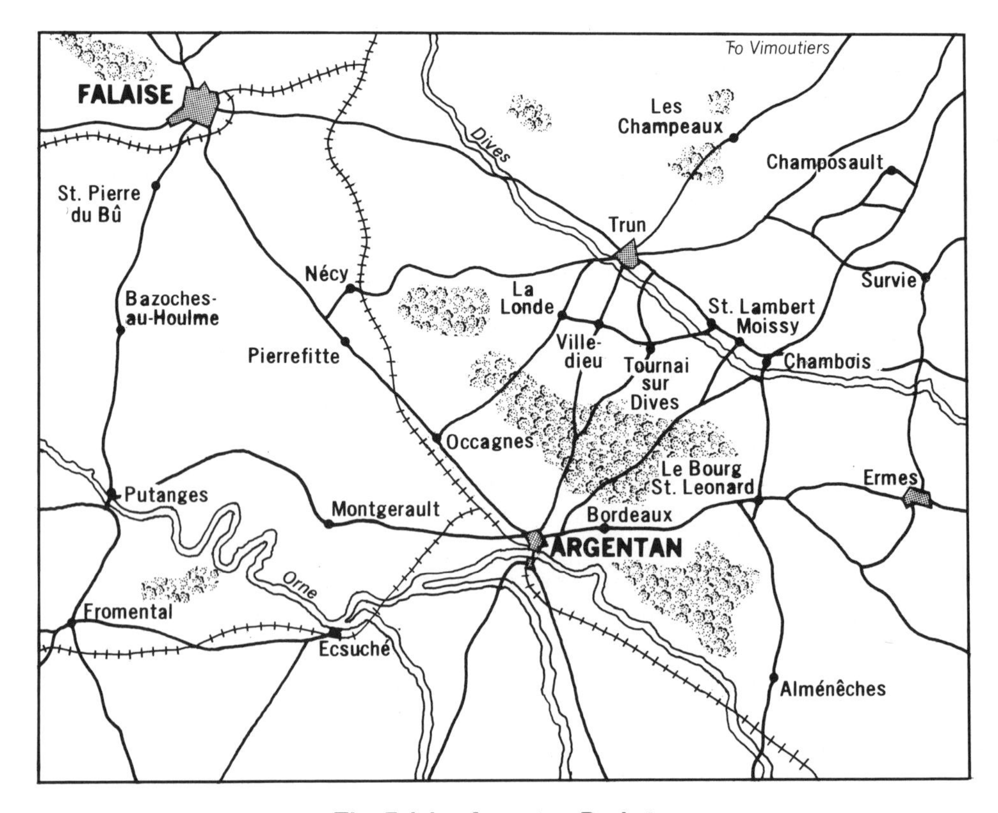

The Falaise-Argentan Pocket

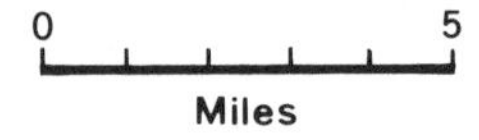

rockets and cannon. When they finished, Spitfires would dive down to strafe the remains. Because the Luftwaffe was not making an appearance over the battlefield, Broadhurst directed 2 TAF wings to operate their aircraft in pairs. Accordingly a "two ship" of Spitfires or Typhoons could return to the gap after being refueled and rearmed without waiting for a larger formation to be ready to return. This maximized the number of support sorties that could be flown, and, indeed, pilots of one Canadian Spitfire wing averaged six sorties per day.[141] Nothing that moved was immune from what one Typhoon pilot recollected as "the biggest shoot-up ever experienced by a rocket Typhoon pilot."[142] Another recalled the flavor of attack operations:

> The show starts like a well-planned ballet: the Typhoons go into echelon while turning, then dive on their prey at full throttle. Rockets whistle, guns bark, engines roar and pilots sweat without noticing it as our missiles smash the Tigers. Petrol tanks explode amid torrents of black smoke. A Typhoon skids away to avoid machine fire. Some horses frightened by the noise gallop wildly in a nearby field.[143]

Nor was it strictly a 2 TAF operation; the AAF was likewise heavily committed. Battlefield communications—from tactical reconnaissance aircraft to ground forces to the TAC advanced headquarters, and to other airborne strike aircraft—had "improved immensely" so that "fighter-bombers began to figure more and more prominently in tank battles and armored thrusts as they were going on."[144] Over the duration of the Falaise fighting, air strikes gradually moved from west of Argentan to north, to east, and finally to east of the Dives River. One strike by P-47's on August 13 gives a graphic indication of the sizes of German forces open to attack:

> That morning 37 P-47 pilots of the 36th Group found 800 to 1,000 enemy vehicles of all types milling about in the pocket west of Argentan. They could see American and British forces racing to choke off the gap. They went to work. Within an hour the Thunderbolts had blown up or burned out between 400 and 500 enemy vehicles. The fighter-bombers kept at it until they ran out of bombs and ammunition. One pilot, with empty gun chambers and bomb shackles, dropped his belly tank on 12 trucks and left them all in flames. All told, on 13 August, XIX TAC fighter-bombers destroyed or damaged more than 1,000 road and rail vehicles, 45 tanks and armored vehicles, and 12 locomotives. Inside the pocket they reduced 10 enemy delaying-action strong points to rubble.[145]

Four days later another Thunderbolt squadron, below-strength, came over a huge traffic jam, radioed for assistance, "and soon the sky was so full of British and American fighter-bombers that they had to form up in queues to make their bomb runs." The next day, 36th Group Thunderbolts spotted another large German formation, marked out by yellow artillery smoke; since the vehicles were in a zone designated as a British responsibility, XIX TAC sat back "disconsolately" while 2 TAF launched a series of strikes that claimed almost 3000 vehicles damaged or destroyed.[146] On August 19, one Spitfire wing put in a claim for 500 vehicles destroyed or damaged in a single day; that same day, another Spitfire wing claimed 700.[147]

Nothing and no one was immune from attack. Oberst Heinz-Gunther Guderian, son of the victor of Sedan, was seriously wounded when his Volkswagen was strafed and set ablaze by an Allied fighter. Generalmajor von Gersdorff was strafed and slightly wounded by a P-38 Lightning at Chambois, and he subsequently reported that "The very strong low flying attacks . . . caused high losses . . . units of the Army were almost entirely destroyed by low flying attacks and artillery." One country road eastward from Moissy earned the grim sobriquet *le Couloir de la Mort*: the Corridor of Death. At night, intruder aircraft attacked river crossings and ferries over the Dives. At least 10,000 German soldiers died, and 50,000 fell prisoner. Nearly 350 tanks and self-propelled guns, nearly 2500 other vehicles, and over 250 artillery pieces had been lost in the northern section alone of the Falaise pocket. Von Gersdorff stated that armored divisions that did withdraw from the gap had "extremely low" strength: for example the 1 SS Panzer had only "weak infantry," and no tanks or artillery; the 2 Panzer had one battalion, no tanks, and no artillery; the 12 SS Panzer had 300 troops and no tanks; the 116 Panzer had two battalions, twelve tanks, and two artillery batteries; and the 21 Panzer had four battalions and ten tanks.[148] As historian Max Hastings has shown in his own studies, such figures were by no means unique; four other SS Panzer divisions could muster no more than fifty tanks among them.[149] (Wehrmacht armored divisions typically possessed an organizational strength of 160 tanks, and approximately 3000 other vehicles.) The carnage of the battlefield was truly incredible, and sickened many fighter-bomber pilots who visited the site. Eisenhower, touring the gap area two days after it closed, encountered "scenes that could only be described by Dante."[150] Perhaps the twisted allegories of Hieronymous Bosch would have been more fitting a choice, for Dante, at least, offered hope.

With the conclusion of the battle of the Falaise gap came the final end to the battle of Normandy. It did not, of course, spell an end to the war, which would rage on for another nine months. But it did confirm that the ultimate defeat of Nazi Germany was in the cards. Though much has been written by critics who should know better about the remarkable ability of the Wehrmacht to rejuvenate and reform itself, and about the "toughening" and "thickening" of German resistance in the weeks and months ahead, not enough attention is paid to the flip side of this: where that strength was coming from. It was coming from the Russian Front, as units were hastily transferred to the West (adding inevitably to the prospects of an eventual Soviet triumph in the East), and from within the critical bone marrow of the Third Reich itself. Hitler and his minions were spending capital they did not have. The thickening of the resistance at the front was the thickening of a crust: a crust that would prove costly as the Allies cut through it in the fall and winter of 1944–45, but which would open up to them the poisonous souffle that was the Nazi heartland in 1945.

Anglo-American Tactical Air Power: A Retrospective View

By the end of the Normandy campaign, all the elements and relationships that would characterize the employment of Anglo- American tactical air power during the rest of the European campaign were in place: forward observers and controllers, occasional airborne controllers, radar strike direction, "on-call" fighter-bombers, armored column cover, night intruders, to name just a few. It had only been thirty-six months—a mere three years—from the disappointment of a Brevity and Battleaxe to the harmony and orchestration of the invasion and breakout. Normandy was neither the victory of a single branch of arms, nor the victory of a single nation. It is the classic example of complex combined arms, multiservice, coalition warfare. In this regard, it is well to remember that the battlefield triumphs of air power were part and parcel of infantry-artillery-armor assaults on the ground. It was true air-land battle.

The effectiveness of the Anglo-American air support in the Normandy campaign—and by extension, through the collapse of Germany as well—is beyond question, attested to alike by airmen, ground commanders (both beneficiaries and victims), soldiers in the field (again on both sides), and prisoners of war. Typical were the comments of a battalion commander in an armored regiment, who reported:

> Our air cover has been excellent and has helped us out of many tight spots. At El Boeuf they knocked out eight German Mark V [Panther] and Mark VI [Tiger]

> tanks that were giving us a great deal of trouble. They also helped us at Tessy-sur-Vire by knocking out tanks. They are on call by any unit down to a platoon, calling through company and battalion, and giving the location of the target. Then the ASPO [Air Support Party Officer] contacts the air cover and gets a strike within a matter of minutes. I have seen the air strike within three minutes after the call was made. We like to know the air is there. We want it all the time.[151]

Two other battalion commanders from the same regiment endorsed his remarks. VII Corps's "Lightning Joe" Collins stated that "we could not possibly have gotten as far as we did, as fast as we did, and with as few casualties, without the wonderful air support that we have consistently had."[152]

As evaluated by Bradley's 12th Army Group air effectiveness committee, fighter-bombers in particular proved valuable for a number of missions, including operations within striking range of artillery. Only when utilized against heavily constructed positions such as casemated guns and pillboxes did they prove "not particularly effective." They actually proved more accurate than long-range heavy artillery, specifically the 240mm howitzer and the 8-in gun or howitzer. Operating with 500-lb general purpose and 260-lb fragmentation bombs, fighter-bombers—particularly the rugged P-47—routinely conducted close-in strikes within 300 to 500 yards of friendly troops. Pure bombers were a different matter. Mediums (such as the B-25, B-26, and even the A-20 and its successor the Douglas A-26 Invader) were viewed as a mixed blessing. While they were not as criticized as the occasionally errant "heavies," commanders felt that they lacked the strong control and communication relationship with medium bomber units that they had with the fighter-bombers. In particular, mediums were seen as too inflexible—they lacked the quickness, ease of response, and availability of the fighters. Heavy bombers were seen as devastatingly effective in the Cobra breakthrough, but were also believed to have inherent disadvantages that the fighter-bombers did not have, namely the need to be concerned about friendly casualties and the consequent necessity of a large safety area between friendly forces and the target area. All of this reinforced a generalized view from the ground that, when air support was needed, it could best be delivered by the fighter-bomber. And it is interesting to note that despite all the brouhaha of the early war years concerning dive bombers, arguments favoring them for battlefield air support were conspicuously absent by the spring of 1945, as were arguments for specialized battlefield attack aircraft. The "attack" airplane was dead; long live the fighter-bomber.[153]

The performance of the fighter-bomber nevertheless needs to be examined in light of the Normandy experience. First, when confronted with dense light antiaircraft fire, fighter-bombers did take losses; IX TAC lost a total of eighty aircraft from July 25 through August 7, 49 percent from flak, 7 percent to enemy aircraft, 24 percent to small-arms fire, and 24 percent from unknown causes. Thus 73 percent—and possibly more nearly 90 percent—of combat losses came from some form of light or heavy ground fire. Undoubtedly the rugged construction and dependability of the P-47's air-cooled engine prevented even further losses, a luxury the liquid-cooled Typhoon lacked.[154] Wolfgang Pickert, a General der Flak-Artillerie in charge of the III Flak Korps, reported that when "fighter-bomber weather" existed, "the movement of large vehicles during the hours of daylight was practically tantamount to their certain loss." Nevertheless, when light antiaircraft forces were present in sufficient strength (a rarity in Normandy), "fighter-bombers had hardly any successes, or only with heavy losses." (Pickert also complained that the flak forces lacked sufficient heavy flak to "ward off" B-26 Marauders on deep interdiction sorties.) III Flak Korps had one regiment in the Falaise pocket during the hectic withdrawal eastward, and by good fortune it had an unlimited supply of ammunition due to its proximity to III Flak Korps ammunition depots in the area. Pickert alleged that the regiment "reported that it had inflicted heavy casualties on the enemy and had put numerous enemy tanks and planes out of action," but such does not seem warranted—as least as regards aircraft claims—from other accounts. Undoubtedly, the firing of antiflak artillery fire immediately prior to air strikes and in some cases during them had a salutary benefit on Allied close air support operations—a reminder of the necessary combined-arms interactions between air and land forces fighting to achieve victory on the battlefield.[155]

An argument that will probably remain unanswered is which aircraft was more effective as a ground-attack airplane—the P-47 or the Typhoon? Actually, the argument centers more on the relative armament employed—the bomb or the rocket. Quesada considered bombs "perfect," but the RAF swore by the rocket-armed Typhoon. Eisenhower, as has been seen, credited the rocket-firing Typhoons with saving the situation at Mortain, and it is difficult to fault that judgment. But Montgomery's 21st Army Group took a different view completely. The

Operations Research Section of the 21st AG reported that after examining 301 heavily armored vehicles (tanks and assault guns) destroyed in Normandy fighting, only *ten* were found to have been destroyed by rockets! When a rocket did hit a vehicle, it did destroy it. Only three out of eighty-seven armored troop carriers (halftracks) were found destroyed by rockets. Cannon and machine gun fire were "very deadly to all except heavily armored vehicles."[156] Muddying the water, of course, is Collins's statement to Quesada about the effectiveness of machine-gun fire against German tanks, and the fact that many Panzer crews simply abandoned their vehicles for a variety of reasons in the gap fighting and withdrew on foot. Then, of course, there are all the claims of the Typhoon pilots, with the graphic descriptions of burning Tigers and shattering rocket impacts. Wherein is the truth?

The truth likely lies between these factors. Pilot claims for ground destruction have always been prone to distortion, if for no other reason than the speed at which an aircraft passes over a target. The flash of exploding cannon shells and the spark of ricochets off a hardened target can all too easily be mistaken for the signs of a "kill," as can the dense smoke from an uncleaned exhaust system pumped into the air from a revving tank engine. Then, there is the problem of vehicle identification. To a plane skimming over the ground at several hundred miles per hour, with the pilot concerned about flak, it is increasingly difficult to distinguish between a genuine tank and, say, a potentially more vulnerable vehicle such as a self-propelled gun carriage or an armed halftrack. They all become tanks, and then all too often an inflation factor elevates them not merely to the status of tanks, but to the most formidable tank of all, which, in 1944, was the Mk VI Tiger. (Even to the Allied infantry and tank crews on the ground this was a difficulty. The Wehrmacht undoubtedly would have liked to have had as many Mk VI Tiger tanks in service as were claimed destroyed in the drive through Europe, but which were, in reality, varieties of Mk IV's and Mk V Panthers.) Having said this, however, all those Typhoon and Thunderbolt pilots can't be wrong, and given the immense number of armored vehicles and other transport left from the beachhead to the Seine, it is not difficult to find fault with the sampling of the 21st AG report. Undoubtedly far more than ten tanks were lost to rocket-firing Typhoons; what the exact number was will likely never be known, even in an approximation. But the accounts from eyewitnesses on the ground to the terror and ferocity of Typhoon attacks—and for that matter, the bomb-toting Thunderbolts—make it clear that not all rockets were misplaced, nor all bombs short or long. What is beyond question, of course, is the devastation wreaked upon other forms of transport by the fighter-bombers. Even allowing some exaggeration and duplicate claims, the sortie claims of the 9th AF and 2 TAF during the Normandy fighting is most impressive (Table 3).[157]

Table 3. Fighter-bomber sortie claims in Normandy

	2 TAF	9 AF	Total
Sorties flown	9,896	2,891	12,787
Claims for motor transport destroyed	3,340	2,520	5,860
Claims for armor destroyed	257	134	391
Total claims	3,597	2,654	6,251
Claims per sortie	0.36	0.92	0.49

No stronger endorsement of the air support of the Allied armies in Normandy can be found than that offered by Omar N. Bradley himself, in a letter to Hap Arnold at the end of September 1944. "I cannot say too much for the very close cooperation we have had between Air and Ground," Bradley wrote. "In my opinion, our close cooperation is better than the Germans ever had in their best days."[158]

Normandy is the classic example of modern combined-arms, mechanized, air-land, coalition warfare. It was a struggle in which the Allies were fortunate to have not merely air superiority, but air supremacy; with it, their task of winning on the ground was that much easier. Where the Allies had won the critical battle for air supremacy was not over the beachhead, but in the several years of air war that had previously gutted the Luftwaffe. To those inclined to minimize the value of air to the Normandy operation, the final word must come from Eisenhower himself.

In June 1944, John S. D. Eisenhower, Ike's son, graduated from West Point—ironically on the same day (June 6) that Allied forces stormed ashore at Normandy. June 24 found him riding through Normandy with his father, observing the aftermath of the invasion:

> The roads we traversed were dusty and crowded. Vehicles moved slowly, bumper to bumper. Fresh out of West Point, with all its courses in conventional procedures, I was offended at this jamming up of traffic. It wasn't according to the book. Leaning over Dad's shoulder, I remarked, "You'd never get away with this if you didn't have air supremacy." I received an impatient snort:
>
> "If I didn't have air supremacy, I wouldn't be here."[159]

15

Battlefield Air Support in the East: The Case of Kursk

When Hitler invaded the Soviet Union in June 1941, he had every expectation of a quick victory. He had a honed military system benefiting from its experiences in the blitzkrieg. In contrast, the Soviets had done poorly in Finland, and Finland could be counted on to assist Nazi Germany when the invasion began. Further, Stalin's purges had devastated the Soviet military: 3 of 5 Marshals perished, as did 13 of 15 army commanders, 57 of 85 corps commanders, and 110 of 195 division commanders. Then, there were the captive peoples, and large numbers of disaffected Soviet citizens, in Latvia, Lithuania, Estonia, and the Ukraine. There were, however, a number of factors that mitigated against his success: the core of weakness of the Wehrmacht, already showing signs of being badly led and being overcommitted in an increasingly widening war; the vast geography of the Soviet Union; and, finally, the racial policies of the Third Reich itself. Subordinating common sense to racial ideology, the Nazi regime ruined for itself any chance of exploiting the considerable anti-Stalin anti-Soviet feeling in the eastern territories by its rigid insistence on treating the population as collective *Untermensch* to be presided over, tyrannized, enslaved, and brutalized by Nazi *Herrenvolk*. Within a year of *Fall Barbarossa* (Operation Barbarossa, the invasion of Russia), partisan activity had become a serious problem for the Germans, functioning in a way analogous to Allied air interdiction of German lines of communication in the west.[160]

The Nazi Blitz and Soviet Reaction

At first, all went well. In particular, the Luftwaffe shattered the Soviet army air force; 1800 Soviet airplanes had been destroyed on the first day alone, 4000 by the end of the first week. On the ground, the German blitzkrieg knifed through Russian defenders, and for a while it appeared like the heyday of German arms—an invasion of France writ large, as it were. The old Bolshevik Semyon Budenny ("a Slavic synthesis of Foch and Patton—with the talents of neither," in Alan Clark's memorable phrase) proved even less successful defending the southwestern front against Hitler's generals than he had been against the Poles two decades before. Summer passed into fall, then winter approached, and the Nazi forces found themselves on the outskirts of Moscow. As Stuka pilot Paul-Werner Hozzel recollected:

> All of us were still sure of victory, even somewhat overbearing, not knowing what kind of winter was ahead of us. Our *Generaloberst* [von Richthofen] hovered in his Fieseler *Storch*, equipped with a radiophone, over the Russians as though they were flocks of sheep, directing single *Stuka* and *Schlachtflieger* units to targets spotted by him.[161]

There were, nevertheless, disquieting signs: the Russian tanks disengaging and heading eastward, hull-down on the horizon, beckoning German forces ever deeper; the never-ending resistance; and the continual appearance of the Soviet army air force, which like the Chinese air force half a world away, refused to be destroyed. Nazi Panzer gunner Karl Fuchs's letters to home offer an interesting mix of optimism and frustration (as well as revealing the malignant spirit of the Third Reich's military). On July 17, less than a month into the invasion, he wrote:

> Russian opposition is weak. . . . Our Air Force, in particular the *Stuka* attack bombers, actively support our efforts. Our comrades from the Air Force are top-notch guys. . . . Yesterday . . . we saw our first women soldiers—Russian women, their hair shorn, in uniform! And these pigs fired on our decent German soldiers from ambush positions.[162]

On November 21, a Soviet tank crew brought the life of this hostile product of the Hitler years to a close. The frustration with which German commanders regarded their Russian opponents as 1941 edged toward 1942 can be found in numerous memoirs and summary reports. One such document refers to Russian ground-attack aviation at the end of 1941, noting that "during the course of the winter battles the Russians employed every last one of their aircraft against ground targets, billets, roads, traffic congestion, bridges, etc. They even used trainers and obsolete planes utterly unfit for combat flying."[163]

As noted earlier, the Soviets had profited tremendously from their experience in Spain and China, but the purges of Stalin had, to a great degree, negated any benefits that might have accrued to their military. Thanks to prewar economic and industrial planning, new military systems were in development that would have a profound effect upon the course of the war itself: the T-34 medium tank (which was superior to every German tank until the advent of the Panther and Tiger); battlefield support rocket artillery (the infamous Katyusha developed by Sergei Korolev); the Ilyushin Il-2 ground-attack aircraft; the Petlyakov Pe-2 dive bomber and attack airplane; new fighters such as the Yak, MiG, and LaGG families; and new bombers such as the Tupolev Tu-2. Of at least equal significance, however, the Soviet military leadership placed increasing emphasis on the self-reliance, morale, and training of the average Soviet soldier.[164]

The course of the war on the Russian front is generally well known. The 1941 offensive surged forward to Moscow, and then halted with the onset of a fierce winter. Zhukov's subsequent offensive, while not resulting in any major strategic reversal in the East, nevertheless thwarted Hitler's plans to seize Moscow, and ended once and for all the myth of German battlefield invincibility. In 1942, hopes for a general Soviet offensive against the Germans were dashed by the ineptly conceived and executed offensive at Kharkov. Germany continued to push eastward, and the year ended in the battle of Stalingrad, which degenerated into savage street fighting before the ultimate collapse and surrender of Generalfeldmarschall Friedrich Paulus's 6 Armee in 1943.

Stalingrad is rightly hailed as the pivotal battle on the Eastern Front. In that battle, the encircling Soviet aerial blockade around Stalingrad included thirty-two fighter, bomber, ground-attack, and "mixed" air divisions, nine of which were ground-attack (Shturmovik) divisions; in June 1941, the total ground-attack strength of the Soviet army air forces consisted of only eleven regiments of obsolete Polikarpov I-15 and I-16 fighters, plus token numbers of newer Sukhoi Su-2 and Ilyushin Il-2 aircraft. (The Soviet air division was subordinate to the air army, and usually consisted of three or four regiments. The Soviet air regiment was the basic "building block" of Soviet military aviation, and consisted of approximately thirty-two aircraft by February 1943. Ground-attack divisions occasionally included a fighter regiment for escort duties.) At

Stalingrad, the Luftwaffe was unable to maintain an air corridor into the beleaguered city, and Russian fighters and flak took a heavy toll of resupply aircraft and even German combat airplanes. The Germans lost air superiority at Stalingrad, and that loss, coupled with reversals on the ground, sealed the fate of Paulus's troops.[165]

Restructuring and Rebuilding the Soviet Army Air Force

In 1942, months before the beginnings of the Stalingrad debacle, the Soviet Supreme High Command *(Shtab glavnogo/verkhovnogo komandovaniya,* or *Stavka)* reorganized the Soviet army air force *(voyenno-vozdushnyye sily,* or *VVS),* and placed it under the command of Gen. A. A. Novikov, one of the three architects (together with Zhukov and Nikolai Voronov) of the salvation of Moscow. Novikov brought a pronounced tactical orientation to the VVS that went beyond even its prewar leanings. Adapting some lessons from the Luftwaffe itself (notably tactical combat formations, and an emphasis on air-ground liaison), he also stressed areas in which the VVS was increasingly showing its own strength: namely, ground-attack aviation. In the Soviet system, the front commander—equivalent to an army group commander in the British and American armies—had direct control over his own "frontal" aviation. In fact, so tight was this control over the air forces that close support aviation was within a so-called "army aviation" command subordinate to the army commander. At lower—corps and division—level, an equivalent "troop aviation" structure existed for the same purpose. The Stavka reorganization of 1942 swept the army and troop aviation components away and, instead, assigned air units to the air commander working with the front commander. It was, in somewhat analogous fashion, similar to the structuring of air power in the Western Desert under Coningham, or the restructuring of American air power after FM 31-35, though without the strategic aviation perspective of either the British or American systems.[166] By the end of 1942, the VVS had adopted the notion of the "aviation offensive" and revealed considerably tightened operational control procedures as well. The air army commander located his command post in close proximity to the front (army) commander, and the command posts of subordinate fighter and Shturmovik aviation forces were located near forward observation posts within no greater than two or three kilometers of the front lines. Infantry and armored corps and divisions had air liaison officers located at their command posts, and these liaison officers kept track of friendly troop locations, and helped direct and control supporting Shturmovik attack flights. In short, then, by the end of 1942, the Soviet air-ground system was showing the same degree of organizational sophistication that was evolving in the Royal Air Force at that time, and it appears that the VVS system was, if anything, appreciably more practical and sophisticated than the contemporaneous American system, hamstrung as it was by FM 31-35. In contrast to its Western counterparts, however, the VVS was an intensely mobile force; air armies (which numbered approximately 1400 aircraft) had as many as 4000 trucks, permitting their withdrawal or advance as the ebb and flow of the land battle dictated.[167]

In part, the German invasion acted as a violent purgative to rid the service of its obsolescent and obsolete designs, and forced the reequipping of the VVS with newer and more suitable aviation technology. Designing aircraft was one thing; producing them was something else. The Soviet Union had a number of excellent designers, such as Sergei Ilyushin, Nikolai Polikarpov, Vladimir Petlyakov, Andrei Tupolev, Alexander Yakovlev, and Semyon Lavochkin, and in the months immediately prior to the outbreak of the Great Patriotic War, they had produced prototypes and the initial production models of some impressive designs that would eventually contribute markedly to the defeat of the Wehrmacht in the East. But the German invasion threatened the production base of the Soviet aircraft industry, and, as a consequence, the Soviet government undertook a massive relocation of industry west of the Volga to Siberia, across the Urals. By February 1942, this move had been completed, and in a surprisingly short time, aircraft began rolling off production lines, not in the hundreds, but in the thousands. Then, of course, there is the largely unrecognized effect of Lend-Lease upon the Soviet air forces. Large numbers of American and British aircraft arrived in the Soviet Union, and while many were kept behind the front for local training and defense purposes, many others—notably the American Bell P-39 Airacobra and Douglas A-20 Havoc—saw extensive service, as revealed from Russian and German memoirs alike.[168]

Four Soviet aircraft were to prove particularly significant in the air war in the East: the Yakovlev and Lavochkin fighter families, Ilyushin's Shturmovik, and Petlyakov's Pe-2. Together these four formed the backbone of the VVS as it grappled with the Wehrmacht. The first two were fighters, and operated primarily in the counter-air role against the Luftwaffe, but with secondary ground-attack roles as well.[169]

The Yakovlev family were aerodynamically clean, low-wing designs fabricated from wood. The initial production model, the Yak-1, first flew in mid-1939, and subsequently spawned many variants, the most important of which were the Yak-1M, Yak-7B, Yak-9B, Yak-9DD, Yak-9T, Yak-9U, and Yak-3. They all had roughly the same appearance, but over time the Yak family progressed markedly in performance. Early Yaks showed a pronounced French design influence and, in fact, had a performance similar to that of the Dewoitine D-520; the late-war Yak-3 closely approximated the later Spitfire and Mustang in performance. Over time, the all-wood structure of the Yak-1 evolved to wood and metal and then, with the Yak-9U, to an all-metal stressed-skin design. The Yak-9B, which appeared in the spring of 1943, reflected the swing-role emphasis of the VVS; the B stood for *Bombovoy* (bomber), and this aircraft was a true fighter-bomber, carrying a 450-kg bomb load in an internal bomb bay. The Yak-9T, which appeared late in 1943, was an antitank aircraft (T = *Tankovo*, tank) carrying a 37mm cannon (though sometimes replaced with a 23mm or 20mm weapon) firing through the propeller hub. By and large, however, the Yak fighters made their reputation as lightweight, fast, and highly maneuverable dogfighters, more than able to hold their own with later German Bf 109 and FW 190 opponents. Seemingly infinitely adaptable, the basic design of the Yak family spawned Soviet Russia's first postwar jet airplane, the Yak-15.

The Lavochkin fighter family was less prolific, but nonetheless suitable to the mix-and-match design approach that has commonly characterized Soviet aircraft design practice. The family began with the LaGG-1 which, like the Yak-1, was of prewar origin. The LaGG-1 and its successor the LaGG-3, as with the Yak, utilized a liquid-cooled engine, featured wooden construction, and had a highly varnished and polished exterior. Of pleasing shape, it nevertheless possessed extraordinarily vicious handling characteristics, leading VVS fighter pilots to aver that LaGG stood for "guaranteed varnished coffin." Replacing the inline engine with a lighter radial engine generated the La-5 variant, which first appeared in early 1942. The better-flying La-5 restored Lavochkin's reputation in the fighter community, and in turn spawned the La-5FN fighter-bomber, which could carry RS 82 rockets or PTAB antiarmor bombs, and the La-7, which featured a more powerful radial engine. Again like the Yak family, the La-5 and La-7 made their name primarily as air-superiority fighters (the La-5 was a particularly dangerous opponent at low altitudes), but with an ability to undertake swing-role ground-attack missions as well.

Both the Yak and Lavochkin fighters flew top cover and escort missions for the Soviet Union's two most important ground-attack aircraft: the Ilyushin Il-2 Shturmovik and the Petlyakov Pe-2 dive bomber. The Il-2, affectionately known to its crewmen as the Ilyusha, earned an altogether different sobriquet from the Wehrmacht: *der Schwarze Tod:* the Black Death. The Il-2 sprang from the design bureau of Sergei Ilyushin, a "tough-minded perfectionist"; it was said that the history of Soviet aircraft instrumentation was etched in the face of Ilyushin, who had test-flown—and crashed—many early Soviet designs.[170] The Il-2's prototype first flew in 1939, and proved seriously underpowered, necessitating a change to a more powerful 1680-hp liquid-cooled Mikulin engine. Thus modified, the new aircraft entered service in the spring of 1941, and 249 had been completed by the time of the Nazi invasion of Russia. The Il-2 had an armored-shell forward fuselage up to 12 mm thick, the aft fuselage was a wooden monocoque, and the wings and tail surfaces were metal as well. The initial Il-2 had 700 kg (1542 pounds) of armor plating, while later models had approximately 950 kg (2092 pounds) of armor; the armored metal monocoque forward fuselage essentially conferred invulnerability to any caliber shell less than 20mm, and even 20mm shells often failed to penetrate the curvacious contours of this aircraft. Thus armored, the Il-2 avoided many of the problems that afflicted other liquid-cooled-engine–powered aircraft used for ground attack, such as the Typhoon and Mustang. With a top speed of 292 mph "on the deck," the Il-2 was no world-beater, but it was faster than contemporary medium bombers, and made the most of low-altitude operations, usually operating in two-plane *para* or six-to-eight-plane *gruppa* formations at altitudes between 30 and 150 feet.[171] The initial Il-2 was a single-seat attack airplane with an awkward canopy design that offered a notoriously poor rear view, and rendered the plane vulnerable to any roving Luftwaffe fighters that might have the temerity and ammunition supply to attack them at low altitude, from above and behind. Ilyushin overcame this weakness by adding a defensive gunner with a 12.7mm (.50 cal.) machine gun. Over time the offensive striking power of the Il-2 improved as well. The initial model Il-2 had two 7.62mm (.30 cal.) machine guns and two 20mm cannon, could carry eight 82mm rockets, and a 400-kg (881-lb) bomb load. The later Il-2m3 switched to harder-hitting 23mm cannon, 132mm RS 132 rockets, a

600-kg (1312-lb) bomb load, and a DAG 10 grenade launcher to dispense grenades in the path of closing German fighters. Then, in 1943, the Il-2m3 Mod appeared (just in time for Kursk), with all of the above, but replacing the 23mm cannon with 37mm cannon for tank-busting sorties. A number of specialized air-dropped weapons had appeared that the Il-2 could carry, particularly the PTAB antiarmor bomb (discussed below). By early 1943, then, the VVS possessed in the Shturmovik an extremely formidable ground-attack aircraft. In the fluid situation of the Eastern Front, it could be expected to operate with a great deal of success.

Petlyakov's Pe-2 represented a significant departure on a number of counts; it had two liquid-cooled engines, could undertake dive bombing, and had performance high enough—including a 361-mph top speed, in later models—so that it could act as a long-range fighter (particularly against German attack and bomber formations), or, on the defensive, evade intercepting German fighters. In some respects, this fine aircraft constituted a Russian equivalent to the De Havilland Mosquito. The Pe-2's prototype flew in 1939, originally to fill a requirement for a high-altitude fighter. Subsequently redesigned as a high-altitude bomber, the Pe-2 design team next had to redesign it as a dive bomber. Perhaps because this aircraft had been designed originally as a fighter and subsequently modified as a bomber, it retained superlative performance characteristics (going the other way—from a bomber to a fighter—works only in the rarest of circumstances). Designed to an ultimate safety factor of 11g, the Pe-2 obviously had the ruggedness to be operated with abandon by its crews, and it possessed agility the more heavily armored Il-2 series lacked. Affectionately known as the Peshka, the Pe-2 served on occasion as a dive bomber, but more frequently flew in sorties similar to Anglo-American fighter-bombers: approaching a target at medium and high altitudes (up to 23,000 feet), and then diving downward, making repeated strafing and bombing runs. Only several dozen were available to the VVS during the opening of Barbarossa, but by mid-1942, the Pe-2, like the Il-2, was rapidly emerging as one of the major types of Soviet aircraft.

By 1943, the Soviet Union was clearly winning the production battle over Nazi Germany. By the summer of that year, the Luftwaffe boasted a total strength of 6000 airplanes (its wartime peak, with 2500 on the Russian Front), while the VVS was already at 8300.[172] More importantly, the quality of these new Soviet planes was every bit as good as that of their German counterparts. Equipment is significant, of course, but as Zhukov shrewdly recognized, the human factor is more decisive. Accompanying the introduction of new aircraft, the VVS had increasing numbers of aircrew and mechanics entering its ranks, and while training standards were not up to those of the West, they were adequate. It would fall to these individuals, many in their 'teens and straight off collective farms, to grapple with the Luftwaffe and ultimately secure the air superiority that would help send Soviet forces to Berlin.

German *Schlachtflugzeug* in the East

German ground-attack operations over Russia generally took the form of three kinds of attacks: concentrated efforts in direct support of the army on the battlefield, and (less frequently) indirect support operations (such as attacks on enemy airfields, troop concentrations, and rail lines); rolling attacks to furnish continuous support of German ground forces by pinning down Soviet forces on the battlefield; and, finally, "free-sweep" attacks to hunt for targets of opportunity while, typically, covering the flanks of an armored breakthough or continually supporting an armored spearhead. Interestingly, this system mirrored closely the Soviets' own division of ground-attack catagories. Formation size depended on the kind of mission; the Luftwaffe's basic fighter and attack aircraft operational unit consisted of the 2-ship *Rotte*; two of these comprised a 4-ship *Schwarm*; four of these formed a 16-ship *Staffel* (squadron); four of these plus a *Stabschwarm* (staff flight) created a 68-ship *Gruppe*; four of these plus a *Stabschwarm* established a 276-ship *Geschwader* (wing). Under the conditions of Eastern Front fighting, units typically had a "best" availability of 75 percent, thus a Geschwader, if at full strength (not always the case), could call upon a little over 200 aircraft. Only rarely would a full Geschwader go on the attack, but attacks by smaller formations up to Gruppe size were common.

Weather and the nature of Russian defenses imposed constraints and dictated certain operational procedures in German ground-attack activities. The Luftwaffe traditionally believed that it could conduct air support operations whenever it had a ceiling of 4500 feet over a battle area, but it also recognized that the average German pilot had mediocre bad-weather flying skills. Therefore, the lower the ceiling, the more restricted were attack operations, until only a unit's most experienced pilots—a true handful—would be out flying. Unlike the Soviets, the Germans usually stressed approaching the battle area as high as practicable so as to avoid the intense light flak that increasingly accompanied Soviet ground formations. If operations had to be conducted under low cloud conditions, however, the Schlacht formations flew as close to the ground as possible to prevent losses from light flak shooting at aircraft clearly silhouetted against the cloud. Over the target, a strike flight would undertake either very steep 60- to 80-degree diving attacks, shallow diving attacks, or near-horizontal runs. Steep dives worked best against dense antiaircraft concentrations. If at all possible, a strike leader would have a portion of his flight attack antiaircraft sites, usually with antipersonnel cluster bombs. Constant operations such as this demonstrated how urgently the Luftwaffe needed to introduce new weapons and tactics to meet a constantly evolving Soviet threat.[173]

As mentioned previously, the German assault into Russia had progressed along the lines of the blitzkrieg in the West, but with one important difference: Germany actually had fewer air resources with which to attack Russia than it had possessed during the invasion of France in 1940. Steady attrition, the Battle of Britain the previous fall, and the failure of the German industrial base (as well as the failure of the German procurement and acquisition leadership) all contributed to a situation in which the Luftwaffe found that it had increasingly marginal, and eventually inadequate, forces with which to fulfill its mandate to support German army operations against the Soviet Union. Weather, too, played a role, as did the almost constant operations. By the Battle of Moscow, for example, the Luftwaffe had vastly reduced in-commission rates for its combat aircraft: only 40 percent of its bombers and 58 percent of its fighters were available for service. Meanwhile, the strength of the VVS steadily rose. This situation came at a time when, increasingly, the Luftwaffe realized that it had to expand the scope of its attack aircraft operations in the East, particularly to confront the growing menace of the T-34 tank. At the end of 1942, the Rechlin *Erprobungstelle* (flight test center) had established a special operational test and evaluation command called the *Versuchskommando für Panzerbekämpfung* (Experimental Command for Antitank Warfare), evaluating in combat a number of modified aircraft and weapons capable of being used against Soviet tanks. This level of interest "at home" reflected and reinforced rising interest in the antitank mission at the front. For example, Luftflotte 4 established a special formation, *Panzerjagdkommando Weiss* (Tank-hunting Command Weiss), in February 1943 to combat Soviet tanks after they had effected a breakthrough. Such were the successes of this unit, headed by Oberstleutnant Otto

Weiss, one of Germany's most prominent Schlachtflieger, during fighting near the Don that Weiss was subsequently appointed *Führer der Panzerjagdstaffeln* (Leader of Tank-hunting Squadrons). A forceful promoter of the antitank aircraft, Weiss eventually became an influential staff officer dealing with Schlachtflugzeug affairs.[174]

Up until mid-1942, the Luftwaffe had had to rely on the elderly (though still surprisingly effective) Henschel Hs 123 attack biplane, the rapidly obsolescing Ju 87 Stuka, and modified bomb-carrying Messerschmitt Bf 109E's as its primary close support airplanes. But in the late spring of 1942, they were joined by an altogether new aircraft: the twin-engine Henschel Hs 129B. In many respects, the small Hs 129 (about the size of France's prewar Breguet 693) was intended to be a German equivalent of the heavily armored Ilyushin Il-2.

The Hs 129B had a pronounced slab-sided appearance, and was of all-metal construction; the pilot sat in an armored "bathtub" of extremely small dimensions, and had a limited view, in part because of the twin-engine layout, but primarily because the canopy itself was small, and had a 75mm-thick armored glass windscreen. Initially, the Hs 129B carried an armament of two 7.9mm machine guns and two 20mm cannon, though eventually 30mm cannon replaced the 20mm weapons on subsequent models. Additionally, the Hs 129 typically carried a small bomb load of two SC-50 bombs or two bomblet dispensers each containing twenty-four 2.2-kg antipersonnel bomblets. More significantly, the Hs 129B had provisions for a single 30mm antitank cannon (and eventually a 37mm or even 75mm cannon) carried in a belly fairing, together with 30 rounds of antitank ammunition.

Unfortunately, the Hs 129B proved a disappointment when it entered service. Only 20 mph faster than the Stuka, it actually had even less maneuverability. Committed initially to Tunisia, the Henschel virtually immediately revealed its greatest weakness, one that would plague it its entire life: its French engines were vulnerable to damage from dust and sand ingestion, often seizing without warning, and could not absorb even the slightest battle damage. This disastrous introduction braced the Luftwaffe for what to expect when it entered large-scale service in Russia. There, intensive maintenance generally kept the Hs 129 in the air, but involved a prohibitive number of manhours, and even so, dust ingestion was a serious problem. At one point, only two Hs 129's of an entire Gruppe were airworthy! By July 1943, in time for Kursk, the Luftwaffe had five Staffeln of Hs 129's in service.[175]

Complementing the Hs 129 were cannon-armed models of the ubiquitous Ju 87. The early experience of the Russian campaign had convinced the Luftwaffe what Coningham's Royal Air Force had learned on the Western Desert: bombing moving tanks is, by and large, an unprofitable, difficult business. For example, four days into the invasion of Russia, an entire wing of Stukas attacked a concentration of sixty Soviet tanks near Grodno, destroying only one of them. Ironically, the success of a few expert pilots who did demonstrate that they could consistently destroy moving tanks with bombs—such as Oberst Hans-Ulrich Rudel—delayed efforts to find alternative solutions such as cannon-armed airplanes or air-to-ground rockets. Eventually, however, the Luftwaffe had no choice, and ultimately elected to employ modified Stukas equipped with antitank guns. The Ju 87 received two 37mm weapons, mounted on two short pylons beneath the wings. The first Ju 87 so modified entered experimental service in 1942, and eventually the Luftwaffe formed special *Panzerjäger-Staffeln* equipped with the production model, the Ju 87G. The "G" had greatly reduced performance compared to the straight dive-bomber version of the Ju 87, and the location of the cannon (immediately outboard of the fixed landing gear) ensured poor lateral maneuverability. The Ju 87's needed heavy escort to conduct gun missions against Soviet tanks, and as the air war increasingly turned in the favor of the Soviets, losses from fighters and flak rose appreciably. In the last two years of the war, Rudel, the leading antitank Stuka pilot of the Luftwaffe, was shot down no less than thirty-six times, losing a leg to enemy flak but returning to combat after recuperation.[176]

This desire to employ heavier and heavier cannon for use against tanks led to some truly incredible modifications of German aircraft, not all of which represented a grasp of reality on the part of their originators. The Hs 129's, for example, ultimately sported 75mm cannon by the fall of 1944, endowing them with even poorer flying qualities than they had possessed previously. In desperation, the Luftwaffe even modified some twin-engine Junkers Ju 88 medium bombers as cannon-carrying antitank airplanes. The latter proved disastrously slow and unmaneuverable. The best answer, of course, would have been the large-scale production and widespread introduction into service of a genuine fighter-bomber, and this the Luftwaffe attempted to do, beginning in late 1942, with the introduction of a specialized armored ground-attack version of the famed FW 190 fighter. The new aircraft, the FW 190F and

FW 190G, were true fighter-bombers, with provisions (depending on the sub-model) for a variety of 20mm and 30mm cannon, and bomb racks for a variety of antipersonnel and high-explosive bombs. Eventually, the FW 190F/G replaced the Ju 87 in daytime service, the remaining Stukas going to the so-called *Nachtschlachtgeschwadern* (night ground-attack wings) by the fall of 1944. But by that time, defeat was around the corner. Incongruously, the ever-growing demand for advanced air superiority models of the FW 190 to counter the Allied air supremacy that was an unpleasant fact of life by 1944–45 resulted in limiting ground-attack fighter-bomber production precisely at the time when it should, if anything, have been expanding—yet another graphic comment on the interrelatedness of air superiority and battlefield air support operations.[177]

Setting the Stage for Kursk: Soviet Air Support Developments, 1941–1943

From the fall of 1942, through the climatic fighting around Stalingrad, the battles in the Kuban, and, finally, the summer battle of Kursk in 1943, Soviet attack aviation steadily advanced in power and strength. Early in the war, the attack formations had consisted primarily of older aircraft such as Polikarpov I-15's and I-16's, and, predictably, German light flak massacred these attackers as effectively as they had the aircraft of the Armée de l'Air and the Royal Air Force during the battle of France. But as 1941 went on, innovations appeared that, in and of themselves, seemed to have little significance. Taken together, however, they spelled eventual bad news for the Wehrmacht. First was the appearance of air-to-ground rockets, courtesy of the prolific work of two Soviet rocketry pioneers, G. Langyemak and I. Kleymyenov, of the Moscow Gas Dynamics Laboratory. These small, seemingly innocuous 82mm solid-fuel rockets had first appeared in the fighting against Japan, used then as air-to-air weapons. At first a rather ineffectual curiosity, they soon became a distinct annoyance and danger to troops and transport. A greater shock was the Il-2, which one German officer termed "a very effective and unpleasant ground-attack plane," noting that

> [it] was invulnerable to rifle and machine-gun fire of any caliber. Its armor also withstood 20mm flak projectiles. It is, therefore, understandable that these ground-attack aircraft were used at danger points, and unceasingly harassed ground troops once they had caught them in a low-level attack. In that manner

> they were able to bring daylight movements of motorized troops to a standstill, and to inflict considerable losses on them with their twin-barreled machine-gun [actually cannon] fire and small fragmentation bombs. Flak guns of 37mm or heavier caliber were of no use in the defense against their hedge-hopping attacks, because they flew too fast for allowing proper aim.[178]

By the summer of 1943, the VVS had completely reequipped its attack regiments with the Il-2, most of which were the two-seat Il-2m3 model.[179] Then, in early 1943, the Soviets introduced another important innovation: the PTAB hollow-charge antiarmor bomb, designed by I. A. Larionov. The PTAB "was highly destructive, light and small, and inexpensive to manufacture," and an Il-2 could carry containers for as many as *200* of them.[180] Similar cluster bomb units proved deadly when used against German troops; the same officer quoted above recalled that

> the bombs would fall within a radius of a hundred meters in such a dense pattern that no living object within the effective beaten zone could escape the splinters. The bombs fell into even the narrowest trenches and, because of their great fragmentation, were very dangerous and greatly feared.[181]

There were some other important changes in Soviet operations as well: the introduction of radar and radio, and the development of a genuine air support system. The Soviets had first used a primitive, indigenously developed air warning radar as early as 1941, and followed this with large numbers of radars supplied from the United States and Great Britain via Lend-Lease. By 1943, RUS-2, a 65–85 megacycle mobile early warning radar, and a fixed-base derivative called "Pegmatit" were operating in support of Frontal Aviation for tactical air control and early warning purposes. Further, by 1943 every new Soviet combat airplane employed radio communications, and a tactical air navigation system dubbed "ZOS" "underwent extensive development."[182]

More significant than any of this was the emergence of a well-thought-out air-ground support system, in place by early 1943, and subsequently refined and utilized up through the collapse of Nazi Germany in May 1945. The support system functioned largely to meet the needs of the "front" commander, on down to army, corps, and division level. The air units were subordinate to an air commander for that portion of the front, but he received his direction and guidance on where to employ his air assets from the ground commander. This system made use of joint air-ground command posts at levels from division up to "front." At the tactical level, the CP had both an Air Liaison Officer and a Ground Liaison Officer, who received reconnaissance and ground observer information, consulted with their associated air and ground forces organizations, and then passed along (or denied) requests for air support. Interestingly, the Soviets seemed little bothered by communications security requirements. A captured Soviet document on air support stresses that radio communications were to be in the open, so that the broadest range of air and land forces could be apprised of a situation. Generally, ALO's communicated to air organizations via radio, and with ground organizations via land line. This was, of course, a serious weakness, and German sources acknowledge the success that Nazi signals intelligence forces had intercepting launch orders for strike flights, enabling the Luftwaffe to be vectored onto Soviet attack and bomber formations, and ground forces to be warned of an impending attack.[183] Perhaps the most interesting feature of the Soviet air support scheme was its dependency upon a so-called Traffic Control Post located at the front no more than eight to ten kilometers from a forward CP. The Traffic Control Post contained an air and ground officer, and was responsible for directing an incoming airstrike, or for cancelling it if unnecessary. As a result, Soviet attack formations would check-in over the post, looking for particular signal panels, smoke markers, or rockets, before undertaking a support strike (interestingly, radio seems not to have been utilized, or perhaps only rarely, for these final "mother may I?" clearances). The Soviet ground commander would establish the post on a distinctive landmark easily recognizable from the air to facilitate its being located by strike flights. While commendable for its simplicity, this, too, was a weakness, as such positions were certain to come under accurate artillery fire as soon as they were detected.[184]

By early 1943, Soviet air support employment strategies stressed combined arms operations. During a breakthrough of a hostile defense, for example, air would complement the activities of ground forces. During the artillery preparation period, air strikes would hit targets beyond the range of friendly artillery. Then, no later than ten minutes prior to an attack, the artillery commander would be informed of the approach path of Shturmovik and light bomber (usually Pe-2 formations) strikes, and would, accordingly, either discontinue fire, or adjust the fire so that "the trajectory of the shells passes 50 to 100 meters lower than the flight of the airplanes."[185] Immediately prior to the infantry and armor assault, attack aircraft would strike at the German main line of resistance with "a concentrated blow" intended to reinforce the effects of artillery fire. To avoid casualties from their own strikes (either from bomb and rocket fragments, or from still-falling Soviet artillery fire), the Il-2's would attack no lower than 500 meters. Meanwhile,

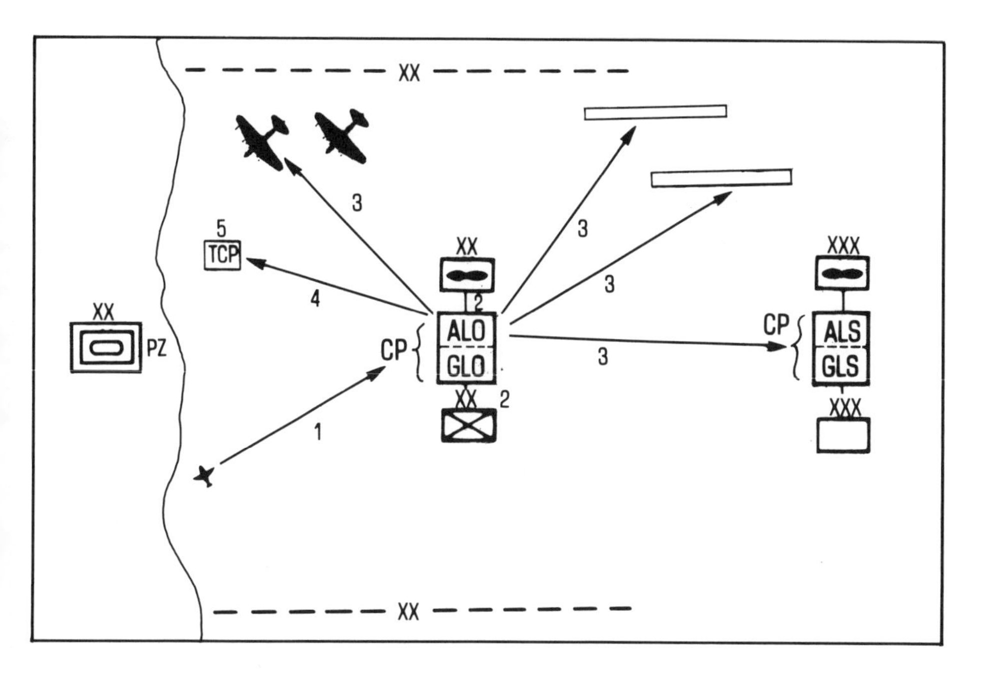

Soviet Air Support System, 1943–1945

1. Reconnaissance aircraft spots advancing Panzer division and radios report to Command Post (CP) of endangered infantry division and an associated VVS air division.
2. CP Air Liaison Officer (ALO) and Ground Liaison Officer (GLO) confer with commanders of infantry and air divisions. Ground commander orders air support strikes.
3. ALO radios airborne Shturmovik flight and orders a strike. Shturmovik leader will communicate with ALO for situation updates, relying for final clearance from Traffic Control Post (TCP) at front before initiating strike. At same time, ALO radios scramble orders for additional strike flights from the air division's forward airstrips. ALO, on behalf of division CP, uses land lines to inform corps CP, with Air Liaison Staff (ALS) and Ground Liaison Staff (GLS), of situation at front, so that army corps and associated VVS air corps are fully aware and prepared to intervene, if necessary.
4. ALO informs frontline TCP (located approximately 8–10 km in front of Division CP, on a site deliberately chosen to be clearly visible from the air) of incoming strike flight.
5. TCP staff (consisting of an air officer and a ground officer) lay out appropriate panel signals for strike flight, or use other means of signal communication, such as rockets or smoke pots.

Soviet artillery would shift to shell German antiaircraft positions, suppressing enemy air defenses while the attack aircraft delivered this brief, concentrated assault. These attacks often took place as close as 200 to 250 meters from friendly troop and tank positions, and, as a consequence, the Soviet procedures stressed use of mosaics, large-scale maps, and suitable signals so that strike aircraft did not inadvertently attack friendly positions. As the assault unfolded and the Russian and German forces came into close combat, Soviet attack and fighter aircraft would remain over the battlefield for as long as two or three hours, being rotated as fuel and ammunition depletion demanded. The fighters ensured that German aviation did not intervene, and the attack aircraft remained on station to "ensure the advance of the friendly infantry and tanks" by "means of their own attacks upon enemy tanks, artillery, mortars, and personnel."

As a rule, Il-2 and Pe-2 formations employed attack circle tactics; whether this represented a heritage going back to the experience Soviet forces had in Spain witnessing the Cadena is difficult to ascertain; most likely it simply resulted from an indigenous appreciation of the advantages that a circular attack pattern could have in constantly keeping an enemy under attack, and suppressing German air defenses.[186] Interestingly, the two communities used different terms to describe this tactic; Ilyusha crews termed it the "circle of death," while Peshka crews dubbed it the *Vertushka* (Merry-go-round). Typically, an Il-2 *gruppa* (a formation of six or eight aircraft, consisting of a four-ship *zveno* and an attached two-ship *para*, or two *zveno*) would approach a target in echelon formation. If antiaircraft fire was thought to be light, the formation would be "on the deck." Otherwise, they would fly at between 2500 and 4000 feet. The Il-2's would shift into a circle, maintaining a separation of approximately 1500 feet between aircraft. The lead aircraft would begin the attack, and as long as the aircraft had ammunition to dispense, the circle would be kept up, one aircraft attacking, one pulling off target, others spaced around the circle, and one rolling in on its attack. Typically an Il-2 pilot would fire his cannon, and once the cannon were striking the target, he would fire any rockets he might have. Then, he would initiate his pullout, dropping any bombs either singly or in salvo. Since some of these Shturmovik attacks could involve upward of thirty-six aircraft, it would not be uncommon to have multiple circles in place over the battle area, some just formed, some breaking up as aircraft left to return to base, and some in progress. At least a portion of the strike flight would be responsible for suppressing German antiaircraft batteries, and Soviet fighters—at least in theory—were supposed to provide top cover to prevent the Luftwaffe from intervening. In fact, one gets an impression from reading German sources that all too often the Soviet fighters abandoned the Il-2's to their fate, and that unnecessary losses of Il-2's occurred because fighter cover had been inadequate or absent. Il-2's also used other attack tactics, depending on the nature of the target and the ground defenses. For example, in heavily defended areas, Il-2's usually executed "blowthrough" tactics, making a single high-speed pass and then withdrawing back across the front. Against precision targets, Il-2's often behaved much like Anglo-American fighter-bombers, diving on the target from at least 2500 feet. When confronting enemy fighters, the Il-2's (following standard VVS practice) would remain at low altitude and weave in a "snake" formation back toward their own lines.

The Pe-2's, having a higher performance than the Il-2's, followed somewhat different tactics, developed largely by Colonel (subsequently Major General) I. S. Polbin, the VVS's greatest exponent of the Peshka, and an individual who, tragically, was to perish in combat in February 1945, on the eve of final victory. Polbin created the Vertushka method of attack for the Pe-2, taking advantage of the Pe-2's abilities as a dive bomber and its fine performance. Unlike the Il-2, the Pe-2 would usually approach a target from at least 10,000 feet, and sometimes as high as 23,000 feet. The leader would then enter a 70-degree diving turn and begin his attack, followed in trail by the others. As he pulled off target, he would break into a circle, and the following Pe-2's would keep this up in Ilyusha fashion, but at higher altitudes, higher approach speeds, and with greater agility. While it is difficult to ascertain with any assurance, it is likely that the Pe-2's higher performance more than compensated for its lack of heavy armor, such as the Il-2 had.

The evolution of the Soviet air support system and the development of specialized tactics to employ strike aircraft such as the Il-2 and Pe-2 in battle boded ill for the Wehrmacht, for it came at a time when Germany's air and ground forces were being stretched increasingly thin to meet the demands of multifront war. The evolution of the Soviet air army gave the VVS the ability to generate concentrated, sustained air power over a battle area, and then to shift that army along the front quickly and expeditiously. It was the air support system that was the critical development, however. In 1944, backed up by muscular VVS air armies fielding reliable aircraft piloted by highly motivated aircrews, this air support system would combine with Soviet land warfare techniques to destroy the Wehrmacht in Russia, sending it back in an increasingly disorganized and demoralized retreat to the Nazi heartland.

Preparations for Citadel

Kursk, the greatest tank battle of the Second World War, was of crucial importance to fighting on the Russian front. After it, Germany had no hope of winning, and the locus of decisive initiative for the Eastern Front war shifted from Berlin to Moscow. It was the first battle in history in which opposing sides employed masses of specialized antiarmor attack aircraft. While their use was not decisive (in the sense that they determined the outcome of the battle) they contributed significantly to local decisiveness in particular engagements. More importantly, however, Kursk demonstrated the first flexing of the VVS's ability to participate as a full-fledged member of a combined-arms military team, cooperating with and assisting the activities of Soviet ground forces, particularly armored forces. It placed an indelible combined-arms offensive stamp on Soviet forces, doctrine, and force structure that remains to the present day, and confirmed the essentially tactical orientation of Soviet military air power.

In early 1943, following the collapse of Stalingrad, it appeared that the Soviet Union would continue to move decisively against German forces. Virtually all the gains made in the German 1942 summer offensives against Russia were swept away. However, operational blunders by the Soviets and the timely generalship of Generalfeldmarschall Erich von Manstein (who had assumed command of the southern portion of the Russian front in mid-February), restored a measure of stability to the front. Previously, Germany had always done well in good weather, but then had suffered badly at the hands of "General Winter." As a result, 1943 would be a year of testing, to see if the Soviets could fight and win in their traditionally worst period, the summer.[187]

Kursk has gained an understandable measure of fame among the major engagements of the Second World War; in popular literature, it is often seen as a last-ditch German effort to destroy the Soviet army and thereby "win" the war in the east. In fact, Kursk represented the acceptance of an Eastern Front defensive strategy on the part of the Germans: the recognition that after Stalingrad, the sweeping blitzkrieg was a thing of the past, and that Germany would have to increasingly take into account what was happening in the rest of the world—which, in early 1943, meant coping with the certain loss of Tunisia, and an inevitable invasion of Italy. The battle of Kursk is more properly understood if it is recognized that Kursk itself ultimately was little more than a pivot point around which the action (German thrust, Soviet counterthrust) rotated.

Preparations for the Kursk offensive, dubbed *Zitadelle* (Citadel), began in early 1943, accompanied by all the best and worst elements of German planning. The best included meticulous attention to detail, forces, coordination, and scheduling. The worst included indecisiveness among senior planners, a divided general staff, direct intervention by Hitler himself into the planning process (as inevitable for German planners as Stalin's was to the Soviets), and overreliance on new tank and weapon technology. Kursk was, in many ways, a personal clash between Hitler and Stalin, for each intervened significantly in the command of their respective military for the battle. For his part, Hitler seems to have vacillated between manic enthusiasm for Citadel and depressive worry over its failure. Victory at Kursk, Hitler decreed (sounding somewhat like a Teutonic John Winthrop), "must have the effect of a beacon seen around the world."[188] On the other hand, when Generaloberst Heinz Guderian (Hitler's newly appointed inspector-general of all armored forces, including tank development) asked *der Führer* "Why do you want to attack in the East at all this year?" Hitler answered "You're quite right. Whenever I think of this attack my stomach turns over."[189]

Planning for Citadel began in early 1943, at the instigation of Hitler himself, who saw in the operation a chance to strengthen the German position in Russia by ending any chance for a Soviet return to the offensive in 1943. Army Group Center would attack the Kursk salient from the north, and Army Group South would attack northward, pinching the salient and essentially destroying or trapping any Russian forces within. Citadel was, in fact, only one of three related operations intended to complement each other. The others were *Habicht* (Hawk) and *Panther.* Habicht would have continued von Manstein's success in the Donets by driving across the Donets and threatening the underbelly of Soviet forces. Panther would have occurred south of Hawk, and would have driven Soviet forces further eastward. The cumulative effect of these three—Citadel in the north, Hawk in the middle, and Panther in the south—would have been to straighten the front over a distance of approximately 450 miles and retain the German initiative in the east, denying the Soviets the opportunity to launch a summer offensive themselves.

One major problem Hitler faced was which operation to order first: Citadel, Hawk, or Panther. Hawk was the least significant of the three, and meaningless unless Citadel itself were undertaken. Panther logically followed Citadel, and so Hitler—committed to Citadel—directed that Hawk precede Citadel, which would, in turn, lead to Panther. In fact, however, the flood schedule of the Donets prohibited the undertaking of Hawk, and so, in April 1943, Hitler abandoned plans for Hawk to precede Citadel, and, instead, decreed that Citadel would take place (with six days notice) sometime after late April, with Panther to follow. Therein followed a tragi-comedy of scheduling conferences, with Citadel falling further and further behind schedule. This was an important point, for in its initial conception, Citadel was to catch the Russians largely unaware and unprepared, with German forces in their prime, and Russian forces still disorganized after the fighting around Kharkov. Any delay played into the hands of the Soviets who were, in any case, already aware of the planned offensive, thanks to their own covert intelligence-gathering sources (confirmed as well by a warning from Britain, based on Enigma communications intercepts). Already, in mid-April, German reconnaissance aircraft spotted increasingly heavy supply traffic into the Kursk salient, from the direction of Moscow and Stalingrad.

Hitler was bothered by these signs, and emphatically warned his senior commanders to beware a Soviet pincer counterattack against the Orel salient above Kursk, lest the pincher become the pinchee. As weather delayed the Citadel offensive even further, Hitler next urged a delay into June so that some new weapons—chiefly the Tiger and Panther tanks, and a large tank destroyer called the Ferdinand—would be ready. Von Manstein, concerned about the inevitability of collapse in Tunisia and the upcoming Italian campaign, urged cancellation of Citadel if it went beyond May. Guderian, having just rebuilt Germany's Panzer forces, was not willing to see them thrown away in a battle that would result in obvious heavy attrition of armored units, and likewise encouraged rethinking the planned offensive, supported in his arguments by Hitler's armaments minister, Albert Speer. Hitler, mesmerized by Kursk, pressed on, despite these objections and later ones by the operations staff of the OKW (*Oberkommando der*

Wehrmacht, the High Command of the Armed Forces). On June 20, he decreed that the Citadel offensive would commence on July 5. By this time, while Germany might have been able to hope for some tactical surprise in actually launching the offensive, any strategic surprise was impossible; both the German and Soviet armies were at near-full strength, promising a protracted, punishing, and intense struggle.

Much of the anticipated success of the Kursk offensive would depend upon the Luftwaffe being able to support the ground forces in timely and effective manner. While that service cannot be said to have done so, its inability stemmed from the strength of the VVS, and not from errors made at Kursk by its own leadership and personnel. Indeed, German ground commanders go out of their way in memoirs and interviews to praise the activities of the Luftwaffe in the battle, one referring to that service's "gallant effort."[190]

Some measure of the potential of German antitank aircraft had been gained from the experiences of Weiss's special command operating with Luftflotte 4 (under the command of close support specialist von Richthofen) during fighting between the Don and Dnepr rivers. In one notable case, a strong force of 250 Russian tanks broke through Italian lines, triggering intensive Luftwaffe bombing missions against them. The bombers "achieved practically no successes whatever," but the Henschel Hs 129's of 4/S.G.1 (Schlachtgeschwader 1's fourth staffel) gunned ten of them in two days: a testimonial to the value of the antitank airplane, but hardly encouraging in terms of the number of tanks destroyed. More encouraging were results from fighting around Orel in February 1943, when German attack aircraft destroyed over sixty tanks in a single day, "annihilating" troop and mechanized formations. On the other hand, III/K.G.1's (the third Gruppe of Kampfgeschwader 1) evaluation of the cumbersome Junkers Ju 88P (a modified medium bomber armed with a 75mm hand-loaded antitank cannon) as a tank destroyer ended catastrophically when *all* of the unit's aircraft were shot down, apparently by ground fire; proof, if any were further needed, that attack aircraft have to be small, quick, and agile. By and large, however, the Luftwaffe's support forces did well in the early months of 1943; concentrated attacks by Luftflotte 4 greatly assisted the Das Reich SS division's assault on Kharkov.[191] The new gun-toting Stukas likewise demonstrated their value when Rudel, the Luftwaffe's *Panzerknacker* (tank cracker) took an experimental detachment to the Kuban bridgehead, but also their terrible vulnerability to new Soviet fighters, particularly the La-5. Rudel pressed home attacks to as low as fifteen to thirty feet, often flying through the blast cloud of the exploding tank he had just attacked.[192]

These operations were crucially important in training and familiarizing the Schlacht units with the potential and problems of their "new" mounts. For example, Henschel pilots initially wasted far too much ammunition in repeated attacks on single tanks, and their mechanics were so unskilled in the maintenance of the cannon that 4/S.G.1 temporarily removed the belly cannon from its Hs 129's, replacing them with small bomb racks! Operational problems with the Henschel continued to limit its usefulness. In contrast to the ever-dependable Shturmovik, the Hs 129 was so unreliable that the II Gruppe of S.G.1 at one point had only six serviceable aircraft. Of the antitank Stuka, former Luftwaffe Generalleutnant Herman Plocher recollected:

> It was too dangerous to fly low with these planes . . . at an altitude of only a few feet over areas which had well-established fronts and strong local ground defenses. The Luftwaffe knew that losses in such situations would be greater than the results which could thereby be achieved. *If the Ju 87G could be used at all, it could only be in areas where the fronts were relatively fluid and the ground situation was constantly changing* [emphasis added].[193]

In May and June 1943, the first strikes and counterstrikes by the Luftwaffe and the VVS occurred against targets in the Kursk area. By this time, what Hitler's staff had feared most was an accomplished fact: Soviet forces were already stronger than those of Germany. Nowhere was this disparity more apparent than in aircraft. The Luftwaffe had approximately 1830 combat aircraft available to support Citadel, 1100 belonging to the VIII Fliegerkorps (assigned to Luftflotte 4, based near Kharkov), and another 730 with the 1 Luftdivision assigned to Luftflotte 6 near Orel. In contrast, the VVS had a strength of approximately 2900 aircraft distributed among five air armies, the 1st, 16th, 2nd, 5th, and 17th. While is it difficult to determine with any precision the actual breakdown according to type, the figures in Table 4 appear to be reasonably accurate (miscellaneous types—reconnaissance, air transport, night harassment, etc.—are not shown).[194]

Table 4. Luftwaffe vs. VVS air strength at Kursk

Category	Luftwaffe	VVS
Fighters	545	1,060
Bombers	500	900
Attack	450	940

Thus, in significant combat types, the VVS had a slightly less than 2:1 advantage in fighters and bombers over the Luftwaffe, and a greater than 2:1 advantage in attack aircraft. But in addition, the VVS's Frontal Aviation could call upon two divisions of *PVO Strany* (Soviet air defense) fighters for additional air superiority aircraft, and upon 500 *ADD* (long-range aviation) bombers as well, helping boost total VVS air strength to approximately 5400 aircraft. When it is remembered that on the eve of Kursk the German air units had (by their own figures) an operational readiness of 75 percent, that the VVS had aircraft that were generally easier to maintain and keep in service than the Luftwaffe (particularly their Shturmoviks), that many Russian fighters were swing-role aircraft assigned to air-to-ground tasks, and that strong partisan activity in May had destroyed so many fuel trains that the Luftwaffe had to severely curtail its anticipated operations, the true magnitude of the Soviet air advantage over the Luftwaffe is even more apparent.[195]

On the German side, Army Group Center and Army Group South, commanded respectively by von Kluge and von Manstein, confronted no less than five Soviet army "fronts" (equivalent to army groups), though only two posed an immediate challenge to the Wehrmacht: the Central Front (Marshal Konstantin Rokossovskiy), and the Voronezh Front (General Leytenant N. F. Vatutin). The total force structure available to the Soviet ground forces is in some dispute, but there is general agreement that the Soviets deployed between 3300 and 3600 tanks, 20,000 cannon and mortars, and over 1,330,000 troops. But these figures do not convey the true significance of Russian preparations, for the Soviets had constructed deeply echeloned in-depth defenses, involving three to five trench lines constructed within two to three miles of the front, secondary trenches constructed running between six to eighteen miles from the front, and, finally, tertiary defensive lines twenty-five miles from the front. Central Front alone had 3000 miles of trenches, and 400,000 mines.[196]

If the Luftwaffe could look to its experiences in early 1943 with its new tank hunters as offering some encouragement for the future, so, too, could the VVS, and with much greater confidence. After all, the equipment disparity that had afflicted the VVS at the beginning of the war was now a thing of the past. The new Yak and Lavochkin fighters, the Il-2's and Pe-2's, new medium bombers such as the Tupolev Tu-2, all were as good as or better than their German counterparts. (In fact, at Kursk, the Germans still were operating some elderly Henschel Hs 123 attack

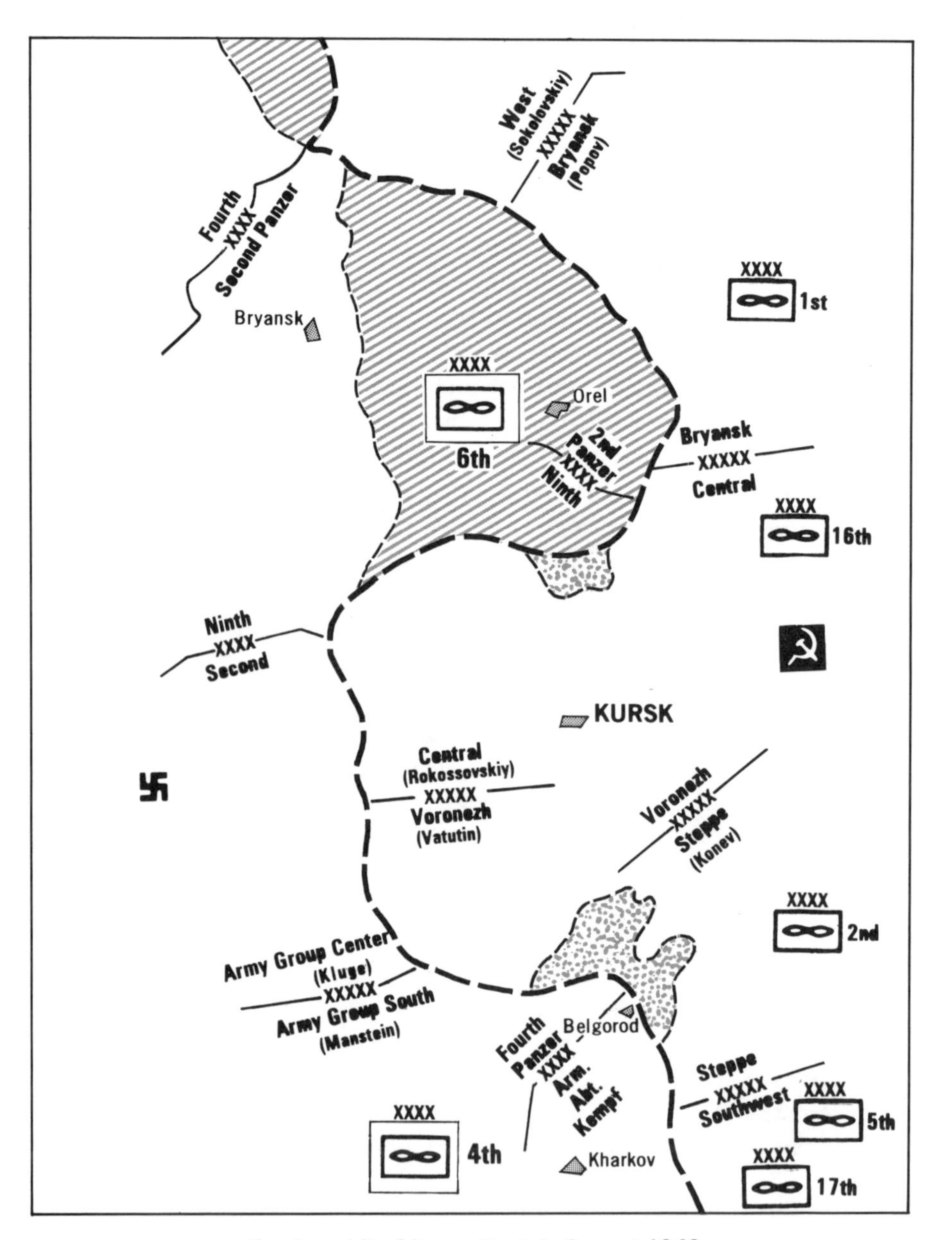

Kursk and its Aftermath, July-August 1943

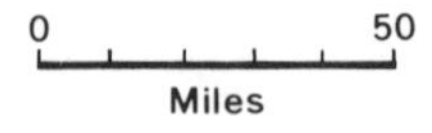

Front as of July 4

Territory captured by Nazi advance

Soviet-controlled territory as of August 18

biplanes.) The pervasive myth of German invincibility had been shattered at Moscow (one German commander ruefully recollected that during the Russian winter offensive of 1941, "Our air support became more and more inadequate, while the Russians all of a sudden employed wings consisting of 100 airplanes, each.").[197] At Stalingrad, the Soviets had destroyed Germany's most formidable eastern army, and the VVS had staved off the Luftwaffe, imposing an air blockade that Germany could not crack. Over the Kuban in early 1943, the VVS matured further, employing new control procedures, particularly large-scale use of radio. With radio, the VVS no longer had to rely on large, mass formations of fighters "controlled" by a leader watched by his pilots. Instead, as with the RAF, AAF, and Luftwaffe, pilots and flight leaders were now free to communicate among themselves on a radio net, and, more importantly, with a ground controller station. Accordingly, the VVS began using more flexible formations, deploying fighters echeloned over the battle front in formations dubbed "Kuban stacks."[198]

Of even greater significance were advances in training and air-ground liaison. In January 1943 the VVS Main Directorate of Frontal Aviation Combat Training established a tactical training program resembling, in some respects, the modern Red Flag or Top Gun programs. Pilots flew intensive air combat maneuvering training missions; additionally, "various methods of attacking small mobile ground targets and guiding aircraft by radio to enemy air and ground-based targets were tested."[199] In the weeks prior to the Kursk offensive, as German intentions became clearer, the Soviet military stepped up liaison between air and ground units. Airmen and flight commanders journeyed to the front to become familiar with ground units and their operation; likewise, ground forces—and particularly armored forces—representatives ventured to VVS bases to learn more about what air could—and could not—do for them.[200] As June ended and July began, all that remained was the final test: that of combat itself.

The Battle of Kursk

Hitler had planned the Kursk offensive to begin in the early morning of July 5, but the Soviet Union stole a march on *der Führer* by undertaking a furious artillery barrage to catch the Germans at their most vulnerable: as their forces were deployed preparatory to the assault. The barrage—which Zhukov subsequently referred to as "the grand 'symphony' of the great battle at the Kursk Bulge"—included tube artillery, battlefield rockets, and "the steady roar of planes overhead."[201] But despite this favorable opportunity, limited intelligence resulted in the barrage having more of an annoyance impact than the terrible destructiveness that the Soviet commanders had hoped. Accordingly, when the German assault in the south got underway several hours later, it made dogged progress against stiff Russian resistance. In particular, the dense network of Soviet antitank emplacements and minefields, coupled with imaginative use of specialized antitank squads operating from trenches in the battlefield itself, held the Panzers up, even though, one for one, the new Tiger and Panther tanks were more than equal to the 76mm-armed T-34's deployed against them.[202]

The Schlachtgeschwader of Luftflotten 4 and 6 supported the German assault as best they could, and they did earn praise from their army colleagues. Nevertheless, the scarcity of German air power in view of the magnitude of the tasks to be done showed early; the crossing of the Donets, for example, was completed without air support because the air had to go elsewhere—to support the battle of the II SS Panzerkorps. Both this unit and another, the XLVIII Panzerkorps, made rapid progress through Vatutin's first defensive line to the right of Belgorod (as seen from the Russian side of the front), but a rainstorm held up the XLVIII until nightfall, and massive Soviet artillery barrages disrupted the II SS Panzer attack. VIII Fliegerkorps had orders to support the attack of II SS Panzer, and now its Stukas did so with a vengeance, pounding Russian artillery positions and clearing the way for II SS Panzer to continue its attack. Frustratingly, the attack bogged down again, this time from a massive mine belt. On the other side of Belgorod, things were not going much better. There, two corps of German forces, having crossed the Donets without air support, now came across a thick network of Russian defenses; "Without air support of their own," Army historian Earl Ziemke notes, "and harried incessantly by Soviet planes, they inched along while their casualties mounted alarmingly."[203]

The second day of the offensive found Russian power virtually undiminished, while German ground forces found themselves locked in increasingly constraining battles that drained their strength; worse, the Luftwaffe had so few resources for the number of tasks it faced, that it had to be shifted from point to point along the front, and the ground units could not expect the kind of continuous, available air support they had always known. Command broke down. Paul-Werner Hozzel, himself a distinguished Stuka pilot, recalled that at Kursk and all subsequent Russian front operations,

we could not break the Russian front, it was too heavy, too massive, too deeply entrenched. So we distributed all units again from army to army. The Geschwader had no possibility of common guidance any more. For example [Hans-Ulrich] Rudel led one group with cannons, and when there was a threat of armor breakthrough at any place, Rudel was called as a fire brigade to prevent the breakthrough. Then another army cried for help and Rudel went there with his special group. He never was in a position to lead the Geschwader.[204]

Worse, the enormously successful partisan campaign to interdict German fuel supplies on the way to the front had resulted in critical fuel shortages; as former Luftwaffe Generalleutnant Hermann Plocher recollected:

In order to have air power available for particular crises and major phases of battle, it became necessary to refuse air support in many situations which would normally have been recognized as critical. Every assigned mission had to be carefully examined to determine whether it was really worth the fuel expenditure. In deciding what type of units to employ, it often became necessary to accept certain disadvantageous conditions because of the possibility of saving a few tons of fuel.[205]

Still, the Schlachtflieger had their share of successes. On July 5, for example, Rudel personally destroyed twelve tanks near Belgorod with his 37mm-armed Ju 87, operating within 1500 to 2000 yards of German tank forces. (It was, in fact, this proximity to friendly forces that encouraged him to take his vulnerable Stuka into the hornets' nest of Russian light flak; other gun Stukas did not do nearly as well.) The most rewarding of their operations, however, came several days later, on July 8, and involved the use of Henschel's *Büchsenöffner* (tin opener): the Hs 129.

Like drink spilled on a carpet, the German offensive in the north and south had quickly become absorbed in the dense Soviet defensive network. Such was even true south, where von Manstein's forces had had some success in forcing past Soviet defenders. Under such conditions, as Plocher has accurately noted, "In many instances air units were the only forces at hand which could deliver really destructive fire to repel Soviet armored attacks."[206] By June 8, II SS Panzer was overextended, with a vulnerable right flank waiting for cover from another tank division. But until that unit arrived, the possibility of a strong Russian attack against the flank caused VIII Fliegerkorps to devote special attention to preventing Soviet forces from exploiting any such opportunity. IV(Pz)/S.G.9 (a Gruppe of sixty-four Hs 129's, though only thirty-six—nine from each of the four Staffeln—participated in the assault), based at Mikoyanovka (approximately twenty miles from the Belgorod front) had sent a patrol along the exposed flank, and detected a large formation of Soviets (estimated at one armored brigade plus accompanying infantry) advancing toward the II SS Panzer. Under the leadership of the Gruppenkommandeur, Hauptmann Bruno "Panzer" Meyer, the entire Gruppe attacked the Soviet force (located east of Gostchevo, approximately twenty minutes from their base); typically one Staffel would be attacking, one returning to base, one en route, and one taking off. Schlacht FW 190's accompanied the Henschel strike force to suppress antiaircraft fire and attack the accompanying infantry with antipersonnel bombs. After approximately an hour's attack, the Russian force had been decimated and stopped in its tracks. While sources disagree on the actual losses the Soviets experienced under Meyer's assault (though forty tanks seems to be a generally accurate figure), there is general agreement that in this case, cannon-armed attack aircraft played a decisive role in influencing the course of the ground battle. In another case two days later—and subsequently described by von Mellenthin—Stukas ably supported the advance of a Panzer division held up by strong Russian resistance near Beresowka.[207]

With the amount of postwar attention that has been focused upon the Luftwaffe, it is easy to accept some of the glib generalities that have appeared in many of the admittedly partisan and defensive accounts that have appeared by ex-Luftwaffe officers and hangers-on. In some of these accounts it is difficult to determine that Germany lost the air war in the East, even at Kursk. In fact, throughout the battle, the VVS succeeded in preventing the Luftwaffe from having a free hand in conducting its battle. German sortie rates fell precipitously. Soviet figures allege a decline in German sorties from 4298 on July 5, to 2100 on July 6, 1162 on July 7, 870 on July 8, and down to 350 on July 9. While these figures may be exaggerated to emphasize the success of the VVS in dealing with the Luftwaffe, this general trend is undoubtedly accurate. As Von Hardesty, a perceptive student of Russian air warfare, has noted, "After two years of war, Kursk exposed a grim reality for the Germans. They lacked the striking power, reserves, and mobility to crush the Soviets, even in a summer offensive."[208] What, then, of the VVS during the Kursk battle?

After a dangerous start, the VVS acquitted itself well at Kursk, particularly in its ability to undertake sustained, continuous operations in the face of intense combat, and accompanying heavy losses. Its command and control structure improved throughout the battle, and it proved an exceptionally tough, resilient opponent. On July 5, the VVS launched a massive air attack intended to destroy the Luftwaffe on its major airfields around Kharkov. Instead, thanks to early warning radar, the Germans were able to meet this assault with two fighter wings, and in the resulting air combat, the VVS took heavy losses that, for a while, severely limited its ability to strike at German forces. German accounts that allege that after this engagement the Luftwaffe had unchallenged air superiority over the battle area are, however, most suspect. So strong were Soviet forces that the losses were quickly made good. For most part, Soviet commanders are bluntly honest in assessing the outcome of this first battle; Zhukov, for example, stated that

> Air Force participation was negligible and, putting it bluntly, ineffective, while dawn strikes at enemy airfields failed to accomplish their purpose, as by that time the enemy already had his aircraft up to support ground troops.
>
> Our aviation inflicted far more telling blows at tactical battle arrays and enemy columns regrouping in the course of fighting.[209]

That night, Stalin (keeping in constant touch with his commanders) asked Rokossovskiy "Have we gained command of the air?" The embattled Marshal responded (no doubt cautiously choosing his words) "Fierce air battles are going on with alternating success." Stalin then demanded that the VVS spare no effort in gaining control of the air.[210]

For a while, the VVS reacted sluggishly following its spectacular morning setback, and, as a result, VIII Fliegerkorp's support aircraft were able to operate on the first day with a relatively free hand. But that quickly disappeared. Russian Shturmoviks and Pe-2's appeared over the battlefield in increasing numbers. On the first day, formations of six and eight attacked German targets near Yasnaya Polyana, Ozerok, and Arkhangelskoye, using the new PTAB munitions to explode German tanks and mechanized transport. On the second day, July 6, Sergei Rudenko's 16th Air Army attacked the German 9 Armee as it assaulted the stoutly defended Ol'khovatka ridge (in the Central Front), using a massive fighter sweep to disperse and tie up German fighter forces so that the attack aircraft and bombers could intervene over the battlefield. Altogether, the operations on the 6th were, as Rudenko reported subsequently, "more orderly . . . with more purpose."[211]

On July 7, the tempo of the air war had shifted, and so had the initiative. As German sortie rates fell, those of the VVS rose appreciably (particularly those of attack aircraft), to the disconcertion of the

Wehrmacht; likewise, it was the VVS that now had the luxury of dominating battle management, of committing units in large-scale strength and according to well-thought-out plans. By this time, as one German commander noted, "powerful Russian flying formations were in the air at all times."[212] Emboldened, Soviet ground-attack aircraft extended their operations deeper over the front, striking as much as fifteen to twenty miles behind the main line of resistance. Nevertheless, it was over the battlefield that they continued to make their major presence felt. One Ilyusha attack cost a German tank division seventy tanks in twenty minutes on July 7, and in an attack against which Meyer's Henschel effort pales, four hours of Il-2 strikes destroyed an estimated 240 tanks out of 300 of another Panzer division. Even allowing for overoptimistic claims and outright exaggeration, it is obvious that the Soviet attack aircraft formations took a heavy toll of German armor and transport. Von Mellenthin recollected that "many tanks fell victim to the Red Air Force—during this battle Russian aircraft operated with remarkable dash in spite of German air superiority."[213] Plocher likewise noted grudgingly that "the performances of some hostile air units, especially among the ground-attack forces, were quite commendable."[214] In short, the VVS had come a long way since 1941.

On July 8, the 9 Armee ground to a halt before Ol'khovatka ridge, and on the 10th, recognizing that the Orel front was wide open to an anticipated Russian counterattack, the Luftwaffe began shifting units northward, further weakening the ability of the German air force to deal with Russian forces near Belgorod. (That same day, Allied forces landed in Sicily, further distracting Hitler and the OKW's attention from what was happening in the brutal battle east of them.) On the llth, by the Germans' own accounts, VVS airmen were able to wander virtually at will over the front. On the ground, German forces received a setback when their air liaison officers (who had done good work so far in the battle) were unable to participate to any great degree in the battle of Prokhorovka, because their command vehicles could not keep up with attacking units. (Prokhorovka, the largest armored battle of all time, eventually was a German "victory," but a Phyrric one; German forces were too exhausted and weakened to exploit it, and, in any case, the Russian counterthrust toward Orel had rightly necessitated a shift in the German effort northward.) Then, on July 12, the other shoe dropped. Supported by massive air and ground action, the Soviet Union launched a counteroffensive by the Bryansk and West Fronts. The embattled 9 Armee now had to reduce its strength before Ol'khovatka and shift some of its forces north to Orel. On July 12, with little real choice (despite a postwar flood of literature suggesting the contrary, including von Manstein's own memoirs), Hitler abandoned Citadel, instructing von Kluge and von Manstein that the progress of the battle itself and the potential of the Allies' invasion of Sicily dictated a retrenchment.[215]

Aftermath

Like a major earthquake, there were numerous aftershocks after Hitler's order, and many bitterly fought actions as the Russians fought to exploit their advantage, and the Germans fought to prevent total collapse of the front. Eventually, German forces withdrew to the so-called Hagen line, licking their wounds and preparing to experience the full fury of the Russian summer offensive. Within a month of Kursk, all of the meager territorial gains made by the Wehrmacht in the Citadel operation had been liberated by the Red Army. That the German forces were able to withdraw as well as they did was, in fact, due in large measure to the Schlachtflieger, who, it must be admitted, were a dedicated and valorous group of airmen. But they had their equals—certainly in dedication and valor—in the airmen and airwomen of the VVS who likewise greatly assisted the ground forces of the Red Army, particularly during the subsequent collapse of the German Army Group Center and the drive onward to Berlin.

It had been Kursk, of course, that had made the crucial difference. Before Kursk, even after Moscow and Stalingrad, it had been the Wehrmacht that called the tune for the Russian bear to dance. After Kursk it was quite another matter. Now the forces of the Third Reich were on the defensive, responding to combined-arms symphonies of Soviet artillery, armor, air power, and infantry. Whatever measure of strength the Wehrmacht had enjoyed because of its well-honed air-ground support network had likewise been lost. From now on, the Luftwaffe would be committed piecemeal, while the VVS would increasingly function more smoothly and professionally. By the time the Red Army assaulted East Prussia, the VVS would be able to mount not a few thousand sorties per day, but upward of *45,000* combat sorties per day by bombers, fighters, attack aircraft, and fighter-bombers. Kursk was the last organized air offensive by the Luftwaffe. Unfortunately for Hitler it was the first organized air offensive by the VVS. After Kursk, the Luftwaffe slid into the abyss on the Russian Front as it already had in the West. Evaluating the impact of Kursk on Soviet military operations, the commander of the 16th Air Army, Marshal Sergei Rudenko, remarked:

> Noteworthy [was] the air armies' close cooperation with the tank armies, operating as the fronts' mobile task forces. Attack and fighter formations were assigned for joint action with the tank armies in sustaining attacks in operational depth. This experience was widely used in all the subsequent operations of the war and holds good to this day.[216]

It must never be forgotten that the Soviet Union inflicted a decisive military defeat upon the armed forces of the Third Reich, just as it must never be forgotten that the Third Reich, far from being a Western European nation that found itself in conflict with the Soviet Union, was instead a truly demonic state posing the most terrible international threat of its time. This might seem an obvious point were it not for the great amount of postwar literature (mostly of a popular nature) that has served to glorify the efforts of the Wehrmacht to subjugate the Russian people, as they had the French, the Dutch, the Norwegians, and others. Perhaps this is an unfortunate by-product of the Cold War—a war that was, after all, not the doing of the West but rather a result of the aggressive intentions of the Soviet leadership in the years since 1945. As time goes by, there is always a fatal tendency to believe "the enemy of my enemy is my friend." Such a line of reasoning often is conveniently suggested or implied in the memoirs of Hitler's military, many of whom argue that, somehow, they were fighting "our" battle over forty years ago. That, of course, is pernicious rubbish: a disingenuous attempt to evade responsibility and ignore the ghastly history of the Third Reich (with its death camps, slave labor, perverted "medical" experimentation, racism, and rampaging militarism) that deserves to be treated with utter contempt. Indeed, were it not for Hitler, most of the problems now facing the West would likely not even exist. The Soviet aircrews who fought—and too often died—over the Russian Front in combat against the Wehrmacht are no less deserving of the Free World's respect than their compatriots of the Royal Air Force, the AAF, the Free French, or any other of the nations that joined together to defeat Hitler. It is a tragic irony that these men and women were themselves servants of a totalitarian state that in time has caused incredible hardship for its own citizens and for millions of others in countries around the world.

EPILOGUE

Where We Have Come, Where We Are, Where We Are Going

The years since 1945 have witnessed many changes in military technology, doctrine, strategy, and war-fighting. Though postwar developments affecting battlefield air support clearly merit detailed examination themselves, some general comments need to be made so that the experience up to 1945 can be placed within the broader context of subsequent development of battlefield support aviation and forces. These postwar developments have generally confirmed trends seen earlier, in the era up to 1945, and thus we can offer the following points—some not necessarily obvious, and many bound to be controversial—from the perspective of more than seventy years of close air support and battlefield air interdiction operations:

1. Armies and air forces traditionally bicker over the nature and control of CAS/BAI operations. Just as air forces traditionally view almost all their missions as contributing to the success of friendly land forces in battle, armies traditionally fear an enemy air force more than they respect their own. They overemphasize the anticipated effectiveness of the enemy air force both in defending its own air space and in projecting its power across the front, and they minimize the ability of their own air force to defend them from enemy attack and to conduct meaningful CAS/BAI against the enemy.

2. We have always done what are now delineated as CAS/BAI operations, since the time of the Great War's "trench strafing" and "ground strafing."

3. BAI operations have always been of more value—as well as more extensive—than CAS operations. CAS is in extremis air support, and its use typically reflects more desperate or peculiar circumstances. In mobile warfare, BAI has been employed more frequently and decisively than CAS. *Thus the desire to achieve a CAS capability should not come at the expense of developing systems suitable for the historically more profitable BAI mission.*

4. With rare exception, the strategic bomber has been of minimal value in battlefield air support. Even when it has proven useful—such as with B-52 Arc Light missions in Southeast Asia—its missions fell far more under the rubric of BAI rather than CAS, and were undertaken in conflicts where both air and ground commanders greatly preferred the flexibility of smaller, more agile, and faster aircraft, notably the fighter-bomber.

5. "Classic" (e.g., non-BAI) air interdiction has proven disappointing, and of questionable value in its impact on battlefield operations, as exemplified by four separate air campaigns: Operation Strangle (Italy, 1944); Operations Strangle and Saturate (Korea, 1951–52); French interdiction efforts against Viet Minh supply lines, 1952–54; and the long and arduous campaign by the United States and its allies against the Ho Chi Minh trail network over a decade later. Air interdiction history indicates that, as a rule, *air interdiction works best only when it is synchronized with ground maneuver warfare.* Under those circumstances, an enemy is forced to maneuver across a battlefront while exposed to simultaneous air and ground threats.

6. The single greatest recurring problem in battlefield air support has been that of effecting timely strikes with satisfactory communications, control, and coordination. Contemporary experience from war gaming and training exercises indicates that this remains a serious challenge to effective battlefield air support operations even in the present day.

7. Troops exposed to air attack experience serious psychological and morale damage that hinders their subsequent combat performance, and, indeed, the shock they suffer seems greater than that experienced by civilian populations exposed to more impersonal bombing.

8. The ground-to-air threat environment has always posed a serious challenge to battlefield air operations, and has now reached a stage where ground defenders have been able, as in Afghanistan, to occasionally inflict "air denial" upon battlefield attackers.

9. The swing-role fighter-bomber has always performed more satisfactorily in the CAS/BAI role than the special-purpose attack airplane. Fighters have a natural dual-role air-to-air and air-to-ground nature, and thinking otherwise can fuel the dangerous belief that there is something inherently desirable in a specialized attack aircraft for battlefield air support.

10. Taken together, CAS and BAI have tremendous beneficial synergy that has proven, on occasion, critically important during specific military operations.

11. CAS/BAI experience from limited wars has only limited relevancy to high-intensity conflict, even as limited wars themselves have but limited relevancy to larger and more intensive conflicts. The benign environment of limited war—benign compared to high-tempo multiple-threat modern war—results in CAS/BAI operations that are more static in nature, and not characterized by the loss-rates, "fog of war," and operational constraints imposed in a high-intensity war where every

significant aspect of military operations is usually up for grabs or in question for much of the time.

12. CAS and BAI effectiveness relate directly to the kind of war being fought. BAI works best in fluid, high-tempo war situations, while CAS is most effective in more static ones. Likewise, both CAS and BAI work best when applied as part of a combined-arms strategy involving "pile-on" war, with the defender or attacker subjected to a multitude of air and ground threats, of which BAI and CAS are two. The combined synergy of CAS/BAI with ground assault is true "airland" war, and has been successfully prosecuted since the time of the battle of Cambrai in 1917.

13. Night CAS/BAI has been the most difficult and frustrating form of CAS/BAI to employ, and has proven less significant than daytime CAS/BAI operations. Even with modern technological advances such as the LANTIRN (Low-Altitude Night Targeting Infrared Navigation) system, realistically we should probably not expect to achieve identical accuracies or overall efficiencies comparable with daytime CAS/BAI employment, though the attempt to achieve such goals is a most laudable one.

Four developments in the postwar years had a profound impact on the evolution and employment of battlefield air support and the development of specific aircraft designs for that mission. The first of these was the experience of limited war, typified by Korea, Indochina, Malaya, Algeria, Southeast Asia, Africa (both Saharan and sub-Saharan) and, most recently, Afghanistan. The second was the development of the attack helicopter. The third was the development of the surface-to-air missile, especially the shoulder-launched SAM such as the SA-7, Blowpipe, Redeye, and Stinger. The fourth was the emergence of the remotely piloted vehicle (RPV) and the precision-guided cruise missile.

Aircraft were used in limited warfare before the First World War and extensively afterward. But what had changed in the post-1945 world was the nature of war in the atomic era. For example, having concluded that future wars would, in all likelihood, involve strategic nuclear warfare, the United States paid little attention to tactical air warfare requirements in the critical early postwar years, learning to its shock in Korea in 1950 that the five years since 1945 had witnessed the disestablishment of American tactical air power.[1] In the limited wars of the 1950s onward, the major powers were unwilling to use their atomic arsenals and unable to do so as well, if for no other reason than there was a lack of high-value targets upon which to employ such weapons. In particular, the "Wars of National Liberation" of the late 1940s onward lacked the kind of profitable targets that had worked so well for allied air power in the Second World War: supply columns, armored formations, and the like. To be meaningful, air power would have had to be applied in massive amounts that were simply lacking. Thus, in Indochina, the French were unable to prevent the Viet Minh from imposing air denial at Dien Bien Phu in 1954. Lacking aircraft themselves, the Communists simply moved in massive amounts of light and heavy flak that devastated supply flights, bomber strikes, and even fighters operating over that lush valley of what is now northwestern Vietnam.[2] (France did markedly better in Algeria toward the end of the 1950s because of larger air forces, use of "airmobile" assaults, and the absence of dense terrain conditions that masked the enemy so well in the tropics.)[3] From the perspective of both air and ground commanders, the United States generally did well in its own Southeast Asian war with battlefield air support, thanks to the infusion of massive numbers of tactical aircraft, high technology, sophisticated sensor and navigation aids (which even permitted CAS strikes by strategic bombers), and large numbers of helicopters.[4] The French had Dien Bien Phu; we, on the other hand, had Khe Sanh. But all of these conflicts were of the sort that were not really amenable to traditional military solutions. They demanded political solutions, and ultimately the will and desires of the local populations proved of far greater significance than any particular military aspect of the struggles.[5]

Ironically, what these conflicts did generate, however, was a return to the special-purpose attack airplane for battlefield air support. In Korea enemy air power was a negligible factor over the actual fighting front. In Indochina, Algeria, Malaya, Southeast Asia, Africa, and Afghanistan it likewise was of no consequence. As a result, air commanders could get away with operating obsolescent, slow, high-payload aircraft (such as the Douglas A-1 Skyraider and B-26K Counter-Invader), modified trainers (such as the North American T-28D Trojan), or even gunships modified from older transports (typified by the Douglas AC-47 "Spooky"). When these aircraft proved to have numerous operating problems because of their age (all were essentially World War II vintage or even pre-World War II designs), calls went forth for newer aircraft to replace them. Thus began a return to the specialized ground-attack airplane.[6]

In the United States this call resulted in adopting the Navy's Ling-Temco-Vought A-7 Corsair II, which was itself inspired by the earlier F-8 Crusader fighter, for the Air Force, and eventually spawned the Fairchild A-10 Thunderbolt II. Overseas it resulted in the birth of a whole series of armed trainer derivatives, such as the British Aerospace Strikemaster, the Italian Macchi MB-339, and the Dassault Alphajet. While useful in 1970s counterinsurgency warfare, these are hardly the sort of aircrft that can survive the modern high-intensity battlefield in the age of the Stinger. Ironically, of course, the same is true of the A-10. Required in the mid-1960s for the war in Southeast Asia, the A-10 did not emerge until the early 1970s, by which time its mission had changed to Warsaw Pact tank-killer. The 1980s have found it increasingly vulnerable, and, indeed, the A-10 may ultimately be replaced by an advanced version of the older A-7—which the A-10 itself was supposed to replace! In further irony, this proposed advanced A-7 essentially represents a return to F-8 performance, therefore affirming the basic need for fighterlike performance in the attack mission. What should be remembered is that the A-10, inspired in part by intense Congressional debates over the whole close air support issue in the 1960s, was designed to give the ground forces exactly what they claimed they needed: an up-to-date heavily armed high-payload long-loiter bomb-dropper, without regard to other issues such as its survivability against anticipated sophisticated air-to-air and ground-to-air threats of the 1980s and early 1990s. A greater irony, of course, is that in mirror-image fashion, the Soviets undertook development of their own equivalent, the Sukhoi Su-25 (NATO code named "Frogfoot"), which saw extensive service in Afghanistan, but which has itself but limited expectations of survival in more conventional conflicts. The lesson of the Second World War seems reaffirmed by current defense trends: the best aircraft for the contemporary tactical air support mission is the genuine fighter-bomber, such as the General Dynamics F-16 and McDonnell-Douglas F/A-18.

The helicopter gunship first made its appearance during the French war against the Front de Libération Nationale (FLN) in Algeria, where it was used for suppressive fire during airmobile assaults by transport helicopters. Subsequently, of course, it made its major mark during fighting in Southeast Asia, which witnessed the introduction of the first designed-from-the-outset attack helicopter, the Bell AH-1 Cobra. By war's end, the Cobra had proven invaluable as a support aircraft and had demonstrated its lethality against attacking tanks as well. The Southeast Asian experience resulted in an equivalent explosion of interest in attack helicopters, both in the United States and abroad. Ironically, America abandoned the ambitious Lockheed AH-56 Cheyenne just at the point where it was becoming a practicable system, and abrogated its lead in attack helicopter development to the Soviet Union, which has shown a particularly aggressive appreciation of such craft, typified by the infamous Hind of Afghani and Nicaraguan fame and the anticipated Havoc and Hokum. The Hughes-McDonnell-Douglas AH-64 Apache is a long-overdue response to this growing Soviet capability, as well as to the increasing demands of modern antiarmor warfare. Attack helicopters will, in all likelihood, not replace fixed-wing aircraft for the air-support mission, but they add their own distinctive capabilities. Already air-to-air combat has taken place between opposing attack helicopters (during the Iran-Iraq War), and the potential of the attack helicopter as a launch platform for air-to-air missiles is one that has already received serious attention around the world. The amount of attention devoted by the U.S. Army and U.S. Air Force to development of joint tactics and the increasingly strong relationship between the Air Force's Tactical Air Command and the Army's Training and Doctrine Command (TRADOC) bespeaks the interest of both parties in effectively utilizing fixed-wing and rotary-wing support systems over the battlefield.[7]

No discussion of vertical flight aircraft would be complete without mention of the potential for genuine vertical and short-takeoff-and-landing aircraft such as the British Aerospace Harrier and its American spinoff, the McDonnell-Douglas AV-8B. One of the critical weaknesses of modern air forces around the world is their dependency upon runways. These runways are vulnerable in the extreme, and it is certainly unrealistic to expect that a nation engaged in a general war could rely for any extended period of time upon its airfields. This problem, of course, is particularly acute with regard to NATO, but it is a problem that faces many other nations as well, including those of the Third World. The V/STOL jet fighter is an idea that has intrigued many individuals and nations, and its story has not been an altogether successful one. For that reason, the Harrier had more than its share of critics, but the performance of the Harrier in the Falklands war fully met the expectations of its supporters. Harriers proved that they could thrive in the air-to-air environment and that they could furnish fast and effective air support to friendly forces as well. For this reason it is to be hoped that defense planners will persevere in their pursuit of V/STOL technology and that they will not fall victim to the unreasoning prejudices of conventional thinking that so far have resulted in an overreliance on large stretches of concrete that constitute ideal targets for a variety of sophisticated precision-guided munitions.

Though the surface-to-air missile as it is known today dates to the rocketry research of the Third Reich and the United States during the Second World War, it was only in the war in Southeast Asia that the SAM came into its own. The SAM threat over North Vietnam severely constrained American strike operations until specific countermeasures—notably electronic warfare aids and specialized anti-SAM Wild Weasel and Iron Hand protective escorts—were available. But the SAM threat took a particularly dangerous course in 1972 with the introduction of the shoulder-mounted Soviet SA-7 system, which complemented the earlier SA-2 during the North Vietnamese spring offensive.[8] The SA-7 quickly proved a serious threat to helicopter gunships, older fixed-wing aircraft such as the A-1, and even unwary jet-propelled strike aircraft. The problems first highlighted in Southeast Asia were dramatically reinforced by the Israeli experience in 1973, when the Syrians were able to deny the Israeli Air Force access over the Golan Heights until the IAF arranged for the Israeli Army to seize the territory. In the 1973 Arab-Israeli War the synergistic interaction of the larger SA-6 SAM, the portable SA-7, and the mobile radar-directed ZSU-23-4 gun carriage posed serious problems for the IAF, already experiencing an attrition rate higher than it had experienced in any other previous Middle Eastern war; 61 of the 109 Israeli aircraft lost were shot down on CAS sorties. More recently, the experiences in Angola and Mozambique, Morocco's Polisario guerrilla war, the Falklands War, and in Afghanistan have confirmed these earlier lessons. The Afghani resistance, in particular, was able to actually inflict a situation of air denial upon Soviet forces, employing Stinger surface-to-air missiles and thereby defeating a combined-arms Soviet airmobile system using fixed and rotary-wing support that had previously worked well.[9]

The precision-guided weapons revolution of the 1960s and 1970s has had a profound impact on modern military operations. By the end of the Vietnam War strike flights using laser-guided bombs were destroying high-value targets without incurring the kinds of punishing losses that had typified earlier air operations with aircraft dropping conventional "dumb" bombs. The remotely piloted vehicle (RPV)—in effect, a small pilotless aircraft controlled from a ground station or from an airborne controller—proved useful for reconnaissance and information gathering. Since that time the sophistication and capabilities of precision-guided weapons and RPV's have increased tremendously. The conventionally armed long-range air and ground-launched cruise missile is an example of a system that combines features of both the precision-guided munition and the RPV. When coupled with emergent weapons and munitions development such as area-denial mines and "skeet" type antiarmor munitions, the long-range cruise missile can be seen as a potential air support weapon in its own right, functioning in situations and against targets that a commander would be reluctant to attack with a more vulnerable fighter-bomber.[10]

Another area of emerging significance is that of so-called low observables or stealth technology. The existence of one such aircraft, the Lockheed F-117A, has already been acknowledged, and, of course, a strategic stealth bomber, the Northrop B-2, is under development. In the intense electronic and sensor environment of modern war, the development of military systems—particularly aircraft—that offer minimal or dramatically reduced detectable "signatures" promises to revolutionize future war. The potential of the stealth attacker may be likened to the deadly Acquired Immune Deficiency Syndrome (AIDS) virus and, indeed, gives "penetration aids" a whole new meaning. The stealth attacker, in analogous fashion, could attack the "T-cells" of the enemy's "immune system" (defensive networks), allowing "opportunistic infections" (conventional strike aircraft) to slip by the weakened or destroyed defenses, and thereby cripple or destroy an enemy's ability to wage war.

It is misleading, however, to think that the future of battlefield air attack is dependent solely on technology. Technology devoid of strategic thought and doctrinal underpinnings is incapable of serving a nation's defense needs. Thus, it is worth noting that, at least in the United States, a considerable amount of thought has been given to the nature of future air-land war. Since 1975, the Air Force's Tactical Air Command and the Army's Training and Doctrine Command (TRADOC) have examined the arena of air-land war, forming a joint study agency, the AirLand Forces Application (ALFA) agency. ALFA has structured three major aspects of current "AirLand Battle" thought: joint air attack team operations (JAAT), joint suppression of enemy air defenses (J-SEAD), and joint attack of second-echelon forces (J-SAK). Traditionally, it has been the "first-echelon" opposing forces—those within approximately fifteen kilometers of the front—that have attracted the most attention from ground commanders and air support units. But the second-echelon forces—which begin at roughly the fifteen kilometer mark and extend back deep into enemy territory—pose considerable risk to defenders as well, particularly given the fluid and fast-moving tempo of modern war. (Experience from the National Training Center at Ft. Irwin, California, supports this; a typical Soviet-style motorized rifle regiment can move six kilometers in fifteen minutes—under battle conditions.) Fortunately, since second-echelon forces may still be configured for travel as they approach the battle area, they likely constitute a "softer" and easier target than the first-echelon forces ahead of them, and thus a more profitable subject for intensive air attack.[11]

The recognition that the modern battlefield is not limited in depth to a few vital kilometers but, rather, extends rearward to great depth, with a multiplicity of threats and complexities, has spawned the recognition that battlefield air attack is more diversified than "merely" close air support and interdiction. Between the interdictor operating to a depth of perhaps hundreds of kilometers and the CAS aircraft or attack helicopter operating right over the front is the province of the battlefield air interdictor—the interdiction of enemy forces not yet in close combat with friendly forces, but whose presence cannot be ignored. This is, admittedly, a hazy area, as it does not fall within the jurisdictional niceties that have previously governed battlefield air support thought—particularly the crisp "dividing line" between having to coordinate air action with the fire and movement of friendly forces (typical of CAS) and the freedom to proceed without regard for the ground situation (typical of interdiction). The vagueness has crept over into the definitions of battlefield air interdiction (BAI) (as of the end of 1984):

> Air interdiction (AI) attacks against land force targets which have a near-term effect on the operations or scheme of maneuver of friendly forces, but are not in close proximity to friendly forces, are referred to as battlefield air interdiction (BAI).[12]

As one Air Force general recently wrote, "Our concept of BAI—what it is, how it is controlled, etc.—is still evolving."[13]

The concern over the role of CAS, BAI, traditional air interdiction, and strategic air warfare is commendable, for it reflects the desire of the air and ground communities to resolve potential problems before finding them out firsthand in the last place that one wishes to have to learn a lesson—in the midst of a modern battlefield. But if air-land warfare is currently embroiled in a debate over CAS/BAI issues, control of air resources, the ideal technology for such missions, and the like, practitioners should perhaps take heart in the lessons to be drawn from history. The men who prosecuted air-land warfare over the Western Front did not use the terms CAS/BAI, but that is what they did; so, too, did the airmen of the Second World War, in North Africa, Europe, and the Soviet Union. *Plus ça change, plus c'est la même chose.* In our attempt to further clarify the role of air power over the battlefield, it would be wrong to so structure battlefield air support that we inadvertently build in a doctrinal and procedural mind-set so rigid as to cause us to constrain even more the flexibility of air power application and to impose even greater demands for command, control, and communication than we currently expect. Above all, it would be wrong to presume that we are the first who have confronted such issues and attempted to address them.

What does all this say about the battlefield of the future? More than ever before, it will demonstrate that warfare is truly a combined-arms operation. The air-land battlefield of the future will be intense, fluid, fast-moving, and deadly. Unlike the scenarios in most future-focused military "techno-thrillers," the outcomes will not be easily predictable, nor will advances in technology compensate for or permit failures in command. Rather than give an opponent time to adjust to any single particular threat, both sides will attempt to maintain and "choreograph" a tempo of warfare forcing the enemy to constantly defend against a multitude of threats: "fast-movers" such as jet fighter-bombers, "pop-up" attacks from gunships, "fearless" RPV attacks against battlefield electronic emitters, stealth aircraft, conventional and rocket artillery, and combined infantry-armor assault. Time will be critical, and a major challenge will be preventing the enemy from gaining the initiative in combat. More than ever, troops will be endangered from "shoot first, ask afterward" friendly fire, resulting in greater training and recognition needs and self-imposed firing discipline. Prosecuting battlefield air support attacks under night and adverse weather conditions will assume greater significance than ever before. Indeed, overall, in this "pile on" war, delivering day and night air support will remain as critical as it has always been, for both defender and attacker.[14]

Undeniably, the growing sophistication of ground-to-air threats has forced a rethinking of contemporary and future air support operations. But the advent of gunpowder did not end cavalry, only the lumbering knight. While the nature of air support changes and evolves, its need will not. There always will be a place in assault for aircraft—and there will always be a need for the brave to fly them.

Glossary

AAEAF, Advanced Allied Expeditionary Air Force
AAS, Army Air Service (U.S.)
AASF, Advanced Air Striking Force (of RAF)
ACC, Army Co-operation Command (RAF, WW II)
AEF, American Expeditionary Force(s) (WW I)
AFC, Australian Flying Corps (WW I)
AFM, U.S. Air Force Museum, Wright-Patterson AFB, Dayton, Ohio
AOC, Air Officer Commanding
ASC, Air Support Command
ASPO, Air Support Party Officer
BAI, battlefield air interdiction
bocage, thick hedgerow country around Normandy
Büchsenöffner, can opener; nickname for the Hs 129
CAF, Chinese Air Force (1937–1945)
CAS, close air support
CBU, cluster bomb unit
CP, command post
Edwards Collection, Glen Edwards Collection, History Office, Air Force Flight Test Center, Edwards AFB, California
Erprobungstelle, flight test center
FAC, forward air controller
FASL, Forward Air Support Links
Fliegerkorps, Air Division
Geschwader, wing (organizational unit)
Groupe Mobile: French colonial counterinsurgency formation comprised of from four to nine battalions, three to six artillery batteries, one or two *escadrons* of *spahis*, and, if possible, an additional cavalry *escadrille*
gruppa, a Soviet aerial formation of six or eight aircraft, consisting of a four-ship *zveno* and an attached two-ship *para*, or two *zveno*

Herrenvolk, "Master Race"; Nazi ideological expression
Heeresflak, "organic" German Army flak (antiaircraft) forces
Heeresgruppe, Army Group (combat command)
IAF, Israeli Air Force
JAAF, Japanese Army Air Force
Jagdbomber, abbr. *Jabo,* fighter-bomber
Jagdgeschwader, fighter wings
Jagdstaffeln, fighter squadrons
JNAF, Japanese Naval Air Force
Kette, a German combat formation of three aircraft
Kriegstagebuch, War Diary, compiled by major German commands
LANTIRN, Low-Altitude Night Targeting Infrared Navigation system (U.S.)
Lotnictwo Wojskowe, Military Aviation (Poland)
Luftflotte, "Air Fleet"; the largest overall German theater air command
Luftstreitkräfte, the Imperial German Air Service
Luftwaffe, Nazi German Air Force
MA, Military Attaché
MEW, microwave early warning radar
MHI, U.S. Military History Institute, Carlisle Barracks, Pennsylvania
NAAF, Northwest African Air Force
NASM, National Air and Space Museum, Washington, D.C.
Nachtschlachtgeschwadern, night ground-attack wings
Nahkampfverband z.b.V., "close battle association for special duties"; predecessor of VIII Fliegerkorps
NATAF, North African TAF
Panzer, tank
Panzerjagdkommando Weiss, Tank-hunting Command Weiss
Panzerkampfgruppe, armored battle group (tank command)
Panzerknacker, "tank cracker"; nickname for antitank pilots and aircraft
Panzerjagdstaffeln, Panzerjäger-Staffeln, tank-hunting squadrons
PVO Strany, Soviet air defense forces
RAAF, Royal Australian Air Force
RAF, Royal Air Force, formed in 1918 by merger of RFC with RNAS
RFC, Royal Flying Corps
RNAS, Royal Naval Air Service
Rotte, 2-ship element; two *Rotte* constitute a *Schwarm,*
RPV, remotely piloted vehicle
SAM, surface-to-air missile
Schlacht, ground attack (lit., slaughter)
Schlachtfliegergruppen, ground-attack flier groups; predecessors of the *Schlachtgeschwader*
Schlachtfliegerschule, ground-attack tactics and training school
Schlachtflugzeug, ground-attack aircraft

Schlachtflieger, ground-attack aircrewmen
Schlachtgeschwader, abbr. *S.G.,* attack wings
Schlachtstaffel, abbr. *Schlasta,* attack squadron; WW I
Schutzstaffel, protection flight; WW I
Schwarm, 4-ship flight; four *Schwarm* constitute a *Staffel*
SHAEF, Supreme Headquarters Allied Expeditionary Force
Shturmovaya Aviatsia, attack aviation
Shturmovik, attack plane; spec., the Ilyushin Il-2
Staffel, 16-ship squadron; four *Staffeln* and one *Stabschwarm* constitute a *Gruppe*
Stabschwarm, staff flight; an additional administrative *Schwarm* attached to a *Gruppe*
Stavka (Shtab glavnogo/verkhovnogo komandovaniya), Soviet Supreme High Command
Stuka, abbr. for *Sturzkampfflugzeug,* "diving battleplane"; spec., the Junkers Ju 87 dive bomber
TAC, Tactical Air Command (U.S.)
TAF, Tactical Air Force (U.K., also Allied, WW II)
TCC, Tactical Control Center
TRADOC, Training and Doctrine Command (U.S. Army)
Untermensch, "Inferior people"; Nazi ideological expression
USAAF, U.S. Army Air Forces, WW II
Versuchskommando für Panzerbekämpfung, Experimental Command for Antitank Warfare; German antitank operational test and evaluation command
V/STOL, vertical/short takeoff and landing
VVS (voyenno-vozdushnyye sily), Soviet army air force
Wehrmacht, German Armed Forces

Source Notes

Major sources and references cited frequently are listed with full publication data in the bibliography.

Part One

1. For additional detail on the state of early aeronautics, see Charles H. Gibbs-Smith, *Aviation: An Historical Survey from Its Origins to the End of World War II* (London: HMSO, 1970) and *The Rebirth of European Aviation, 1896–1908—A Study of the Wright Brothers' Influence* (London: HMSO, 1974); Hallion, *The Wright Brothers, Test Pilots,* and *Rise of the Fighter Aircraft, 1914–1918;* Goldberg; and Christienne and Lissarrague.

2. Anthony Robinson, *Aerial Warfare* (New York: Galahad Books, 1982), p. 127; John R. Cuneo, *The German Air Weapon, 1970–1914* (Harrisburg, Pa.: Military Service Publishing Co., 1942), p. 264; Melli, pp. 55, 75, 132; Humphreys, pp. 145–146.

3. Christienne and Lissarrague, pp. 39–40, 49–50; Goldberg, p. 10; Hinds, pp. 194–201.

4. Robert C. Ehrhart, "Ideas and the Warrior," *Air University Review* 37, no. 6 (September–October 1986): 104; quote is from Col. Louis de Grandmaison, cited in Cyril Falls, ed., *The Great War* (New York: Capricorn Press, 1959), p. 35. The literature of the First World War is voluminous; for three useful books that explore the mind-set of higher command in that war, see Corelli Barnett, *The Swordbearers: Supreme Command in the First World War* (New York: William Morrow, 1964); Alistair Horne, *The Price of Glory: Verdun 1916* (New York: Harper & Row, 1967); and Leon Wolff, *In Flanders Fields: The 1917 Campaign* (New York: Time Inc., 1958).

5. For information concerning the differing strategies of the Allies and Germany during the first war in the air, see Hallion, *Rise of the Fighter Aircraft,* pp. 21–22, 27–33, 112, 125, and 139; additionally, Christienne and Lissarrague offer useful French perspectives, while Boyle's biography *Trenchard: Man of Vision* and Hoeppner's autobiographical *Deutschlands Krieg in der Luft* offer useful British and German perspectives.

6. For discussion of the generations of World War I fighter aircraft development, see Hallion, *Rise of the Fighter Aircraft,* especially pp. 44–46, 112–118, and 131–132.

7. For details on these two aircraft, see Munson, *Bombers, Patrol and Reconnaissance Aircraft, 1914–1919,* pp. 137–138, 144–145; and Bruce and Noel, *The Breguet 14,* pp. 3–16.

8. Georg Paul Neumann, ed., *Die Deutschen Luftstreitkräfte im Weltkriege* (Berlin: E. S. Mittler & Sohn, 1920), p. 472.

9. H. Jones, *The War in the Air, III,* pp. 371–373, 378–379.

10. H. Jones, *The War in the Air, IV,* p. 129; see also pp. 130, 163–166, and 175–180, and Baring, pp. 244–250.

11. Lee, *Open Cockpit,* pp. 139, 141; see also pp. 132–138; H. Jones, *War in the Air, IV,* pp. 233–239, 243, and 257; for additional information, see Lee, *No Parachute,* pp. 161–194.
12. H. Jones, *War in the Air, IV,* pp. 245–246.
13. W. M. Lamberton, *Fighter Aircraft of the 1914–1918 War* (Fallbrook, Calif.: Aero Publishers, 1964), pp. 132, 136; Neumann, ed., pp. 74, 90–93; Nami, pp. 155–172; Gray and Thetford, pp. xii–xv, 154–157; Jones, "Appendix 12: Employment of Battle Flights," in H. Jones, *The War in the Air, Appendices;* H. Jones, *War in the Air, IV,* pp. 251–259.
14. For example, see Stuart Cloete, *A Victorian Son: An Autobiography, 1987–1922* (New York: John Day, 1973), p. 288.
15. Ludendorff, p. 463.
16. W. Shaw Sparrow, *The Fifth Army in March 1918* (London: John Lane, 1921), p. 22. The Fifth Army commander, General Sir Hubert Gough, has written a frank account of the collapse; see *The Fifth Army,* pp. 260–329.
17. Sparrow, p. 23; H. Jones, *War in the Air, IV,* pp. 290–291, 299–301, 311, 315–317, 342–343, 376; H. Jones, *War in the Air, VI,* p. 507; Hohmann, pp. 249, 253–258.
18. Sparrow, p. 23.
19. Ludendorff, p. 483.
20. Jones, "Appendix 22: Protection Against Enemy Aeroplanes," in H. Jones, *War in the Air, Appendices,* pp. 113–114.
21. Bill Lambert, *Combat Report* (London: Corgi Books, 1975), p. 157.
22. Ecole Superieure de Guerre, Cours d'Aeronautique, *Notes sur l'Emploi de l'Aeronautique par le Commandement* (1928–1929), pp. 4, 18–21; copy in the MHI archives. While of significance to the subsequent development of independent air forces elsewhere around the world—notably the U.S. Air Force—the creation of the RAF per se falls outside the scope of this study. Fortunately, a large body of memoirs and historical studies examining the establishment of the RAF exists.
23. Christienne and Lissarrague, p. 128.
24. General Voisin, *La Doctrine de l'Aviation française de combat* (Paris: Editions Berger-Levrault, 1932), pp. 112–119, 128–130, 135–140.
25. See Battle Orders No. 3, HQ Air Service First Army, September 13, 1918, reprinted in Maurer, pp. 365–367.
26. Quoted in H. A. Toulmin, Jr., *Air Service American Expeditionary Force, 1918* (New York: Van Nostrand, 1927), p. 368.
27. Christian A. Bach and Henry Noble Hall, *The Fourth Division: Its Services and Achievements in the World War* (unknown: "Issued by the Division," 1920), p. 173.
28. Mitchell, *Memoirs,* pp. 259–260.
29. Ibid., p. 258.
30. This account is based on information from the following works: Department of General Instruction, *History: The Palestine Campaign* (Ft. Riley, Kansas: The Cavalry School, 1922–23), copy in the MHI archives; Falls, *Armageddon;* Savage; Sutherland; Cutlack; H. Jones, *Sir Walter Raleigh* and *War in the Air, VI;* The Palestine News; and Sanders.
31. Sanders, p. 352. For additional information, see Savage, pp. 296–297.
32. H. Jones, *War in the Air, VI,* pp. 218–219; Cutlack, pp. 152–154; Sutherland, pp. 249–250.
33. H. Jones, *War in the Air, VI,* pp. 219–223; Cutlack, pp. 155–158; Sutherland, pp. 250–253. Sutherland offers a useful perspective as he flew as one of the attacking pilots in virtually all of these actions.
34. Cutlack, pp. 159–161. For additional details see H. Jones, *War in the Air, VI,* pp. 224–227; and Sutherland, pp. 253–259.
35. Sanders, pp. 364–365. For additional details, see H. Jones, *Sir Walter Raleigh,* pp. 43–45; information on the Mills bomb "dispenser" is from Sutherland, pp. 242–243, 256

(though it allegedly worked, it seems not to have been repeated); Falls, *Armageddon*, pp. 62–63; The Palestine News has an excellent daily account of the fighting, together with superb maps, especially plates 43–49.

36. Falls, *Armageddon*, p. 62.

37. Cutlack, pp. 165–166. See also H. Jones, *War in the Air, VI*, pp. 231–232.

38. Cutlack, pp. 166–167; H. Jones, *War in the Air, VI*, pp. 232–235; The Palestine News, plates 47–51.

39. Quoted in a conversation with the author by Professor John H. Morrow, January 20, 1988, at the National Air and Space Museum, Smithsonian Institution.

40. Mitchell, *Memoirs*, p. 236.

41. Ibid., p. 261. For a description and table of organization of the radio equipment available to American air squadrons attached to the A.E.F., see U.S. War Department, Equipment Manual Division, Supply Section, U.S. Air Service, *Initial Equipment Manual for Service Squadrons in the Air Service American Expeditionary Forces* (Washington, D.C.: War Department, 12 September 1918), pp. 78–89.

42. For example, see Sunderland, pp. 1–3.

43. Greenhous, pp. 92–93.

Part Two

1. Giulio Douhet, *The Command of the Air* (Washington, D.C.: Office of Air Force History, 1983 ed.); see, for example, p. 126.

2. Mitchell's earlier interest—before he devoted increasing attention to strategic bombing—can be found in his "Tactical Application of Military Aeronautics," written in 1921, pp. 4–5; a copy of this document is in the MHI archives. See also Greer, pp. 12, 66–67, 87–88, and 121–123. For an incisive critical review of Douhet's thought, see Michael J. Eula, "Giulio Douhet and Strategic Air Force Operations," *Air University Review* 37, no. 6 (September–October 1986): 94–99.

3. Sources for technical information on these aircraft were: Munson, *Fighters Between the Wars, 1919–1939, Fighters, Attack, and Training Aircraft, 1914–1919*, and *Bombers Between the Wars, 1919–1939*; Thompson; Wagner, *American Combat Planes*; Cynk, *The P.Z.L. P-23 Karas*; Shores, *Martin Maryland & Baltimore Variants*; American Aviation Historical Society, *The Curtiss Shrike*; and Smith, *The Junkers Ju 87A & B*.

4. See Murray's *Luftwaffe*, p. 15.

5. For details on the Il-2, see Liss, *The Ilyushin Il-2*, pp. 3–16; Hardesty has good information on its operational usage.

6. Holley.

7. Greer, pp. 37–38.

8. Air Corps Field Manual FM 1-5, *Employment of Aviation of the Army* (Washington, D.C.: GPO, 1940), p. 42.

9. *The Employment of Air Forces with the Army in the Field* (London: HMSO, 1938), p. 38; copy in the MHI library.

10. FM 1-5, p. 22.

11. Air Corps Field Manual FM 1-10, *Tactics and Technique of Air Attack* (Washington, D.C.: GPO, 1940), pp. 114–119.

12. Air Corps Tactical School, *Bombardment Aviation* (Maxwell Field: ACTS, January 1, 1938); and Air Corps Tactical School, *Air Force: The Employment of Combat Aviation* (Maxwell Field: ACTS, April 1, 1939). Copies of both of these are in the MHI archives.

13. Gabel, pp. 310–313, 327–329.

14. United States Marine Corps, *Small Wars Manual, United States Marine Corps, 1940* (Washington, D.C.: GPO, 1940), section 9-23.

15. Ibid., section 9-28.
16. Ibid., section 9-29.
17. See, for example, Ira Jones, pp. 107–108.
18. Jerzy B. Cynk, "The Truth," p. 176.
19. Murray, *Wings Over Poland*, p. 131.
20. Karolevitz and Fenn, pp. 169–171.
21. Murray, *Wings Over Poland*, p. 244.
22. Karolevitz and Fenn, p. 145.
23. Murray, *Wings Over Poland*, p. 244.
24. Cynk, "The Truth," p. 177.
25. Ibid., pp. 177–181.
26. General Staff, *Official Account*, p. 49; this is the best reference for overall combat operations. See Molesworth for an interesting memoir on the war.
27. General Staff, *Official Account*, pp. 34–35, 49, 56–57, 80–81, 102–103, 118–121; Molesworth, pp. 107–108, 116, 120–121, 145, 147, 148–149, 160–169.
28. General Staff, *Official Account*, p. 133.
29. Jardine, pp. 262–267.
30. Ibid., p. viii.
31. Ibid., p. 280; see also p. 278.
32. Saunders, *Per Ardua*, pp. 287–288. The best overall source on British air control efforts in the Middle East is Elmer Bermon Scovill, *The Royal Air Force, the Middle East, and Disarmament, 1919–1934* (Michigan State University: Ph.D. Thesis, 1972); a copy of the Scovill thesis is in the MHI library.
33. Glubb, pp. 317–318; see also 69, 84–85, 90, 110, 130–131, 144, 146–147, 194–203, 226, 234–235, 239–242, 270, 291. For an example of the self-congratulatory nature of Trenchard's and Churchill's view of air control, see letters between Churchill and Austen Chamberlain, and Trenchard to Churchill, reprinted in Martin Guilbert, *Winston S. Churchill*, vol. 5: Companion Part 1, Documents, *The Exchequer Years, 1922–1929* (Boston: Houghton Mifflin Co., 1981), pp. 392–393, 553–555. For a pilot's-eye view (prejudices and all) of air control, see Ira Jones, passim. Slessor, *The Central Blue*, pp. 45–75, has a good—though partisan—view of the air control notion.
34. Glubb, pp. 236–242.
35. "H.N.T.," "Ziza, 1924," *Royal Air Force Quarterly* 4 (1933): 261–268. For another account of raiding activities and the RAF, see "Raiding—Iraq 1926–27," in the same volume, 245–252.
36. Slessor, *Air Power and Armies*, p. 98.
37. Ibid., p. 101.
38. Ibid., pp. 101–102.
39. Slessor, *Central Blue*, pp. 120–131.
40. Ibid., pp. 656–660.
41. Ibid., pp. 659–660.
42. Fleming, p. 65. I have adopted the spelling al-Karim from Pennell.
43. Fleming, pp. 56–70, 139–142, 175, 202–203; Pennell, pp. 69–72, 99–103, 150–151, 168–169.
44. Fleming, pp. 204–205, 213, 225, 263–297, 351, 361–369, 378–380; Pennell, pp. 182–191, 198–199, 218, 253.
45. Information on the strength of French aviation forces in Morocco is from the previously cited Christienne and Lissarrague, 231–234. There is some confusion in French sources on the role of aircraft as artillery spotters. Christienne and Lissarrague state that artillerymen were responsible for control of their own shot. However, Colonel Fabre (cited below) emphasizes the role of artillery air control in mobile group operations. Perhaps both are correct from their own perspective; Christienne and Lissarrague seem to be emphasiz-

ing the control of truly long-range artillery, while Fabre is talking primarily about the control of artillery over much shorter ranges during tactical operations against the *Riffi.* For additional information on Moroccan air war, see Paul Armengaud's article.

46. Fabre, pp. 31–46, 51–81; p. 82 has a good schematic presentation of a mobile group operating amid hostile forces.

47. Ibid., p. 222.

48. Ibid., p. 197.

49. Ibid., p. 147.

50. Ibid., p. 195.

51. Ibid., pp. 49, 83, 147–148, 187, 193–197.

52. The best summary of the Sandino War is Macaulay's *The Sandino Affair,* which is sympathetic to the rebel leader (Macaulay himself fought for Fidel Castro before Castro took over Cuba).

53. Ibid., p. 118.

54. Interview with Maj. Gen. Ross E. Rowell, USMC, Oct. 24, 1946, in the files of the Marine Corps Historical Center, Washington Navy Yard, Washington, D.C. I wish to thank Danny J. Crawford, Head, Reference Section, for making this interview available to me, and for facilitating my research on Marine close air support. The single best history on the early development of Marine aviation is Robert Sherrod's *History of Marine Corps Aviation in World War II,* pp. 1–23.

55. Rowell interview.

56. Macaulay, pp. 81; see also Brainard, pp. 30–31. Rowell has a detailed account of Marine air ops in Nicaragua in "Aircraft in Bush Warfare."

57. Macaulay, pp. 166–167.

58. Brainard, pp. 32–34.

59. Letter, Ridgway to USA Chief of Infantry, October 30, 1928, Ridgway Papers, Box 2, MHI archives.

60. Brainard, pp. 28–29; Macaulay, p. 239.

Part Three

1. Zook, pp. 24, 50, 52, 62, 67, 68–69, 84, 92, 94, 101–102, 110, 113, 118, 120, 126, 130, 137, 146, 148–149, 159, 163, 165, 168, 182, 196, 210, 218, 238, and 240.

2. Ibid., pp. 62, 84, 110.

3. Ibid., pp. 24, 67–68, 84, 150.

4. Ibid., pp. 24, 238. See also Rauch, pp. 207–213. See also Pablo Max Ynsfran, ed., *The Epic of the Chaco: Marshal Estigarribia's Memoirs of The Chaco War, 1932–1935,* vol. 8 of The University of Texas Institute of Latin-American Studies, *Latin-American Studies* (New York: Greenwood Press, 1969 ed.), p. 14. For an example of another Latin war in which air power played a generally ineffectual role, see Jackson Flores, Jr., "The Brazilian Air War," *AIR Enthusiast Thirty-five* (January–April 1988): 64–73, about the Brazilian Civil War of 1932.

5. Zook, p. 24, 165; Ynsfran, pp. 141, 165; Rauch, passim. The mercenary account is Wewege-Smith's *Gran Chaco Adventure.*

6. Zook, pp. 92–102, 120; Rauch, p. 209.

7. Wewege-Smith, p. 237; see also pp. 97, 133, 141, 149, and 166–167. See Estigarribia's comments on air recon in Ynsfran, p. 141.

8. Zook, p. 238.

9. Peter Young with Michael Calvert, *A Dictionary of Battles: 1816–1976* (New York: Mayflower Books, 1978), p. 110.

10. De Bono, pp. 13, 24, 63, 72.

11. For Meccozzi's views, see his article in *Revista Aeronautica,* pp. 193–201. The best English-language survey of Italian aircraft development in the 1930s is Thompson's, esp. pp. 31–42, 81–89, 143–147, 196–201, and 270–272.

12. De Bono, p. 111; Del Boca, pp. 91–92.

13. De Bono, p. 177; see also pp. 24, 63, 72, 111, 113.

14. Intelligence Branch, *Certain Studies,* pp. 19, 23, 33, 42–43, and 52. See also Fiske, p. 279. In addition, see De Bono, pp. 240–241; and Del Boca, pp. 75, 78, 107–109, 120–121, 155, and 177. Pedro A. del Valle's *Roman Eagles Over Ethiopia* is an interesting memoir by a Marine officer who accompanied the Italian forces as an official observer.

15. Intelligence Branch, *Certain Studies,* pp. 56; see also Fiske, p. 56.

16. Intelligence Branch, *Certain Studies,* p. 30; Fiske, pp. 221, 278; De Bono, p. 292.

17. Intelligence Branch, *Certain Studies,* pp. 3, 41, 52; Fiske, pp. 40, 244.

18. Intelligence Branch, *Certain Studies,* pp. 37–38.

19. Ibid., p. 38.

20. Badoglio, *La guerra d'Etiopia,* p. 215; see also pp. 96, 101, 104–105, 111–113, 169, 221, 223, 226, 233, 236, 241, 243, and 245. Additionally, see Intelligence Branch, *Certain Studies,* p. 39, and Del Boca, pp. 134–135.

21. Intelligence Branch, *Certain Studies,* p. 48; see also p. 47.

22. Banfill, p. 21.

23. Intelligence Branch, *Certain Studies,* p. 50.

24. Thomas's *The Spanish Civil War* is the most authoritative and balanced history of this emotion-laden conflict; see pp. 133, 144, 206, 211, 229–230, 233–234, 253, 297, 304, 316, 382–388, 403–404, 459, 461–466, 513–514, 557–559, 566, 610–615. Memoirs from the war are, not unexpectedly, tinged with partisan flavor. See, for example, Bessie; Wintringham; Bessie and Prago; Tinker; and Larios. Leftist memoirs, in particular, are often polemical in tone, occasionally hysteric, and self-righteous. The International Brigades are treated in Rosenstone's largely uncritical *Crusade of the Left* and Cecil Eby's more balanced and sobering *Between the Bullet and the Lie.* The international brigadesmen in Spain seem to have been a strange lot, prone to violence, intrigue, self-righteousness, and murderous bickering; one wonders how truly effective and democratically inclined (except in their own eyes) they were. For example, see Rosenstone, p. 142, 306–313, and Eby, pp. 66, 118–119, 270–272, 313.

25. Again, the best military survey of the Spanish war is Thomas's *The Spanish Civil War.* The best political and military reference to the war is James W. Cortada's edited (and massive) *Historical Dictionary of the Spanish Civil War, 1936–1939* (Westport, Conn.: Greenwood Press, 1982). The "Aldershot" reference is from B. H. Liddell Hart, *The German Generals Talk* (New York: William Morrow, 1948), p. 92. The best single study of warfare in Spain is Proctor's; see pp. 45, 56, 80, 86, 89, 103, 105, 113, 118–119, 149, 153, 155, 165, 172–173, 175, 178–183, 191–194, 223–227, 242–246, 251–258.

26. MA Valencia, "Special Report on Spanish Government Air Force," Feb. 21, 1937, pp. 15–16 in B. Jones, *Spanish War,* vol. 1. The air war in Spain has received spotty treatment. The only detailed study is that of Larrazabal, but it is so seriously marred by a "trees rather than forest" approach to the subject as to be virtually unreadable in any coherent way. The best source on the German Legion Condor is Proctor's previously cited work. Christopher Shores's *Spanish Civil War Air Forces* is a generally useful introduction, but short on analysis. The same can be said of Emiliani, Ghergo, and Vigna's *Spagna 1936–1939.*

27. For some of these points, see Murray, *Luftwaffe,* pp. 16–18; Murray's study is the definitive history of the German air force in the Second World War.

28. Thomas, p. 206.

29. Cot, pp. 351–355; see also Elstob, pp. 59–60.

30. Thomas, pp. 229–234, 297–304; Shores, *Spanish Civil War Air Forces*, pp. 5–10; Proctor, p. 56; Green and Swanborough, pp. 9–29.

31. *Legion Condor*, p. 67. Tinker and Larios offer good perspectives on air-to-air combat between the Fiat and the Polikarpov fighters; see also Liss, *The Polikarpov I-16*, pp. 9–10; Cattaneo, *The Fiat C.R. 32*, pp. 5–7; Green and Swanborough, pp. 12–21; and Larrazabal, pp. 50–51.

32. MA Paris, "Major Military Operations: Lessons of the Spanish War," Report 24,469-W, Aug. 20, 1938, p. 2, and extracts from Maj. Wanty, Belgian Army, "A Year of War in Spain: Facts and Lessons," *La Revue d'Infanterie*, both in B. Jones, *Spanish War*, vol. 3. See also Thomas, pp. 253–254, and Shores, *Spanish Civil War Air Forces*, p. 17.

33. Larios, pp. 54–55.

34. Thomas, pp. 316–317; Emiliani, Ghergo, and Vigna, introduction, pp. 1–5.

35. MA Valencia, "Special Report on Spanish Government Air Force," p. 10 (see also p. 5); see also MA Valencia, "Spain Aviation: Subject—Organizational Training, Tactics Employed," Report 6530, April 25, 1937, pp. 8–9, and MA Paris, "Spain (combat): Subject—Major Military Operations, Air Combat Operations, Air Equipment & Tactics," Report 23,265-W, March 10, 1937, p. 1, all in B. Jones, *Spanish War*, vol. 1.

36. There are useful comments on this in Kipp's *Barbarossa*, pp. 27–30; see also Mikhailow, "Tactical Employment of Bombardment Aviation," and "Tactical Employment of Pursuit Aviation." The best German look at the activities of the Legion Condor is Karl Drum's unpublished ms. I wish to thank Mr. Lloyd Cornett and the staff of the USAFHRC for making a microfilm copy of this ms. available for my research. See also Schulz, p. 2.

37. Larios, pp. 86–87.

38. MA Valencia, "Special Report," p. 14, in B. Jones, *Spanish War*, vol. 1.

39. Rosenstone, pp. 40–41; Shores, *Spanish Civil War Air Forces*, p. 19; Proctor, p. 105.

40. The account of the battle of Guadalajara is based on Garriga, pp. 109–180; Nassaes, pp. 75–106; Coverdale, pp. 212–260; Tinker, pp. 120–157; Elstob, pp. 127–131; MA Paris, "Spain (Combat Aviation): Subject—Major Military Operations, Air Combat Operations, Operations in Spain," Report 23,461-W (June 2, 1937), pp. 1–4, in B. Jones, *Spanish War*, vol. 1; Arthur Ehrhardt, "Airplanes Versus Ground Motor: Lessons Gained from the Battle of Guadalajara," an Army War College translation of an article from *Wissen und Wehr*, July 1938, in B. Jones, *Spanish War*, vol. 3; Arthur H. Landis, "American Flyers in Spanish Skies," in Bessie and Prago, pp. 118–131; Larrazabal, pp. 127–128; Proctor, pp. 111–115; Thomas, pp. 383–388; Green and Swanborough, p. 16; Mikhailow, "Tactical Employment of Pursuit Aviation," pp. 7–8; and Martinez, pp. 45–55.

41. For the northern campaign and Brunete battle, see OACS-USAFE, *Fighter Operations*, pp. 46–47; MA Paris, "Major Military Operations: Lessons of the Spanish War," pp. 1–2, in B. Jones, *Spanish War*, vol. 3; Miksche, pp. 74–75 (Miksche was a Major in the Loyalist forces during the war); Proctor, pp. 118–154; Thomas, pp. 403–404, 461–466; Rosenstone, pp. 183–186; *Legion Condor*, pp. 67–74; Wintringham, p. 241; Eby, p. 139. There is a good account of Legion Condor ground-attack operations by 3/J/88 in Adolf Galland's largely autobiographical *Die Ersten und die Letzten*, pp. 46–54. (It should be noted that most English-language editions of the Galland work are abridged and do not include this section on his Spanish Civil War service.)

42. This discussion of Teruel is based on Thomas, pp. 504–516; Rosenstone, p. 231; Shores, *Spanish Civil War Air Forces*, p. 37; *Legion Condor*, pp. 64–74; Galland, pp. 31–33; Proctor, pp. 175–183; Larios, pp. 151–161; Extracts from report of visiting RAF Air Staff officers to Republican Spain, Feb. 1938, pp. 16–17, in B. Jones, *Spanish War*, vol. 2; there is an excellent eyewitness account by a military observer of the actions of Nationalist close air support operations in Reilly, pp. 515–518.

43. Joe Gordon and Leo Gordon, "Seven Letters from Spain," and Ben Iceland, "The Big Retreat," in Bessie and Prago, pp. 201 and 227. See also Proctor, p. 191–194; Miksche, p. 75; Rosenstone, pp. 276–296.

44. This account of the Ebro is based on Robert Goldston, *The Civil War in Spain* (New York: Fawcett Publications, 1969), pp. 178–182; Thomas, pp. 544–561; Tom Wintringham and J. N. Blashford-Snell, p. 163; Proctor, pp. 223–227; Miksche, pp. 75–79; Shores, *Spanish Civil War Air Forces*, p. 40; Eby, pp. 286–295; Larios, p. 223; D.A.N. (Alvah Bessie), "120 a Minute," in Bessie and Prago, p. 283; and Bessie, pp. 241, 255, 292–296.

45. Proctor, p. 227.

46. Miksche, pp. 77–78.

47. Larios, pp. 223–224.

48. Miksche, p. 73.

49. Wintringham and Blashford-Snell, p. 165.

50. Maurice Duval, *Les Espagnols et la Guerre d'Espagne* (Paris: Librairie Plon, 1939), p. 218. See also pp. 194–218, and his *Les Leçons de la Guerre d'Espagne*, pp. 153–154, 180–182, 229.

51. Cot, pp. 274, 284–285, 287, 302–305, 314, 324, 330–333, 335; Doughty, pp. 181–182; Duval, *Les Leçons*, p. 229. The bizarre governmental failings of France in the interwar years are brilliantly treated in Alistair Horne's *To Lose a Battle.*

52. Drum, pp. 200–201.

53. Ibid., pp. 192–195; Schulz, p. 2; OACS-USAFE, *Fighter Operations*, pp. 15, 46–57.

54. Drum, pp. 198–202; OACS-USAFE, *Fighter Operations*, pp. 17, 47.

55. Schulz, pp. 2–3; OACS-USAFE, *Fighter Operations*, pp. 47–48.

56. A. Bogdanow and M. Scherbakow, "Tactical Employment of Aviation in Offensive Situations," *Krasnaya Zwesda* (June 21, 1938), pp. 5–6 of an Army War College translation in B. Jones, *Spanish War*, vol. 2. Additionally, for a view of Soviet lessons from Spain, see the previously cited Mikhailow works, as well as the following works translated by the AWC and in B. Jones, *Spanish War*, vol. 2: G. Gagarin, "Aviation in Defensive Actions," *Krasnaya Zwesda* (June 1938); G. Gagarin, "Aviation in Modern Combat," *Krasnaya Zwesda* (April 26, 1938); and p. Mikhailow, "Tactical Employment of Attack Aviation," *Krasnaya Zwesda* (April 4, 1938). See also the following AWC translations in B. Jones, *Spanish War*, vol. 3: K. Sinyeokov, "The Role of Light-Bombing Aviation in Modern Combat as Judged from Experience of the War in Spain," *Voyennaya Mysl*, 7 (1938): 136–140; G. Sibirtsew, "Aviation in General Combat: Experience of the Spanish Civil War," *Krasnaya Zwesda* (August 24, 1938); and p. Mikhailow, "Republican Aviation on the Aragon Front," *Krasnaya Zwesda* (June 15, 1938). In addition, see Kipp, *Barbarossa*, pp. 3–13, 27.

57. Quoted in Kipp, *Barbarossa*, p. 27.

58. Banfill, p. 26.

59. B. Q. Jones, "Observations on Air Force Employment, Reports on Air Combat Operations in China," and attached excerpts from lecture by H. H. Arnold, in B. Jones, *Notes on the Chinese War.*

60. For example, see A. Alimow, "Distant Aerial Bombing Raids (Experience of the Chinese Conflict)," *Krasnaya Zwesda* (July 5, 1938), pp. 1–4 of an AWC translation in B. Jones, *Notes on the Chinese War.*

61. Takejiro Shiba, *Air Operations in the China Area, July 1937–August 1945*, Japanese Monograph No. 76 of the series *War in Asia and the Pacific, 1937–1949* (Tokyo: Headquarters, U.S. Army Forces Far East and Eighth U.S. Army (Rear), Dec. 10, 1956), p. 16. For additional information of the Japanese experience in China, see Monograph No. 70, *China Area Operations Record (July 1937–November 1941)*, Monograph 178, *North China Area Operations Record, July 1937–May 1941*, Monograph No. 179, *Central China Area Operations Record, 1937–1941*, and Monograph 180, *South China Area Operations*

Record (1937–1941), within the same series.

62. Ibid., pp. 16–17. Details on Japanese aircraft types can be found in Francillon.

63. A good summary history of the CAF is Hooton's "Air War Over China," pp. 7–24, 62–63. For an interesting perspective of the international pilots, see Caidin, *The Ragged, Rugged Warriors*, pp. 42–109; and William Labussiere, "The Truth About the 14th Squadron: Foreign Volunteers in China," *Journal of the American Aviation Historical Society* 32, no. 4 (Winter 1987): 284–293. As in Spain, these volunteers seem on the whole to have been garrulous, difficult to control, and overrated. See the following attaché reports: MA Tokyo, Report 9349 (April 30, 1938), and MA Tokyo, Report 9649 (March 23, 1938), in B. Jones, *Notes on the Chinese War*. For a scathing commentary on the practices of aircraft salesmen (from all nations) in China, see Lt. Cmdr. Ralph Ofstie, USN, "Aviation in the Sino-Japanese War," (Jan. 1938), p. 2, in B. Jones, *Notes on the Chinese War*. Ofstie was the former assistant naval attaché for air in Tokyo.

64. Hooton, pp. 10–11. The Shrike information is from Rust and Jefferies, pp. 10–11. There is an interesting look at Japanese air operations during the early phase of the China war in Shuhsi Hsu, pp. 1–67; despite the admittedly partisan nature of the book, its account of Japan's casual brutality in civilian bombardment is undoubtedly accurate and supported by postwar analysis including testimony at the War Crimes Trials held in Tokyo.

65. For an example of how Japan's inability to secure air superiority impacted her development of new aircraft, see Francillon, p. 380.

66. Hooton, pp. 12, 19. There is an excellent memoir of one Soviet airman's service in China; see Slyusarev, pp. 247–284. For the Soviet view of Chinese air operations, see the following AWC translations, all in B. Jones, *Notes on the Chinese War:* A. Alimow, "Bombardment Aviation in the Chinese Conflict," *Krasnaya Zwesda* (April 1, 1938); A. Sviridow, "Some Aspects of the Japanese Operations in China," *Krasnaya Zwesda* (April 24, 1938); and A. Sukhow, "Tactical Employment of Japanese Aviation," *Krasnaya Zwesda* (March 20, 1938). There is a good example of foreign reaction to the Chinese pamphlet raids in the report of the military attaché to China, Report 9663 (May 21, 1938), in B. Jones, *Notes on the Chinese War*.

67. "Lt. Col. Sibata," "Modern Types of Aircraft and Aerial Warfare," translated by the AWC from a Russian translation of a Japanese article that appeared in *Voyenny Zarubezhnik* and in B. Jones, *Notes on the Chinese War*, p. 5. For examples of attaché reports that make mention of air-ground cooperation and attack, see MA Tokyo, Report 9540 (Aug. 27, 1938); MA Tokyo, Report 9483 (July 21, 1938); MA Tokyo, Report 9441 (June 21, 1938); MA Tokyo, Report 9421 (June 9, 1938); MA Tokyo, Report 9377 (May 13, 1938); Evans F. Carlson, USMC, "Observations of the Sino-Japanese Armies During the Battle of Taierchwang (Shantung)" April 3–10, 1938; MA Tokyo, Report 9329 (April 12, 1938); MA Tokyo, Report 9309 (March 30, 1938); MA Tokyo, Report 9303 (March 25, 1938); MA Tokyo, Report 9285 (March 18, 1938); MA Tokyo, Report 9241 (Feb. 17, 1938); G-2 Report 9900, "Japan (Aviation-Military): Subject—Air Combat Operations; 'Operations of Japanese Army Air Corps During China Fighting' "; MID Summary (Nov. 26, 1937), p. 16525; MID Summary (Dec. 24, 1937), pp. 16580–16581; "Notes on Visit to Intelligence Section, CCAC, of Harry T. Rowland, 1st Lt., Air Reserve," (Jan. 6, 1938); MA Tokyo, Report 9050 (Oct. 29, 1937); MA Report 234 (Office of Chief of Coast Artillery, Aug. 27, 1937); and MA China, Report 9586 (Aug. 20, 1937); all in B. Jones, *Notes on the Chinese War*.

68. Hsu and Chang, p. 199.

69. MA Report 234, pp. 8–9.

70. See Sviridow, p. 2, and Sukhow, pp. 3–4, in B. Jones, *Notes on the Chinese War*.

71. Slyusarev, pp. 279–280; and the following reports from B. Jones, *Notes on the Chinese War:* G-2 Report 9900, p. 8; Rowland "Notes on Visit...", p. 5; MA China, Report

9586, p. 1. See also Hsu and Chang, pp. 266–268, 270–271; Shiba, pp. 49–69; Caidin, *The Ragged, Rugged Warriors*, pp. 62–64, 87–90. There is a good account of the Wuhan operation, including the effect of Japanese attacks upon Chinese troops, in the memoir of a leading Soviet advisor to the Chinese army; see Aleksandr Ya. Kalyagin, *Along Alien Roads*, a volume in the series *Occasional Papers of the East Asian Institute* (New York: Columbia University East Asian Institute, 1983), pp. 127–203, 233–235.

72. The following discussion of the Changkufeng/Khasan fighting is based on the only incisive examination of the conflict, Alvin D. Coox's *The Anatomy of a Small War*, particularly pp. 15, 177, 189–196, 205, 209, 230, 265, 361, and 371–374. Coox's work has an extensive—indeed exhaustive—bibliography on the "incident." There is an interesting study of Japanese tactics written after Changkufeng and prior to Nomonhan, prepared by then-Captain Maxwell D. Taylor of the Field Artillery. This is MA Tokyo, Report 9755 (April 1, 1939) entitled "Japan (Combat)—Subject: Tactical Doctrine of the Japanese Army" and it contains an incisive analysis of Japanese combat experience in China and Changkufeng. This study is in the MHI library. For another reference on Japan's China experience—but based primarily upon journalistic accounts—see Robert Foster Hewett, Jr.'s "The Military Aspects of the Sino-Japanese War, 1937–1938," an unpublished Master's Thesis from Stanford University, June 1953. This, too, is in the MHI library.

73. The best overall source on the fighting at Nomonhan is Coox's two-volume *Nomonhan: Japan Against Russia, 1939*, an exhaustive account which emphasizes the Japanese side of the conflict; I found pp. 397, 520, 683–689, 883, and 1069 especially useful. The best balanced tactical analysis of the Nomonhan fighting is Drea's *Nomonhan: Japanese-Soviet Tactical Combat, 1939*. A useful Soviet viewpoint is given in Georgi Zhukov's *Memoirs*, pp. 147–189.

74. Drea, pp. 69, 99.

75. Ibid., p. 34.

76. Zhukov, *Memoirs*, pp. 150, 169; Coox, *Nomonhan*, pp. 683–684; Liss, *The Polikarpov I-16*, p. 10.

77. Zhukov, *Memoirs*, pp. 152; Drea, pp. 72–75; Coox, *Nomonhan*, pp. 397, 520.

78. Zhukov, *Memoirs*, p. 160.

79. Ibid., p. 161.

80. Ibid., pp. 168–171, 184–189; Drea, pp. 76, 86; Coox, *Nomonhan*, pp. 683–689, 779–841, 881–883, 1069–1070.

Part Four

1. The discussion of German work in the CAS/BAI area is based on material in sources cited previously in Part Three, notably Schulz; OACS-USAFE, *Fighter Operations*; and Murray, *Luftwaffe*. In addition, I have used Baumbach; Pegg; and Deichmann's semi-memoir semi-history. See also the following, from the USAF Historical Research Center, Maxwell AFB, Ala. (Ref. no. K113.3019-3): Adolf Galland and Hubertus Hitschold, "Entwicklung der Schlachtflieger" (1 Feb. 1957); Hitschold, "Die Schlachtfliegerei in der deutschen Luftwaffe" (12 Feb. 1957); Bruno Meyer, "Einsatz von Schlachtfliegern" (24 March 1953); and Ernst Kupfer, "Schlachtflieger, Panzerjagd, und Störflugzeug im Jahre 1943" (10 Sept. 1943).

2. This discussion of Anglo-American air support is based on a variety of sources, including Greer; see also W. H. Phinizy, "Background Material for a Tactical Air Study," RAND Research Memorandum RM-669 (Santa Monica, Calif.: The Rand Corporation, Oct. 1950), copy in the MHI library; Jacobs, "Air Support for the British Army, 1939–1943," pp. 174–182; Observers Report, VIII Air Support Command, "Air Operations in Support of Ground Forces in North West Africa, 15 March–5 April 1943," (U.S. Army Air

Forces, VIII Air Support Command, 1943), copy in the MHI library; Headquarters, 9th U.S. Air Force, "Direct Air Support in the Libyan Desert," (U.S. Army Air Forces, Office of the Commanding General, 9th Air Force), copy in the MHI library; USAAF 42nd Bomb Wing, "42nd Bomb Wing in Air-Ground Operations," (1944), in Air Force Museum "L1 Italy 1943–1944" file, Wright-Patterson AFB, Ohio; USAAF 14th Air Force, "Air Support of Ground Troops, Salween River Area, Central China Area, July–Aug.–Sept. 1944," (USAAF: 24th Statistical Control Unit in Conjunction with the 14th AF, Nov. 20, 1944), copy in the MHI library; Jack P. Monroe, Jr., *The Central Pacific in World War II: The Fast Carrier Task Force and Close Air Support*, no. 6013 in the *Professional Study* series (Maxwell AFB, Ala.: Air War College, April 1976); 3rd Marine Division, "Standing Operating Procedure for the Employment of Close Support Aircraft," (Washington, D.C.: Intelligence Section, Division of Aviation, Headquarters USMC, Feb. 20, 1945) and John M. McJennett, "Report on Air Support in the Pacific," (HQ Marine Air Support Control Units, Amphibious Forces, Pacific Fleet, June 9, 1945), both in the files of the Marine Corps Historical Center, Washington, D.C.; AAF Evaluation Board in the ETO, *The Effectiveness of Third Phase Tactical Air Operations in the European Theater, 5 May 1944–8 May 1945* (hereafter cited as AAF, *Effectiveness*); Allied Forces, 21st Army Group, "Some Notes on the Use of Air Power in Support of Land Operations and Direct Air Support," (Holland, 21st Army Group, Dec. 1944), copy in the MHI library; Steadman; Mortensen; Slessor, *Central Blue;* Tedder; Sallagar; Kohn and Harahan, *Air Superiority* and *Air Interdiction;* Terraine; Shores, *2 TAF;* and Momyer.

3. For Soviet practice, see Kipp, *Barbarossa,* and Steadman. See also the following documents which are in the MHI library: Hans von Greiffenberg, "Tactics Employed by Russian Planes in Lowflying Attacks on Ground Targets," Campaign Information No. 8a, MS C-021 (European Command, Historical Division, Sept. 6, 1948); Soviet Army, "Instructions for Air Force and Ground Troops on Combined Operations," (1944), translation of Soviet document, Registry Number F-3733, I.D. Number 744308, Catalog Number DA-G2 TR F-3733; Hellmuth Reinhardt, "Project #42: *Ground Support;* Replies to Questions Concerning MS #B-791: Collaboration between Army and Luftwaffe Forces During Combat" (European Command: Historical Division, n.d.); Hoeffding; Parotkin et al.; Pahenyanik; and Ratley. Relevant Soviet sources are those by Kozhevnikov and Timokhovich. Finally, the best overall survey of the Soviet air war on the Eastern Front is Von Hardesty's *Red Phoenix,* which balances nicely the official Soviet history; the official Soviet history is available in a translation edited by Ray Wagner entitled *The Soviet Air Force in World War II.*

4. Mellenthin, p. 7.

5. See Banfill, p. 22.

6. Ibid., p. 21. For other accounts of German air operations in Poland, see Mellenthin, pp. 5–9; Murray, *Luftwaffe,* pp. 28–32; and Pegg, pp. 5–7. Two interesting German accounts heavily flavored with self-serving bombast are H. Eichelbaum's edited *Schlag auf Schlag: Die deutsche Luftwaffe in Polen,* vol. 1 of the *Adler Bucherei* series (Berlin: Wehrmacht Presse Verlag Reif & Co., 1939); and Josef Gabler, ed., *Mit Bomben und MGs über Polen* (Gutersloh: Verlag C. Bertelsmann, 1940). A more useful German view can be found in a collection of unit accounts assembled by the *Generalstab des Heeres* and entitled *Kampferlebnisse aus dem Feldzuge in Polen, 1939* (Berlin: E. S. Mittler & Sohn, 1941), esp. pp. 37–41, and 76–86. An excellent English-language summary is that of the Department of Military Art and Engineering, U.S. Military Academy, West Point, entitled *The Campaign in Poland, 1939* (West Point, N.Y.: USMA, 1945); copy in the MHI library.

7. German antiaircraft organization is from U.S. War Department, Military Intelligence Service, *German Antiaircraft Artillery,* Special Series No. 10, MIS 461 (Washington, D.C.: War Department, Feb. 8, 1943), pp. 2–13, 18–19; see also Ernst Neuman, *Handbuch für den Flakartilleristen (Der Kanonier) Waffen und Ausbildung der Flakbatterie—8,8*

cm Flak und 2 cm Flak (Berlin: Verlag Offene Worte, 1941), pp. 65–74, 104–107 (copy in the MHI library). The history of German antiaircraft forces is covered in Horst-Adalbert Koch's *Flak: Die Geschichte der Deutschen Flakartillerie, 1935–1945* (Bad Nauheim, FRG: Verlag Hans-Henning Podzun, 1954); for the Polish campaign, see pp. 35–37; for an overview of the operational environment in which German antiaircraft units operated in Poland, see V. Minden, "Antiaircraft Artillery in the Polish Campaign, Judged by the Experiences of an Antiaircraft *Abteilung*," a translation of an article that appeared in *Der Truppendienst*, May 24, 1940, and in the MHI library. Kreis, pp. 32–34.

8. See Cynk, *The P.Z.L. P-23 Karas*, p. 10.

9. Liss, *The P.Z.L. P-11*, p. 10.

10. Cynk, *P.Z.L. P.37 Los*, pp. 137–139.

11. Banfill, p. 24.

12. For example, see Minden, pp. 1–3, on inability of flak units to keep up with the advancing columns.

13. S. Karpinski et al. and The Committee of the Polish Air Force Memorial Book, *Destiny Can Wait: The Polish Air Force in the Second World War* (London: William Heinemann Ltd., 1949), pp. 13–14.

14. The best single history of the French campaign is Alistair Horne's *To Lose a Battle*; I have also drawn upon Mellenthin, pp. 14–30; Murray, *Luftwaffe*, pp. 38–43; and Guderian, pp. 89–98. Kirkland, pp. 101–117, gives an excellent analysis of the underlying causes of France's disastrous wartime military aviation performance.

15. Guderian, p. 98.

16. Ibid., p. 105.

17. This account of the Meuse crossing is based on Guderian, pp. 97–113; Mellenthin, pp. 15–21; and Horne, *To Lose a Battle*, pp. 269–318.

18. Guderian, p. 101.

19. Horne, *To Lose a Battle*, p. 289.

20. Bloch, pp. 54–55.

21. Horne captures these activities very well; see *To Lose a Battle*, pp. 293–304.

22. Richards, pp. 108–109. These formations were, of course, reinforced following the attack upon France. For a good view of the RAF's war in France see Terraine, pp. 118–165.

23. Salesse, p. 90; Richards, pp. 113–119; see also J. de L. Wooldridge, *Low Attack: The Story of Two Royal Air Force Squadrons from May, 1940, until May, 1943* (London: Sampson Low, Marston & Co., Ltd., 1944), pp. 1–4; Horne, *To Lose a Battle*, pp. 239, 247–248; Danel, p. 11; and Pacquier et al., p. 25. For an account of German flak forces in the French campaign, see Georg von Puttkamer, *Flakkorps "I" im Westen* (Berlin: Volk und Reich Verlag, 1943), particularly pp. 31–33, 37–38, and 155–157.

24. Salesse, pp. 95–98; Richards, pp. 120–121; Horne, *To Lose a Battle*, pp. 332–335; Pacquier et al., pp. 25–27; Wooldridge, p. 2.

25. Guderian, p. 100.

26. Ibid., p. 105.

27. Ibid., p. 113.

28. Kirkland, passim; Bloch, p. 40.

29. Kirkland, pp. 102–103; Cot, pp. 274–335; for the history of the French aircraft industry, see Emmanuel Chadeau, "Government, Industry, and the Nation: The Growth of Aeronautical Technology in France (1900–1950), *Aerospace Historian* 35, no. 1 (March 1988): 26–44. I also acknowledge information received in a conversation with M. Henri Ziegler of *Aerospatiale* at the honors and awards banquet, Society of Experimental Test Pilots, Beverly Hilton Hotel, Beverly Hills, Calif., Sept. 1985.

30. For example, see Kipp, *Barbarossa*, pp. 11–13; Jacobs, "Air Support for the British Army, 1939–1943," pp. 174–176; Huston, pp. 166–168; and Gabel, pp. 63–67.

31. See the Schulz memoir piece, pp. 6–30.

32. See OACS-USAFE, *Fighter Operations*, pp. 13, 16, and Kupfer, passim.
33. Cot, p. 274.
34. Jacobs, "Air Support for the British Army, 1939–1943," pp. 174–176; Gabel, passim; Huston, pp. 166–168.
35. Gabel, pp. 204–205, 310–314; Mortensen, pp. 18–19.
36. Dudgeon, pp. 500–503, 548–552, 604–608; quote is from p. 552 (Dudgeon was a flight instructor at the field). See also Terraine, pp. 339–340; Tedder, pp. 83–90; Richards, pp. 313–319; and Lee, *Special Duties*, pp. 114–116.
37. Philip Guedalla, *Middle East 1940–1942: A Study in Air Power* (London: Hodder and Stoughton Ltd., 1944), pp. 105–112; Terraine, pp. 321–323.
38. This memo is quoted in Tedder, p. 169, Terraine, p. 347, and, in slightly different form, in Jacobs, "Air Support for the British Army, 1939–1943," p. 180.
39. For details see Richards and Saunders, pp. 160–162; and Terraine, pp. 345–348.
40. Richards and Saunders, pp. 171–189; Tedder, pp. 191–207; Terraine, pp. 352–361; a good account of the leadership failures implicit in the Crusader operation can be found in Corelli Barnett's *The Desert Generals*, pp. 78–111.
41. Information on the Hurricane—one of the most underappreciated aircraft of the Second World War—can be found in Mason, pp. 66–95, 103–106.
42. The delay figure is from Terraine, p. 359.
43. Richards and Saunders, p. 187.
44. Liddell Hart, *Rommel Papers*, pp. 191–224; Barnett, *The Desert Generals*, pp. 125–134; David Irving, *The Trail of the Fox* (New York: Avon, 1978 ed.), pp. 201–214; Richards and Saunders, pp. 198–202; Lee, *Special Duties*, pp. 156–157.
45. Richards and Saunders, pp. 201, 236; for technical details on the Hurricane IID, see Mason, pp. 85–86, 107–108.
46. Tedder, p. 317; and Terraine, p. 376.
47. Liddell Hart, *Rommel Papers*, p. 245.
48. Ibid., p. 282. For Tedder's view of Auchinleck's dismissal, see Tedder, p. 314; for Montgomery's view of air power at Alan Halfa, see B. L. Montgomery, *El Alamein to the River Sangro* (London: Hutchinson & Co., n.d.), p. 9.
49. Liddell Hart, *Rommel Papers*, p. 283.
50. Ibid., pp. 283–284. Rommel's use of "heavy bomber" refers in reality to Allied twin-engine medium bomber operations. The Germans did not have equivalents to American or British four-engine strategic bombers, and thus considered their own twin-engine designs as "heavy" bombers.
51. Ibid., p. 284.
52. Ibid., pp. 285–286.
53. USAAF, Ninth Air Force, *Direct Air Support in the Libyan Desert, 19 October–8 November 1942* (9AF HQ, c. Nov. 1942), pp. 5–16, copy in the MHI library; E. N. Tucker and P. M. J. McGregor, *Per Noctem Per Diem: The Story of 24 Squadron, South African Air Force* (Cape Town: Cape Times Ltd. and the 24 Squadron Album Committee, 1962 ed.), pp. 70–77; Tedder, pp. 353–361; Terraine, pp. 383–388.
54. Quoted in Jacobs, "Air Support for the British Army, 1939–1943," p. 180.
55. B. L. Montgomery, "Some Notes on High Command in War" (Tripoli: 8th Army, Jan. 1943); for an example of a later edition, see "Some Notes on the Use of Air Power in Support of Land Operations and Direct Air Support" (Holland: 21 Army Group, Dec. 1944 ed.); copies in the MHI library. For background on this pamphlet, see Tedder, p. 397, and Jacobs, "Air Support for the British Army, 1939–1943," pp. 180–181.
56. U.S. War Department, Field Manual 31-35, "Aviation in Support of Ground Forces" (Washington, D.C.: GPO, April 9, 1942), passim; copy in the MHI library.
57. Sunderland, p. 10; USAAF HQ, *Army Air Forces in the War Against Japan, 1941–1942*, pp. 126–129; Zimmerman, pp. 65, 73, 81, 83, 89; Sherrod, pp. 82, 89–90; Miller,

p. 96; and Dial, pp. 6–7. John Howard McEniry, Jr.'s *A Marine Dive Bomber Pilot at Guadalcanal* (Tuscaloosa: University of Alabama Press, 1987) is an interesting memoir of this period.

58. Eichelberger, p. 57; USAAF HQ, *Army Air Forces in the War Against Japan*, p. 140; Craven and Cate, vol. 1, pp. 410, 425, 478; Sunderland, pp. 10–11; McCarthy, pp. 368, 562; Dexter, pp. 97, 429, 466, 511, 548, 551–552, 575, 585–586, 589, 599, 642, 649, 651–652, 682–683, 691, 693–694, 707–708, 711, 729, 766, 778; Rentz, pp. 83–85; 90; Sherrod, pp. 190–192. For details on the Whirraway and Boomerang, see Profile Publications Research Staff, *The Commonwealth Whirraway*, pp. 6–12; and Pentland, pp. 7–9.

59. Letter, DDE to Hartle, Jan. 15, 1943, in Alfred D. Chandler, Jr., ed., et al., *The Papers of Dwight David Eisenhower, The War Years, 2* (Baltimore: The Johns Hopkins Press, 1970), pp. 904–905.

60. Bradley and Blair, p. 128.

61. This discussion of Kasserine is based on Robert H. Ferrell, *The Eisenhower Diaries* (New York: W. W. Norton & Co., 1981), entry for Feb. 25, 1943, p. 90; Paul McDonald Robinett, *Armored Command: The Personal Story of a Commander of the 13th Armored Regiment, of CCB, 1st Armored Division, and of the Armored School During World War II* (Washington, D.C.: McGregor & Werner, Inc., in association with the United States Armor Association, 1958), pp. 160–194; Blumenson's *Kasserine Pass*, pp. 60–62, 85–86, 108, 110, 126–127, 141–142, 147, 160–162, 198, 204, 206, 217, 251, 266, 281–282; Howe, pp. 401–481; Mortensen, pp. 70–72; Craven and Cate, vol. 2, pp. 153–160; Liddell Hart, *Rommel Papers*, pp. 397–408; Bradley, pp. 23–25, 36–37, 41; Bradley and Blair, pp. 127–128; and Eisenhower, *Crusade in Europe*, pp. 142–147.

62. There is a fascinating collection of contradictory messages, reports, memos, transcripts, and excerpts from published accounts that relate to the air side of the Kasserine debacle; see James L. Cash, *Air Power in North Africa* (bound photo-duplicated compilation, USAF, Oct. 9, 1951), copy in the MHI library. See also Raff for some illuminating comments on air support in North Africa. For the official unit account of Kasserine operations, see USAAF XII Air Support Command, *Report on Operations Conducted by XII Air Support Command, USAAF, Tunisia, 13 Jan 43 to 9 Apr 43*, in the archives of the USAF Historical Research Center, Maxwell AFB, Alabama.

63. For Coningham's immediate actions in Kasserine, see Craven and Cate, vol. 2, p. 157; see also Tedder, pp. 397–398, and Eisenhower, *Crusade in Europe*, pp. 149–150. A good account of how the Coningham philosophy imparted itself to AAF operations virtually immediately can be found in USAAF, VIII Air Support Command, *Air Operations in Support of Ground Forces in North West Africa (15 March–5 April 1943) Observer's Report* (VIII Air Support Command HQ, c. April 1943), pp. 4–33, copy in the MHI library. Howe, pp. 492–495, offers the Army view of the reorganization. For an example of the anti-Coningham feeling among some American ground commanders, see Bradley and Blair, p. 143. For Coningham's own account of the evolution of Allied battlefield support, see his article, "The Development of Tactical Air Forces," pp. 211–227.

64. There is an excellent oral history interview of Lt. Gen. Elwood Quesada, USAF (ret.) by Steve Long and Ralph Stephenson, undertaken as part of the Army Military History Institute's Senior Officers' Debriefing Program; a transcript of this interview is in the MHI archives. Quesada's own interpretation on the significance of the North African experience can be found in his essay "Tactical Air Power," *Air University Quarterly Review* 1, no. 4 (Spring 1948): 37–45, esp. p. 40. See also his comments in Kohn and Harahan's *Air Superiority*, pp. 34–35; and John Schlight's "Elwood R. Quesada: Tac Air Comes of Age," in *Makers of the United States Air Force*, John L. Frisbee, ed., a *Project Warrior* study (Washington, D.C.: Office of Air Force History, 1987), pp. 177–204. For information on Cochran, see Lowell Thomas, *Back to Mandalay* (New York: The Greystone Press, 1951), pp. 32–41; USAF, "Unit History of the First Air Commando Force," a typescript ms. in the

"L2 Group 1st Air Commando" file, archives of the Air Force Museum, Wright-Patterson AFB, Ohio; W. J. Sciutti, "The First Air Commando Group: August 1943–May 1944," *Journal of the American Aviation Historical Society* 13, no. 3 (Fall 1968): 178–185; and Luigi Rossetto, "The First Air Commando," *Aerospace Historian* 29, no. 1 (March 1982): 2–12. See also Kohn and Harahan, *Air Superiority,* pp. 4–5, 29–36; Momyer, pp. 256–258.

65. U.S. War Department, Field Manual FM 100-20, passim; Mortensen, pp. 78–79; Greenfield, *Army Ground Forces,* pp. 46–49. See also Greenfield, *American Strategy in World War II,* pp. 101–112. Two very useful survey studies treat FM 100-20 and its subsequent impact: Huston, "Tactical Use of Air Power," pp. 166–191, and Jacobs, "Tactical Air Doctrine," pp. 35–49.

66. U.S. War Department, Field Manual FM 100-20, pp. 1, 10–11.

67. Greenfield, *Army Ground Forces,* p. 47.

68. U.S. War Department, Field Manual 1-5, "Employment of Aviation of the Army" (Washington, D.C.: GPO, Jan. 18, 1943), pp. 26–27, 42–44; U.S. War Department, Field Manual 31-35, "Air Ground Operations" (Washington, D.C.: GPO, Aug. 1946), passim.

69. Greenfield, *Army Ground Forces,* pp. 49–50.

70. See U.S. War Department, General Staff, *To Bizerte with the II Corps, 23 April 1943–13 May 1943* (Washington, D.C.: War Department Military Intelligence Division, Nov. 25, 1943), pp. 21, 36; Bradley and Blair, pp. 146–147; Tedder, pp. 410–411; Albert Kesselring, "Final Commentaries on the 'Campaign in North-Africa, 1941–1945,'" Manuscript #C-075 (Werl, Germany: Historical Division, European Command, Dec.1949), pp. 53–64, copy in the MHI library.

71. Quesada interview, p. 20. There is a remarkable compilation of pro-air support statements contained within a report by Omar Bradley and a special study group for the postwar United States Strategic Bombing Survey: Bradley and Air Effects Committee, pp. 198–221.

72. Wagner, *American Combat Planes,* pp. 84–85, and Ernie Pyle's essay "Dive Bombers," in *Brave Men* (New York: Henry Holt, 1944), pp. 108–124.

73. Bradley, pp. 150–152; see also Garland and Smyth, pp. 342, 403.

74. Garland and Smyth, pp. 421–422; Craven and Cate, vol. 2, pp. 486–487. For a leading German's reaction to Allied operations in Sicily and the constraints imposed upon the Axis by Allied air operations, see Albert Kesselring, *Kesselring: A Soldier's Record* (New York: William Morrow, 1954), pp. 189–200.

75. Quote is from a written statement by Hitschold during interrogation at Latimer House, England, Oct. 24, 1945, reprinted in volume 2 (appendix), pt. 2, of the previously cited OACS-USAFE, *Fighter Operations.* See also volume 1 of this study (pp. 56–57) for details on the checkered career of S.G. 4. For additional information on ground attack units in the Italian fighting, see Pegg, pp. 20–23. Hitschold's career is treated in Brutting, pp. 131–135.

76. Terraine, pp. 595–596; Craven and Cate, vol. 2, p. 545; Colgan, p. 145.

77. USAAF XII TAC, *XII TAC Tactical Operations,* pp. 11–12, 50–53,

78. Slessor quote from memo reprinted in Terraine, p. 587.

79. USAAF 42nd Bomb Wing, "42nd Bomb Wing and the Anzio Beachhead," in *Air-Ground Cooperation,* p. 3.

80. Ibid., p. 4.

81. USAAF XII TAC, *XII TAC Tactical Operations,* pp. 53–54.

82. Ibid., p. 4.

83. Sallagar, p. xiii. For a view differing with Sallagar on the degree of supply denial and tactical paralysis, see Mark, pp. 176–184; Senger und Etterlin's recollection is from Senger und Etterlin, *Neither Fear nor Hope* (New York: Dutton, 1964), p. 224. See also Bingham's "Ground Maneuver and Air Interdiction." For a comparison with Korean *Strangle* opera-

tions, see Futrell, pp. 324–325, 441–442, 451, 471–473; and Hallion, *Naval Air War in Korea*, pp. 102–103.

84. USAAF, 42nd Bomb Wing, "42nd Wing and Cassino Town: Preparation, Training, and Execution," in *Air-Ground Cooperation*, p. 5; Craven and Cate, vol. 3, pp. 367–370, 373–384.

85. Quotes are from diary extracts reprinted in Sallagar, pp. 67–69; the original diary is in the collections of the U.S. National Archives and Records Service, Washington, D.C., Microfilm T-312, Roll 91. For additional information on how the German forces viewed the Allied air campaign in Italy, see *German Version of the History of the Italian Campaign*, a bound undated photostated manuscript in the MHI library, pp. 28, 52–53, 95, 101, 105–106, 108, 110, and 182a; this document is cataloged as D763 I 8G47.

86. Eaker quote is from a pamphlet issued by the Mediterranean Allied Air Forces (MAAF) entitled "Air Power in the Mediterranean" and issued in February 1945, quoted in Slessor, *Central Blue*, p. 566.

87. Eisenhower, *Supreme Commander's Dispatch*, p. i.

88. Again, the best discussion of the decline of the Luftwaffe is to be found in Murray, *Luftwaffe*, pp. 143–242.

89. Ibid., pp. 223–232; Josef Priller, *Geschichte eines Jagdgeschwaders: das J.G. 26 (Schlageter) von 1937 bis 1945* (Heidelberg: Kurt Vowinckel Verlag, 1956), pp. 242–243; Craven and Cate, vol. 3, pp. 30–66.

90. Craven and Cate, vol. 3, p. 171.

91. Ruge, pp. 15, 26, 38, 64, 66, 143, 152, 159, 167, and 172.

92. AAF, *Effectiveness*, pp. 271–282; Craven and Cate, vol. 3, pp. 107–137, 203, 243–244.

93. AAF, *Effectiveness*, p. 271.

94. Terraine, pp. 619; see also pp. 607–619.

95. Ibid., pp. 632–634, 636–637; Saunders, *Royal Air Force 1939–1945*, pp. 412–415; Shores, *2 TAF*, pp. 9–10.

96. Eisenhower, *Supreme Commander's Dispatch*, pp. 9–10; for general information on the attacks prior to D-day, see Craven and Cate, vol. 3, pp. 138–181, and Saunders, *Royal Air Force 1939–1945*, pp. 79–102.

97. Eisenhower, *Supreme Commander's Dispatch*, pp. 21–23; Craven and Cate, vol. 3, pp. 190–193.

98. AAF, *Effectiveness*, pp. 282–293, 370–373.

99. Quesada interview, pp. 35–37; Quesada's account differs from Bradley's (see *A Soldier's Story*, pp. 337–338), but Quesada is insistent that his version is the correct one, and since it is an "air" matter, I have deferred to his judgment.

100. Bradley, p. 250.

101. Quesada interview, pp. 35–37.

102. U.S. 12th Army Group, *Battle Experiences*, no. 38 (Sept. 3, 1944), pp. 1–2.

103. Bradley and Air Effects Committee, pp. 41–42. See also Collins, pp. 248–249.

104. Quoted in Greenhous, p. 92.

105. Quoted in Brian Holden Reid, *J. F. C. Fuller: Military Thinker* (New York: St. Martin's Press, 1987), p. 173.

106. AAF, *Effectiveness*, pp. 305–321; three useful pilot memoirs on P-47 and Typhoon operations are those by Colgan, Golley, and Demoulin.

107. Quesada interview.

108. *AAF, Effectiveness*, pp. 258–259; Quesada interview.

109. Hastings, *Das Reich*, p. 218. Hastings is curiously ambivalent toward air power at Normandy, particularly in his subsequent work *Overlord*, pp. 266–276, in which he refers to the "historical cliché" and the "half-truth" that air power was "decisive in making victory possible" (p. 266). While his critical comments focus (and rightly so) primarily on

strategic air power, he is, I feel, altogether too critical of Coningham, and simplistic in his treatment of the Coningham-Montgomery relationship. While it may indeed be a "cliché" to state that air power was decisive in the Normandy victory, the undoubted "whole" truth of the statement is beyond question, and borne out not merely by statements from the "air" community, but from the "land" community as well.

110. Saunders, *Royal Air Force 1939–1945,* pp. 118–121; Bradley and Air Effects Committee, p. 45; AAF, *Effectiveness,* pp. 384–385.

111. Gen. Eberbach, "Report on the Fighting of *Panzergruppe West* (Fifth *Panzer* Army) from 3 Jul–9 Aug 44," Manuscript B-840 (Bamberg, Germany: Historical Division, Headquarters, United States Army, Europe, 1 Jun 1948), p. 3, copy in the MHI library. See also Saunders, *Royal Air Force 1939–1945,* p. 122; Murray, *Luftwaffe,* p. 267; and Ellis, p. 258.

112. Ruge, p. 181.

113. Liddell Hart, *Rommel Papers,* p. 491.

114. Ruge, p. 186.

115. Ibid., p. 187.

116. Ibid., p. 209.

117. Ibid., p. 218.

118. Ibid., p. 227.

119. Ibid., p. 229; Saunders, *Royal Air Force 1939–1945,* p. 121.

120. Bradley, pp. 341–342.

121. Leigh-Mallory, p. 263; this is a reprint of Leigh-Mallory's report to Eisenhower which he submitted shortly before his death in an aircraft accident in November 1944.

122. "Report of Rommel and von Kluge [to Hitler]," an annex to a translation of a captured German document entitled "German Air Force Operations against Great Britain," reprinted in *Royal Air Force Quarterly* 19, no. 1 (January 1948): 60–61.

123. Bradley, p. 249. The air strikes on Caen are well covered in Alexander McKee's *Last Round against Rommel: The Battle of the Normandy Beachhead* (New York: The New American Library, 1964 ed.), pp. 190–203, 226–241.

124. For an overall survey of Cobra see Blumenson, *Breakout and Pursuit,* pp. 228–246; see also AAF, *Effectiveness,* pp. 85–94; and Sullivan, pp. 97–110. For an interesting account of the far more deadly Bomber Command operation against Falaise, see Brian Horrocks with Eversley Belfield and H. Essame, *Corps Commander* (New York: Charles Scribner's Sons, 1977), pp. 44–45.

125. Bradley and Blair, pp. 279.

126. Eisenhower, *Supreme Commander's Dispatch,* pp. 38–39.

127. Sullivan, pp. 107–108.

128. Bradley, p. 349.

129. Collins, p. 241.

130. Ibid., pp. 242–249; Blumenson, *Breakout and Pursuit,* pp. 247–335; Bradley, pp. 350–370; Eisenhower, *Supreme Commander's Dispatch,* pp. 39–42.

131. Bayerlein, pp. 1–4.

132. Blumenson, *Breakout and Pursuit,* p. 329.

133. Eisenhower, *Supreme Commander's Dispatch,* pp. 43–51 has an excellent summary of operations.

134. See, for example, F. W. Winterbotham, *The Ultra Secret* (New York: Dell Publishing Co., 1975 ed.), pp. 218–219; and Murray, *Luftwaffe,* p. 273.

135. Bradley and Blair, pp. 291–293; Bennett, pp. 115; see also pp. 111–115. The best account of what Ultra meant to air warfare can be found in Diane T. Putney's *Ultra and the Army Air Forces in World War II,* a volume in the *USAF Warrior Studies* series (Washington, D.C.: Office of Air Force History, 1987), which is an extensively annotated interview with U.S. Supreme Court Justice Lewis F. Powell, Jr., who served as the Ultra

officer on General Carl Spaatz's Strategic Air Forces staff. An extensive collection of Ultra reports and studies is maintained in the MHI library.

136. Golley, p. 136; see also pp. 122–144. For the actions of commanders and an overall view of the battle, see Bradley and Blair, pp. 292–295; Hans Speidel, *Invasion 1944* (New York: Paperback Library, Inc., 1968 ed.), p. 123; Blumenson, *Breakout and Pursuit,* pp. 457–505; Eisenhower, *Supreme Commander's Dispatch,* pp. 43–44; Collins, pp. 249–255; Terraine, p. 660; and Saunders, *Royal Air Force 1939–1945,* pp. 132–133.

137. Eisenhower, *Supreme Commander's Dispatch,* pp. 43–44.

138. Gersdorff, "The Argentan-Falaise Pocket," p. 1. See also Gersdorff, "Northern France," pp. 12–20, and Pickert, pp. 24–26.

139. Bradley, p. 377.

140. Saunders, *Royal Air Force 1939–1945,* p. 136.

141. Johnson, pp. 235–236.

142. Golley, p. 150.

143. Demoulin, p. 133.

144. AAF, *Effectiveness,* p. 121.

145. Ibid.

146. Ibid. See also Charles R. Johnson, *The History of the Hell-Hawks* (Anaheim, Calif.: South Coast Typesetting, 1975), pp. 189–202, a privately printed daily record of the 365th Fighter-Bomber Group (copy in the library of Air Force Systems Command, Andrews AFB, Md.).

147. Johnson, p. 237.

148. Gersdorff, "Northern France," pp. 33–34; Gersdorff, "The Argentan-Falaise Pocket," pp. 5–6; see also Hastings, *Overlord,* pp. 307–308, and John T. Bookman and Stephen T. Powers, *The March to Victory: A Guide to World War II Battles and Battlefields from London to the Rhine* (New York: Harper & Row, 1986), pp. 145–146.

149. Hastings, *Overlord,* p. 313; see note 109] where his figures have differed from Gersdorff's, I have accepted Gersdorff's.

150. Eisenhower, *Crusade in Europe,* p. 279. See also Ronnie L. Brownlee and William J. Mullen III, *Changing an Army: An Oral History of General William E. DePuy, USA Retired* (Carlisle Barracks, Pa.: U.S. Military History Institute, n.d.), p. 49.

151. U.S. 12th Army Group, *Battle Experiences,* no. 45 (Sept. 16, 1944), last page.

152. Bradley and Air Effects Committee, p. 201.

153. Ibid., pp. 26–45, 201–205, 212–215.

154. AAF, *Effectiveness,* p. 89.

155. Pickert, pp. 24–27.

156. Quesada interview; AAF *Effectiveness,* pp, 320–321.

157. AAF, *Effectiveness,* p. 321.

158. Letter, Lt. Gen. O. M. Bradley to Gen. H. H. Arnold, "Correspondence with Major Historical Figures, 1936–1960" box, "Bradley, World War II Correspondence, 1942–45" file, Omar N. Bradley Papers, archives of the MHI.

159. John S. D. Eisenhower, *Strictly Personal* (Garden City, N.Y.: Doubleday, 1974), p. 72. Two additional studies that discuss the combined-arms nature of Normandy operations that are particularly useful are: Wolfert's "From Acts to Cobra: Evolution of Close Air Support Doctrine in World War Two," and Michael D. Doubler's "Busting the *Bocage:* American Combined Arms Operations in France, 6 June–31 July 1944" (West Point, N.Y.: USMA Dept. of History, 1985), soon to be published by the Combat Studies Institute of the U.S. Army Command and General Staff College, Ft. Leavenworth, Kansas.

160. There is profusion of books on the Eastern Front war. My personal favorites are Alan Clark's *Barbarossa: The Russian-German Conflict, 1941–45* and Earl F. Ziemke's *Stalingrad to Berlin: The German Defeat in the East.* Two others deserving mention are

Alexander Werth's *Russia at War: 1941–1945* and Albert Seaton's *The Russo-German War, 1941–45.*

161. Comments by Paul-Werner Hozzel in "Conversations with a Stuka Pilot," p. 69.

162. Horst Fuchs Richardson with Dennis E. Showalter, eds., *Sieg Heil! War Letters of Tank Gunner Karl Fuchs, 1937–1941* (Hamden, Conn.: Archon Books, 1987), p. 119. See also pp. 157–158.

163. Hans Dessloch, "The Winter Battle of Rzhev, Vyasma, and Yuknov, 1941–42," Manuscript D-137 (n.p.: Historical Division, Headquarters, United States Army, Europe, May 1947), p. 5, copy in the MHI library.

164. From a captured Soviet document reprinted in Dessloch, p. 2.

165. I have drawn on material from Hardesty, pp. 11–22 and 113; Steadman, pp. 11–12; and Kozhevnikov, pp. 82–103. I have also benefited from consulting Timokhovich, pp. 64–77; I wish to thank Bill Stacey of the FTD History Office for making this document available to me. For organizational structure of Soviet air units, I have relied on U.S. War Department, Technical Manual TM 30-430, sections XI-6 to XI-10.

166. Hardesty, pp. 83–88; see also Kipp, "Soviet 'Tactical' Aviation," pp. 10–12.

167. Hardesty, pp. 83–88; Kipp, "Soviet 'Tactical' Aviation," pp. 10–12; see also Steadman, pp. 11–12, and U.S. War Department, Technical Manual TM 30-430, p. XI-9.

168. See A. S. Yakovlev, *Fifty Years of Soviet Aircraft Construction,* NASA TTF 627 (Jerusalem: Israel Program for Scientific Translation, 1970), pp. 48–49 (an interesting if self-serving memoir); the best account of Lend-Lease support to the Soviet air forces is Richard C. Lukas, *Eagles East: The Army Air Forces and the Soviet Union: 1941–1945* (Tallahassee: Florida State University Press, 1970); for Soviet views of American aircraft, see pp. 230–232.

169. The following discussion is based largely upon technical information in the following works: Liss, *The Yak 9 Series,* pp. 3–7; Liss, *The Lavochkin La 5 & 7,* pp. 3–9; Liss, *The Ilyushin Il-2,* pp. 3–11; and Passingham and Klepacki, *Petlyakov Pe-2 and Variants.*

170. Ilyushin anecdote is from Richard E. Stockwell, *Soviet Air Power* (New York: Pageant Press, Inc., 1956), p. 8.

171. In addition to Liss, *The Ilyushin Il-2,* see U.S. War Department, Technical Manual TM 30-430, pp. XI-41 to XI-45; and Greenwood, p. 86.

172. Greenwood, p. 97.

173. See OACS-USAFE, *Fighter Operations,* pp. 62–78.

174. Murray, *Luftwaffe,* p. 91; Brutting, p. 140.

175. See Green, pp. 390–392; Smith, *The Henschel Hs 129,* pp. 3–6; and OACS-USAFE, *Fighter Operations,* p. 55.

176. Rudel accident figure from Hozzel conversation at NWC. Rudel's own experiences in flying the cannon-armed Ju 87 are presented in his book *Stuka Pilot,* passim, and there is a good summation of his career in Brutting, pp. 67–97. See Pegg, p. 15. For Ju 87G technical details, see Green, p. 443. See also OACS-USAFE, *Fighter Operations,* pp. 17, 54. For the increasing vulnerability of the Ju 87, see Kupfer.

177. OACS-USAFE, *Fighter Operations,* pp. 52, 55; Green, pp. 208–211; the single best account of the development and employment of the ground-attack "190" can be found in Spenser's *Focke Wulf Fw 190.*

178. Raus et al., p. 171.

179. Kozhevnikov, p. 127.

180. Ibid.; see also Rudenko, p. 194.

181. Raus et al., p. 173.

182. Kozhevnikov, pp. 127–128; see also U.S. War Department, Technical Manual TM 30-430, p. XI-127; and "Soviet RUS-Type Radar Equipment," *Air Intelligence Digest* 4, no. 1 (January 1951): pp. 15–17, copy in the files of the Office of Air Force History, Bolling AFB,

Washington, D.C.

183. This information is from a translated Soviet document on air support catalogued as DA-G2 TR F-3733, "Russia. Army. Instructions for air force and ground troops on combined operations. (1944)" [n.d.], passim, in the MHI library.

184. Ibid., p. 3.

185. Ibid., pp. 9–10.

186. This discussion of tactics is based on information in U.S. War Department, Technical Manual TM 30-430, pp. XI-41–XI-47; Hardesty, pp. 172–173; and Passingham and Klepacki, pp. 123–126.

187. The following discussion of the battle of Kursk is based on the following sources: S. M. Shtemenko, *The Soviet General Staff at War, 1941–1945* (Moscow: Progress Publishers, 1970), pp. 159–173; Ziemke, pp. 118–142; Kozhevnikov, pp. 125–144; Parotkin et al., passim; Clark; pp. 277–366; Zhukov, *Memoirs*, pp. 427–466; Mellenthin, pp. 258–283; Jukes, passim; Busse et al., passim; Werth, pp. 679–685; Wagner, *The Soviet Air Force in World War II*, pp. 164–187; Hardesty, pp. 149–179 offers the single best Western analysis of the role of the Soviet army air force in the battle. See also Albert Seaton, *Stalin as Military Commander* (New York: Praeger Publishers, 1976), pp. 178–187; Alexander Boyd, *The Soviet Air Force since 1918* (New York: Stein and Day, 1977), pp. 170–178; Erich von Manstein, *Lost Victories* (Chicago: Henry Regnery Co., 1958), pp. 443–449; "Paul Carell" [Paul Karl Schmidt], *Scorched Earth: The Russian-German War, 1943–1944* (Boston: Little, Brown and Company, 1970), pp. 19–95 (this work is good for the atmosphere of the battle but otherwise must be used with caution); Plocher, pp. 66–105; Seaton, *The Russo-German War, 1941–45*, pp. 353–368; Guderian, pp. 302–312; Caidin, *The Tigers are Burning*, passim; and Georgi Zhukov, *Marshal Zhukov's Greatest Battles*, pp. 203–257.

188. Ziemke, p. 128.

189. Guderian, pp. 308–309.

190. Busse et al., p. 3.

191. Plocher, pp. 18–22, 56; Pegg, p. 35.

192. Plocher, pp. 39–41; Rudel, pp. 81–93.

193. Plocher, p. 41. Weaknesses of Hs 129 are from Pegg, p. 19.

194. The determination of these figures is difficult because of disagreement between records, confusion in the size of various units, and possible inflation for various reasons by all parties. I have accepted Von Hardesty's figures from *Red Phoenix* as being as definitive as we are likely to see on the VVS. Further, I wish to acknowledge the contribution that Von Hardesty made to my understanding and comprehension of the Eastern Front war. For German figures, I have relied on a mix of information from Plocher, pp. 76–83, and Busse et al., pp. 149–150, 163–165, plus my own computation based on Luftwaffe force structure practices.

195. Plocher, p. 81; Busse et al., p. 195; Hardesty, p. 153; Greenwood, p. 99.

196. Ziemke, pp. 133–135; Caidin, *The Tigers are Burning*, p. 103.

197. Hans von Greiffenberg et al., "The Battle of Moscow, 1941–42," manuscript T-28 *in* Detwiler, Burdick, and Rohwer, p. 210.

198. Kozhevnikov, pp. 115–116.

199. Ibid., p. 144.

200. Wagner, *The Soviet Air Force in World War II*, p. 169.

201. Zhukov, *Memoirs*, pp. 455–456.

202. Ibid.

203. Ziemke, p. 135.

204. Hozzel conversation.

205. Plocher, p. 81.

206. Ibid., p. 95.

207. Meyer's attack appears in many different sources. The best overall study is the paper by Capt. Lonnie O. Ratley, III, USAF, passim. See also Plocher, pp. 95–96, and Mellenthin, p. 273.

208. Hardesty, p. 164; see also p. 163. In addition, see Rudenko, p. 193, and Wagner, *The Soviet Air Force in World War II*, p. 172.

209. Zhukov, *Memoirs*, p. 456.

210. Rudenko, p. 191.

211. Ibid., p. 192; Wagner, *The Soviet Air Force in World War II*, p. 170; Greenwood, p. 100.

212. Busse et al., p. 176

213. Mellenthin, p. 267; his conclusion that Germany had air superiority is not borne out from other sources. See also Hardesty, p. 169.

214. Plocher, p. 87.

215. I have drawn on the Busse et al., Ziemke, Plocher, Clark, and Manstein accounts.

216. Rudenko, pp. 194–195. See also Kozhevnikov, pp. 152–179. For insights into the subsequent Russian air campaign against Germany, see Hardesty, passim; Greenwood, passim; Raus et al., pp. 172–176; Wagner, *The Soviet Air Force in World War II*, passim; and Hoeffding, passim. There is a very interesting account of Soviet close support activities in late 1944 in Pahenyanik.

Epilogue

1. See, for example, Futrell, pp. 78–83, 106–107, 704–705; and Hallion, *Naval Air War in Korea*, pp. 41–54; see also Gerald J. Higgins, *Air Support in the Korean Campaign*, a bound letter from the chief of the Army Air Support Center to the Chief of Army Field Forces, Dec. 1, 1950, and William A. Gunn, *A Study of the Effectiveness of Air Support Operations in Korea*, FEC Project No. 2 (Tokyo: Operations Research Office, General Headquarters, Far East Command, Jan. 1952 ed.), copies of both in the MHI library; U.S. Congress, Senate, Armed Services Committee, Subcommittee on the Air Force, *Study of Airpower*, pt. 7 (Washington, D.C.: GPO, 1956); and U.S. Congress, Senate, Armed Services Committee, Subcommittee on the Air Force, *Airpower* (Washington, D.C.: GPO, 1957).

2. See Bernard Fall's two excellent books, *Hell in a Very Small Place* and *Street without Joy*; see also volume 2 of the *Lessons from the Indo-China War*, particularly Part 4, a document prepared by the French Supreme Command, Far East, in 1955; an English translation of this work is available at the MHI library, with the call number DS 550 E5513, v.2. G. J. M. Chaussin, "Lessons of the War in Indochina," *Interavia* 7 (1952): 670–75. Finally, see "Lt. Colonel de Fouquieres," "Aviation Militaire Française: l'Armée de l'Air en Indochine de 1945 à 1954," *Forces Aeriennes Françaises* (July 1954): 142–150. Yolande Simon of the RAND Corporation is currently completing a study of French air operations in Indochina, using French archival sources.

3. See, for example, Peterson, Reinhardt, and Conger; document ARMA F8 R-627-59, "France. Army. General Staff. Helicopter Operations in Algeria," in the MHI library; Aerospace Studies Institute, *Guerrilla Warfare and Airpower in Algeria*; and Calmel, pp. 924–927.

4. There is an extensive body of literature on air support operations in Southeast Asia. The following, then, should be simply considered as introductory works to what is a much broader topic. For particular information see Kenneth D. Mertel, "Tactical Airpower in Support of Airmobile Operations in Vietnam," student essay, U.S. Army War College, Jan. 17, 1968; Maurice E. Seaver, Jr., "Tactical Air: The Vietnam Verification," Professional Study 4445, Air War College, April 1971; Moyers S. Shore II, *The Battle for Khe Sanh*

(Washington, D.C. Historical Branch, G-3 Division, HQ USMC, 1969); U.S. Congress, House, Committee on Armed Services, *Close Air Support: Report of the Special Subcommittee on Tactical Air Support* (Washington, D.C.: GPO, 1966); U.S. Congress, House, Committee on Armed Services, *Close Air Support: Hearing Before the Special Subcommittee on Tactical Air Support* (Washington, D.C.: GPO, 1966); Donald J. Mrozek, *Air Power and the Ground War in Vietnam: Ideas and Actions* (Maxwell AFB, Ala.: Air University Press, Jan. 1988); Momyer; Baxter et al.; Ballard; Nalty; Thayer; BDM Corporation, *Operational Analysis;* and Tolson. See in particular John Schlight's excellent *The War in South Vietnam,* which has a thorough discussion of the problems and experience of air-to-ground operations during a critical period of the Vietnam war.

5. For information on some of these other conflicts of the 1940s to 1960s, see Aerospace Studies Institute, *The Employment of Airpower in the Greek Guerrilla War* and *The Accomplishments of Airpower in the Malayan Emergency (1948–1960);* Waters, pp. 96–100; Wagoner; and Mets. Bruce Hoffman of the RAND research staff is currently completing a study on the evolution of British air power in "peripheral conflict."

6. See, for example, R. Schamberg, *Tactical Aircraft for Limited War(U),* Memorandum RM-3534-PR (Santa Monica, Calif.: The RAND Corporation, March 1963); see also George M. Watson, Jr., *The A-10 Close Air Support Aircraft: From Development to Production, 1970–1976* (Andrews AFB, Md.: History Office, HQ Air Force Systems Command, n.d.); there is an interesting collection of material on this issue in the form of periodic reports of the Air Force Scientific Advisory Board, particularly with regard to the genesis of the A-10. This collection is in the office of the Scientific Advisory Board, The Pentagon, Arlington, Va.

7. See the previously cited Tolson, *Airmobility;* see also Everett W. Duvall, "The Use of the Helicopter in a Combat Role," Student Individual Study, U.S. Army War College, March 15, 1954; Orrin A. Tracy, "The Role of the Helicopter in Modern Warfare," Student Individual Study, U.S. Army War College, March 15, 1955; Frederick E. Charron, "Does the Army Need an Organic Air-to-Ground Support Capability?" Special Study 306-62, U.S. Air Force Air Command and Staff College, Air University, May 7, 1962; Frank T. Taddonio, "What Can We Learn From a War We Lost? The Relevancy of the Vietnam Experience for Today's Assault Helicopter Doctrine," Study 86-2190, School of Advanced Military Studies, U.S. Army Command and General Staff College, Dec. 2, 1985; U.S. Army Combat Developments Command, *Aerial Fire Support Analysis(U),* ACN 16150, Coordination draft, 27 April 1970 (copy in MHI library); U.S. Army, Office of the Adjutant General, "Special Operational Report—Lessons Learned on the AH-1G Employment, Hqs., 307th Combat Aviation Battalion," May 24, 1968, filed as USARV 307-CABN LL AH-1G, MHI library; U.S. Army, Field Manual FM 1-110, "Armed Helicopter Employment" (Washington, D.C.: HQ USA, July 1966); Stanton, pp. 74–86. For contemporary joint Army–Air Force work, see the periodic *Air Land Bulletin* issued by the Air Land Forces Application (ALFA) Agency, HQ USAF Tactical Air Command and HQ Army TRADOC, Ft. Monroe, Va.; and Davis. For a good overall survey of attack helicopter development, see Wheeler, passim.

8. See, for example, Mann, *The 1972 Invasion of Military Region I.*

9. See Mets; Dean; Gaines and Potgieter, pp. 22–30; Aviation Week & Space Technology editorial team, *Both Sides of the Suez;* Jeffrey Ethell and Alfred Price, *Air War South Atlantic* (New York: Macmillan Publishing Company, 1983); I wish to acknowledge a student paper on the Afghanistan air war prepared by Lt. Col. James Gallivan as part of my course on battlefield air power. There is an interesting account of the American role in supplying Stinger missiles to Afghanistan—and their subsequent use—in Fred Barnes's "Victory in Afghanistan: The Inside Story," *Reader's Digest* (December 1988): 87–93. Soviet military press accounts have been unusually frank about the success of the Stingers, particularly against "Crows": the Su 25 Frogfoot. For the longstanding Iran-Iraq

war, see Ronald E. Berquist, *The Role of Airpower in the Iran-Iraq War;* CAS was used only in "extremely dire situations" (p. 59), with BAI being far more typical. See also Segal.

10. See Kenneth P. Werrell, *The Evolution of the Cruise Missile* (Maxwell AFB, Ala.: Air University Press, Sept. 1985).

11. See McPeak, pp. 68–69.

12. Ibid., p. 70.

13. Ibid.

14. There is an exhaustive body of literature pertaining to the present and future of air-land warfare. Some useful perspectives may be found in the following: Office of the Chief of Public Affairs, *AirLand Battle* (Ft. Monroe, Va.: HQ TRADOC, n.d.), copy in the library of the Army War College catalogued as U260 .A3; and Seymour J. Deitchman, *Military Power and the Advance of Technology: General Purpose Military Forces for the 1980s and Beyond* (Boulder, Colo.: Westview Press, 1983). Lt. Col. Price T. Bingham of the Air Force Center for Aerospace Doctrine, Research, and Education (CADRE) has been a prolific writer of works dealing with future battlefield air support; see his "Dedicated, Fixed-Wing Close Air Support—a Bad Idea," "Ground Maneuver and Air Interdiction in the Operational Art," pp. 16–31, and "Air Power and the Defeat of a Warsaw Pact Offensive: Taking a Different Approach to Air Interdiction in NATO." I am grateful for his sending me two unpublished papers as well: "Air Power and the Close-In Battle: The Need for Doctrinal Change," and "Vertical Landing Aircraft and Operational Art." See also Benjamin S. Lambeth, "The Outlook for Tactical Airpower in the Decade Ahead," Study P-7260 (Santa Monica, Calif.: The RAND Corporation, September 1986); James H. Simms, "The AirLand Battle: Doctrine and Policy Failures," Student Research Study, National War College, February 1986; Alton C. Whitley, Jr., "Doctrinal Challenges for Tacair in AirLand Battle," Student Research Study, National War College, February 1986; Michael S. Snyder, "Rehearsal and the Attack Helicopter Battalion," and Philip J. Dermer, "JAAT Eagle 87," both in *U.S. Army Aviation Digest* 1-88-8 (August 1988): 3–8 and 32–36; Richard D. Newton, "A Question of Doctrine?" *Airpower Journal* 2, no. 3 (Fall 1988): 17–22; Hammes, pp. 50–55; Stroud, pp. 16–20; Weinraub, pp. 22–30; Saint and Yates, "Close Operations," pp. 2–15; Saint and Yates, "Deep Operations," pp. 2–9; Chapman, pp. 42–51; and Garden, pp. 149–167.

Selected Bibliography

Major sources are listed below and identified in the source notes by author and, if needed for clarity, a short title. Other sources are given in full in the notes.

AAF Evaluation Board in the ETO, *The Effectiveness of Third Phase Tactical Air Operations in the European Theater, 5 May 1944–8 May 1945.* Dayton, Ohio: Wright Field, Feb. 1946. Copy in the MHI library.

Aerospace Studies Institute. *The Accomplishments of Airpower in the Malayan Emergency (1948–1960).* Project No. AU-411-62-ASI. Maxwell AFB, Ala.: Air University, May 1963.

———. *The Employment of Airpower in the Greek Guerrilla War, 1947–1949.* Project No. AU-411-62-ASI. Maxwell AFB, Ala.: Air University, Dec. 1964.

———. *Guerrilla Warfare and Airpower in Algeria.* Project No. AU-411-62-ASI. Maxwell AFB, Ala.: Air University, March 1965.

American Aviation Historical Society. *The Curtiss Shrike.* Leatherhead, Eng.: Profile Publications, 1966.

Armengaud, Paul. "Les enseignements de la guerre Marocaine (1925–1926) en matiere d'aviation." *Revue Militaire Française* (January–March, April–June 1927).

Aviation Week & Space Technology editorial team. *Both Sides of the Suez: Airpower in the Mideast.* New York: McGraw-Hill, 1973.

Badoglio, Pietro. *La guerra d'Etiopia.* Milan: A. Mondadori, 1937.

Ballard, Jack S. *Development and Employment of Fixed-Wing Gunships, 1962–1972.* A volume in the series *The United States Air Force in Southeast Asia.* Washington, D.C.: Office of Air Force History, 1982.

Banfill, Charles Y. "Air Weapons." In *Report of Committee No. 4, Course at the Army War College, 1939–1940: Analytical Studies.* Washington, D.C.: Army War College, Feb. 23, 1940. Copy in the MHI archives.

Baring, Maurice. *R.F.C. H.Q., 1914–1918.* London: Bell and Sons, 1920.

Barnett, Corelli. *The Desert Generals.* New York: Berkeley Publishing Co., 1962 ed.

Baumbach, Werner. *The Life and Death of the Luftwaffe.* New York: Ballentine Books, 1967 ed.

Baxter, Walter H. III, et al. *An Analysis of Tactical Airpower in Support of the United States Army in South Vietnam.* USAWC Group Research Paper 70-865. Carlisle Barracks, Pa.: Army War College, March 9, 1970.

Bayerlein, Fritz. "Panzer-Lehr Division (24–25 Jul 44)." Manuscript A-902. N.p.: Historical Division, Headquarters, U.S. Army, Europe, n.d.). Copy in the MHI library.

BDM Corporation. *Operational Analysis.* Book 1 of *Conduct of the War,* Vol. 6 of *A Study of Strategic Lessons Learned in Vietnam.* BDM/W-76-128-TR-Vol-6-1. McLean, Va.: BDM Corporation, 9 May 1980.

Bennett, Ralph. *Ultra in the West: The Normandy Campaign 1944–45.* New York: Charles Scribner's Sons, 1979.

Berquist, Ronald E. *The Role of Airpower in the Iran-Iraq War.* Maxwell AFB, Ala.: Air University Press, Dec. 1988.

Bessie, Alvah. *Men in Battle.* New York: Pinnacle Books, 1977 ed.

Bessie, Alvah, and Albert Prago, eds. *Our Fight: Writings by Veterans of the Abraham Lincoln Brigade, Spain, 1936–1939.* New York: Monthly Review Press, 1987.

Bingham, Price T. "Dedicated, Fixed-Wing Close Air Support—a Bad Idea." *Armed Forces Journal International* (September 1987).

———. "Air Power and the Defeat of a Warsaw Pact Offensive: Taking a Different Approach to Air Interdiction in NATO." CADRE report AU-ARI-CP-87-2. Maxwell AFB, Ala.: Air University Press, March 1987.

———. "Ground Maneuver and Air Interdiction in the Operational Art," *Parameters: The U.S. Army War College Quarterly* 19, no. 1 (March 1989).

Bloch, Marc. *Strange Defeat: A Statement of Evidence Written in 1940.* London: Oxford University Press, 1949.

Blumenson, Martin. *Breakout and Pursuit,* a volume in the series *United States Army in World War II.* Washington, D.C.: Office of the Chief of Military History, 1961.

———. *Kasserine Pass.* Boston: Houghton Mifflin Co., 1967.

Boyle, Andrew. *Trenchard: Man of Vision.* New York: W. W. Norton 1962.

Bradley, Omar N. *A Soldier's Story.* New York: Holt, Rinehart and Winston.

Bradley, Omar N., and Air Effects Committee, 12th Army Group. *Effect of Air Power on Military Operations, Western Europe.* Wiesbaden, Germany: U.S. 12th Army Group Air Branches of G-3 and G-2, July 15, 1945. Copy in the MHI library.

Bradley, Omar N., and Clay Blair. *A General's Life: An Autobiography.* New York: Simon and Schuster, 1983.

Brainard, E. H. "Marine Corps Aviation." *The Marine Corps Gazette* 13, no. 1 (March 1928).

Bruce, J. M., and Jean Noel. *The Breguet 14,* Profile Publication 157. Leatherhead, Surrey: Profile Publications, 1967.

Brutting, Georg. *Das waren die deutschen Stuka-Asse, 1939–1945.* Stuttgart: Motorbuch Verlag, 1977.

Busse, Theodor, et al. "Operation 'Citadel' (the Battle of Kursk, July 1943)." Manuscript T-26 *in* Detwiler, Burdick, and Rohwer. Copy of the original T-26 ms. in the MHI library.

Caidin, Martin. *The Ragged, Rugged Warriors.* New York: Ballantine Books, 1967.

———. *The Tigers are Burning.* New York: Hawthorn Books, Inc., 1974.

Calmel, J. "Aviation Militaire Française: Activite de l'Armée de l'Air en Algerie." *Forces Aeriennes Françaises* (May 1957).

Cattaneo, Gianni. *The Fiat C.R. 32.* Leatherhead, Eng.: Profile Publications, 1965.

Chapman, Robert M. Jr. "Technology, Air Power, and the Modern Theater Battlefield." *Airpower Journal* 2, no. 2 (Summer 1988).

Christienne, Charles, and Pierre Lissarrague. *A History of French Military Aviation.* Washington, D.C.: Smithsonian Institution Press, 1986.

Clark, Alan. *Barbarossa: The Russian-German Conflict, 1941–45.* New York: William Morrow, 1965.

Colgan, Bill. *World War II Fighter Bomber Pilot.* Blue Ridge Summit, Pa.: TAB Books, Inc., 1985.

Collins, J. Lawton. *Lightning Joe: An Autobiography.* Baton Rouge: Louisiana State University Press, 1979.

Coningham, Arthur. "The Development of Tactical Air Forces." *Journal of the United Services Institute* 9 (1946).

Coox, Alvin D. *The Anatomy of a Small War: The Soviet-Japanese Struggle for Changkufeng/Khasan, 1938.* Number 13 of the series *Contributions in Military History.* Westport, Conn.: Greenwood Press, 1977.

———. *Nomonhan: Japan Against Russia, 1939.* In two volumes. Stanford: Stanford University Press, 1985.

Cot, Pierre. *Triumph of Treason.* New York: Ziff-Davis Publishing Co., 1944.

Coverdale, John F. *Italian Intervention in the Spanish Civil War.* Princeton: Princeton University Press, 1975.

Craven, Wesley Frank, and James Lea Cate, eds. *The Army Air Forces in World War II.* Vol. 1, *Plans and Early Operations, January 1939 to August 1942.* Chicago: University of Chicago Press, 1948.

———. *The Army Air Forces in World War II.* Vol. 2, *Europe: Torch to Pointblank.* Chicago: University of Chicago Press, 1949.

———. *The Army Air Forces in World War II.* Vol. 3, *Europe: Argument to V-E Day.* Chicago: University of Chicago Press, 1951.

Cutlack, F. M. *The Australian Flying Corps in the Western and Eastern Theatres of War, 1914–1918.* Sydney: Angus and Robertson, Ltd., 1942 ed.

Cynk, J. B. *The P.Z.L. P-23 Karas.* Leatherhead, Eng.: Profile Publications, 1966.

Cynk, Jerzy B. *P.Z.L. P.37 Los.* Windsor, Eng.: Hylton Lacy Publishers/Profile Publications, Oct. 1973.

———. "The Truth About the Operational Doctrine of the Polish Air Force: A Rebuttal." *Aerospace Historian* 25, no. 3 (September 1978).

Danel, Raymond. *The Lioré et Olivier LeO 45 Series.* Leatherhead, Eng.: Profile Publications, 1967.

Davis, Richard G. *The 31 Initiatives: A Study in Air Force–Army Cooperation.* Washington, D.C.: Office of Air Force History, 1987.

De Bono, Emilio. *Anno XIIII: The Conquest of an Empire.* London: The Cresset Press Ltd., 1937.

Dean, David J. *The Air Force Role in Low-Intensity Conflict.* Maxwell AFB, Ala.: Air University Press, Oct. 1986.

Deichmann, Paul. *German Air Force Operations in Support of the Army.* No. 163 in the *USAF Historical Studies* series. Maxwell AFB, Ala.: Air University, 1962.

Del Boca, Angelo. *The Ethiopian War, 1935–1941.* Chicago: University of Chicago Press, 1969.

Demoulin, Charles. *Firebirds! Flying the Typhoon in Action.* Washington, D.C.: Smithsonian Institution Press, 1988.

Detwiler, Donald S., Charles B. Burdick, and Jurgen Rohwer. *World War II German Military Studies.* Vol. 16. New York: Garland Publishing, Inc., 1979.

Dexter, David. *Australia in the War of 1939–1945.* Series One (Army), Vol. 6, *The New Guinea Offensives.* Canberra: Australian War Memorial, 1961.

Dial, Jay Frank. *The Bell P-39 Airacobra.* Leatherhead, Eng.: Profile Publications, 1967.

Doughty, Robert Allan. *The Seeds of Disaster: The Development of French Army Doctrine, 1919–1939.* Hamden, Conn.: Archon Books, 1985.

Drea, Edward J. *Nomonhan: Japanese-Soviet Tactical Combat, 1939.* Number 2 of the *Leavenworth Papers* series. Ft. Leavenworth, Kans.: Combat Studies Institute of the U.S. Army Command and General Staff College, Jan. 1981.

Drum, Karl. *The German Luftwaffe in the Spanish Civil War.* Maxwell AFB, Ala.: USAF Air University, 1956. Unpublished manuscript, a copy of which is on file at the Air Force Historical Research Center, Maxwell AFB.

Dudgeon, Tony. "The Habbaniya Campaign." *Aeroplane Monthly* 9, nos. 9–11 (September–November 1981).

Duval, Maurice. *Les Leçons de la Guerre d'Espagne.* Paris: Librairie Plon, 1938.

Eby, Cecil. *Between the Bullet and the Lie: American Volunteers in the Spanish Civil War.* New York: Holt, Rinehart and Winston, 1969.

Eichelberger, Robert L. *History of the Buna Campaign, December 1, 1942–January 25, 1943.* Bound report, n.d. Copy in the MHI library.

Eisenhower, Dwight D. *Crusade in Europe.* Garden City, N.Y.: Doubleday, 1948.

———. *Supreme Commander's Dispatch for Operations in Northwest Europe, 6 June 1944–8 May 1945.* SHAEF, n.d. Document catalogued as D756 A24c c.2 in the MHI library.

Ellis, Lionel F. *Victory in the West.* Vol. 1, *The Battle of Normandy.* London: HMSO, 1962.

Elstob, Peter. *Condor Legion.* New York: Ballantine Books, 1973.

Emiliani, Angelo, Guiseppe F. Ghergo, and Achille Vigna. *Spagna 1936–1939: l'aviazione legionaria.* A volume in the series *Immagini e storia dell'aeronautica Italiana 1935–1945.* Milan: Intergest, 1973.

Fabre, Colonel [*sic*]. *La Tactique au Maroc.* Paris: Charles-Lavauzelle & Cie, 1931.

Fall, Bernard B. *Hell in a Very Small Place: The Siege of Dien Bien Phu.* New York: Vintage Books, 1966.

———. *Street without Joy: Indochina at War, 1946–54.* Harrisburg, Pa.: The Stackpole Company, 1961.

Falls, Cyril. *Armageddon: 1918.* Philadelphia: Lippincott, 1964.

Fiske, Norman E. *Report of Military Observer with Italian Armies in East Africa.* Reports 1–10. Washington, D.C.: Geographic Branch, Military Intelligence Division, War Department General Staff, 1936. Copy in the MHI library.

Fleming, Shannon. *Primo de Rivera and Abd-El Krim: The Struggle in Spanish Morocco, 1923–27.* Ann Arbor: University Microfilms, 1975.

Francillon, René J. *Japanese Aircraft of the Pacific War.* New York: Funk & Wagnalls, 1970.

Futrell, R. Frank. *The United States Air Force in Korea, 1950–1953.* Washington, D.C.: Office of Air Force History, 1983 ed.

Gabel, Christopher. *The U.S. Army GHQ Maneuvers of 1941.* Ann Arbor: University Microfilms International, 1981.

Gaines, Mike, and Herman Potgieter. "Bush War." *Flight International* (March 1, 1986).

Galland, Adolf. *Die Ersten und die Letzten: Die Jagdflieger im zweiten Weltkrieg.* Darmstadt: Franz Schneekluth, 1953.

Garden, Timothy. "The Air-Land Battle." In *War in the Third Dimension: Essays in Contemporary Air Power,* edited by R. A. Mason. London: Brassey's Defence Publishers, 1986.

Garland, Albert N., and Howard McGaw Smyth, assisted by Martin Blumenson. *Sicily and the Surrender of Italy.* A volume in the series *United States Army in World War II.* Washington, D.C.: Office of the Chief of Military History, 1965.

Garriga, Ramon. *Guadalajara y sus consecuencias.* Madrid: G. Del Toro, 1974.

General Staff Branch, Army Headquarters India. *The Third Afghan War, 1919: Official Account.* Calcutta: Government of India Central Publication Branch, 1926.

Gersdorff, Rudolf-Christoph von. "The Argentan-Falaise Pocket." Manuscript A-919. N.p.: Historical Division, Headquarters, U.S. Army, Europe, 1954. Copy in the MHI library.

———. "Northern France: vol. 5 (Fifth *Panzer* Army (25 Jul–25 Aug 44))." Manuscript B-726. N.p.: Historical Division, Headquarters, U.S. Army, Europe, n.d.

Glubb, John Bagot. *War in the Desert: An RAF Frontier Campaign.* New York: W. W. Norton, 1961.

Goldberg, Alfred, ed. *A History of the United States Air Force, 1907–1957.* Princeton: Van Nostrand, 1957.

Golley, John. *The Day of the Typhoon: Flying with the RAF Tankbusters in Normandy.* Wellinborough, Eng.: Patrick Stephens, 1986.

Gough, Hubert. *The Fifth Army.* London: Hodder and Stoughton, 1931.

Gray, Peter, and Owen Thetford, *German Aircraft of the First World War.* London: Putnam, 1962.

Green, William. *Warplanes of the Third Reich.* Garden City, N.Y., Doubleday, 1970.

Green, William, and Gordon Swanborough. "Of *Chaika* and *Chato* . . . Polikarpov's Fighting Biplanes." *AIR Enthusiast Eleven* (November 1979–February 1980).

Greenfield, Kent Roberts. *Army Ground Forces and the Air-Ground Battle Team Including Organic Light Aviation.* Study No. 35 of the series *Studies in the History of Army Ground Forces.* Washington, D.C.: Historical Section, Army Ground Forces, 1948.

———. *American Strategy in World War II: A Reconsideration.* Baltimore: The Johns Hopkins Press, 1963.

Greenhous, Brereton. "Close Support Aircraft in World War I: The Counter Anti-Tank Role." *Aerospace Historian* 21, no. 2 (Summer 1974).

Greenwood, John T. "The Great Patriotic War." In *Soviet Aviation and Air Power: A Historical View,* edited by Robin Higham and Jacob W. Kipp. London: Brassey's, 1978.

Greer, Thomas H. *The Development of Air Doctrine in the Army Air Arm, 1917–1941.* Washington, D.C.: Office of Air Force History, 1985 ed.

Guderian, Heinz. *Panzer Leader.* New York: E. P. Dutton & Co., Inc., 1952.

Hallion, Richard P. *The Naval Air War in Korea.* Baltimore: Nautical & Aviation Publishing Co., 1986.

———. *Rise of the Fighter Aircraft, 1914–1918.* Annapolis: Nautical and Aviation Publishing Co., 1984.

———. *Test Pilots: The Frontiersmen of Flight.* Smithsonian Institution Press, 1988.

———. *The Wright Brothers: Heirs of Prometheus.* Washington, D.C.: Smithsonian Institution Press, 1978.

Hammes, Thomas X. "Rethinking Air Interdiction." *U.S. Naval Institute Proceedings* 113/12/1018 (December 1987).

Hardesty, Von. *Red Phoenix: The Rise of Soviet Air Power, 1941–1945.* Washington, D.C.: Smithsonian Institution Press, 1982.

Hastings, Max. *Das Reich: The March of the 2nd SS Panzer Division through France.* New York: Jove Books, 1983 ed.

———. *Overlord: D-day and the Battle for Normandy.* New York: Simon and Schuster, 1984.

Hinds, James R. "Bombs Over Mexico." *Aerospace Historian* 31, no. 3 (September 1984).

Hoeffding, Oleg. *Soviet Interdiction Operations, 1941–1945.* RAND Report R-556-PR. Santa Monica, Calif.: The RAND Corporation, Nov. 1970. Copy in the MHI library.

Hoeppner, Ernst von. *Deutschlands Krieg in der Luft.* Leipzig: Koehler, 1921.

Hohmann, Johannes. "Wir Schlachtflieger." In *Unsere Luftstreitkräfte, 1914–1918,* ed. Walter von Eberhardt. Berlin: Vaterlandischer Verlag C. A. Weller, 1930.

Holley, I. B. *Buying Aircraft: Material Procurement for the Army Air Forces* (a volume in the *United States Army in World War II, Special Studies* series). Washington, D.C.: Office of the Chief of Military History, U.S. Army, 1964.

Hooton, Edward R. "Air War Over China." *AIR Enthusiast Thirty-four* (September–December 1987).

Horne, Alistair. *To Lose a Battle: France, 1940.* Boston: Little, Brown and Company, 1969.

Howe, George F. *Northwest Africa: Seizing the Initiative in the West.* A volume in the series *United States Army in World War II.* Washington, D.C.: Office of the Chief of Military History, 1957.

Hozzel, Paul-Werner. "Conversations with a Stuka Pilot." An interview at the National War College, Nov. 1978, transcribed and published by the Tactical Technology Center, Battelle Columbus Laboratories, 1978.

Hsu Long-hsuen and Chang Ming-kai, compilers. *History of the Sino-Japanese War (1937–1945).* Taipei, Taiwan: Chung Wu Publishing Co., 1972 ed.

Humphreys, F. E. "The Wright Flyer and Its Possible Uses in War," *Journal of the United States Artillery* 33 (March–April 1910).

Huston, James A. "Tactical Use of Air Power in World War II: The Army Experience." *Military Review* 31, no. 7 (July 1952).

Intelligence Branch, Military Intelligence Division, War Department General Staff, *Certain Studies on and Deductions from Operations of Italian Army in East Africa, October 1935–May 1936*

Jacobs, W. A. "Air Support for the British Army, 1939–1943." *Military Affairs* 46, no. 4 (December 1982).

Jacobs, William A. "Tactical Air Doctrine and AAF Close Air Support in the European Theater, 1944–1945." *Aerospace Historian* 27, no. 1 (March 1980).

Jardine, Douglas. *The Mad Mullah of Somaliland.* London: Herbert Jenkins, Ltd., 1923.

Johnson, J. E. *Wing Leader.* New York: Ballantine Books, 1967 ed.

Jones, Byron Q., compiler. *Notes on the Chinese War.* Washington, D.C.: War Plans Division, Army War College, Dec. 27, 1940. A typescript compilation in the MHI library.

———. *Spanish War: Reports and Articles.* In three volumes. An undated typescript document in the MHI library.

Jones, H. A. *Sir Walter Raleigh and the Air History: A Personal Recollection.* London: Edward Arnold & Co., 1922.

———. *The War in the Air, III.* Oxford: Clarendon Press, 1931.

———. *The War in the Air, IV.* Oxford: Clarendon Press, 1934.

———. *The War in the Air, VI.* Oxford: Clarendon Press, 1937.

———. *The War in the Air, Appendices.* Oxford: Clarendon Press, 1937.

Jones, Ira. *An Air Fighter's Scrap-Book.* London: Nicholson & Watson, 1938, pp. 107–108.

Jukes, Geoffrey. *Kursk: The Clash of Armour.* New York: Ballantine Books, 1968.

Karolevitz, Robert F., and Ross S. Fenn. *Flight of Eagles: The Story of the American Kosciuszko Squadron in the Polish-Russian War, 1919–1920.* Sioux Falls, S.Dak.: Brevet Press, Inc. 1974.

Kipp, Jacob W. *Barbarossa, Soviet Covering Forces and the Initial Period of War; Military History and Airland Battle.* Ft. Leavenworth, Kans.: Soviet Army Studies Office, n.d.

———. "Soviet 'Tactical' Aviation in the Postwar Period: Technological Change, Organizational Innovation, and Doctrinal Continuity." *Airpower Journal* 2, no. 1 (Spring 1988).

Kirkland, Faris R. "The French Air Force in 1940: Was It Defeated by the Luftwaffe or by Politics?" *Air University Review* 36, no. 6 (September-October 1985).

Kohn, Richard H., and Joseph P. Harahan. *Air Interdiction in World War II, Korea, and Vietnam.* Washington, D.C.: Office of Air Force History, 1986.

———. *Air Superiority in World War II and Korea.* Washington, D.C.: Office of Air Force History, 1983.

Kozhevnikov, M. N. *The Command and Staff of the Soviet Army Air Force in the Great Patriotic War, 1941–1945.* A volume in the *Soviet Military Thought* series. Washington, D.C.: GPO in cooperation with the All-Union Copyright Agency of the USSR, n.d.

Kreis, John F. *Air Warfare and Air Base Air Defense, 1919–1973.* A volume in the USAF *Special Studies* series. Washington, D.C.: GPO, 1988.

Kupfer, Ernst. "Schlachtflieger, Panzerjagd, und Störflugzeug im Jahre 1943." Ref. no. K113.3019-3. Maxwell AFB, Ala.: USAF Historical Research Center, 10 Sept. 1943.

Larios, José [Fernandez de Villavincencio]. *Combat Over Spain: Memoirs of a Nationalist Fighter Pilot, 1936–1939.* New York: Macmillan, 1966.

Larrazabal, Jesus Salas. *Air War over Spain.* London: Ian Allan, Ltd., 1969.

Lee, Arthur Gould. *No Parachute: A Fighter Pilot in World War I.* New York: Pocket Books, 1971.

———. *Open Cockpit: A Pilot of the Royal Flying Corps.* London: Jarrolds, 1969.

Lee, Arthur S. Gould. *Special Duties: Reminiscences of a Royal Air Force Staff Officer in the Balkans, Turkey and the Middle East.* London: Sampson Low, Marston & Co., Ltd., 1946.

Legion Condor, Deutsche Kämpfen in Spanien. Berlin: Wilhelm Limpert Verlag, 1939.

Leigh-Mallory, Trafford. "Air Operations by the A.E.A.F. in N.W. Europe from 15th November, 1943, to 30th September, 1944." *Royal Air Force Quarterly* 18, no. 4 (October 1947).

Liddell Hart, B. H., ed., with the assistance of Lucie-Maria Rommel, Manfred Rommel, and General Fritz Bayerlein. *The Rommel Papers.* New York: Harcourt, Brace and Co., 1953.

Liss, Witold. *The Ilyushin Il-2.* Leatherhead, Eng.: Profile Publications, 1966.

———. *The Lavochkin La 5 & 7.* Leatherhead, Eng.: Profile Publications, 1967.

———. *The Polikarpov I-16.* Leatherhead, Eng.: Profile Publications, 1966.

———. *The P.Z.L. P-11.* Leatherhead, Eng.: Profile Publications, 1966.

———. *The Yak 9 Series.* Leatherhead, Eng.: Profile Publications, 1967.

Ludendorff, Erich. *Meine Kriegserinnerungen, 1914–1918.* Berlin: E. S. Mittler & Sohn, 1919.

Macaulay, Neill. *The Sandino Affair.* Durham, N.C.: Duke University Press, 1985 ed.

Mann, David K. *The 1972 Invasion of Military Region I: Fall of Quang Tri and Defense of Hue.* Hickam AFB, Hawaii: HQ PACAF, Directorate of Operations Analysis, CHECO/Corona Harvest Division, March 15, 1973).

Mark, Eduard. "A New Look at Operation Strangle." *Military Affairs* 52, no. 4 (October 1988).

Martinez, Luis Garcia. "Los Katiuskas," *AIR Enthusiast Thirty-two* (December 1986–April 1987): 45–55.

Mason, Francis K. *The Hawker Hurricane.* A volume in the *Macdonald Aircraft Monographs* series. Garden City, N.Y.: Doubleday, 1962.

Maurer, Maurer, ed. *The U.S. Air Service in World War I.* Vol. 3, *The Battle of St. Mihiel.* Washington, D.C.: Office of Air Force History, 1979.

McCarthy, Dudley. *Australia in the War of 1939–1945.* Series One (Army), Vol. 5, *South-West Pacific Area—First Year, Kokoda to Wau.* Canberra: Australian War Memorial, 1959.

McPeak, Merrill A. "TACAIR Missions and the Fire Support Coordination Line." *Air University Review* 36, no. 6 (September–October 1985).

Meccozzi, Amedeo. "Origini e svilupo dell' aviazione d'assalto." *Revista Aeronautica* 11, no. 2 (February 1935).

Mellenthin, F. W. von. *Panzer Battles: A Study of The Employment of Armor in the Second World War.* New York: Ballantine Books, 1971 ed.

Melli, B. *La guerra Italo-Turca.* Rome: Enrico Voghera Editore, 1914.

Mets, David R. *Land-Based Air Power in Third World Crises.* Maxwell AFB, Ala.: Air University Press, July 1986.

Mikhailow, P. "Tactical Employment of Bombardment Aviation," *Krasnaya Zwesda* (Jan. 17, 1938). Translation in the MHI library.

———. "Tactical Employment of Pursuit Aviation." *Krasnaya Zwesda* (Feb. 8, 1938). Translation in the MHI library.

Miksche, F. O. *Attack: A Study of Blitzkrieg Tactics.* New York: Random House, 1942.

Miller, Thomas G., Jr. *The Cactus Air Force.* New York: Bantam Books, 1981 ed.

Mitchell, William. *Memoirs of World War I: From Start to Finish of Our Greatest War.* New York: Random House, 1960 ed.

Molesworth, G. N. *Afghanistan 1919: An Account of Operations in the Third Afghan War.* New York: Asia Publishing House, 1962.

Momyer, William W. *Airpower in Three Wars.* Washington, D.C.: HQ USAF, 1978.

Mortensen, Daniel R. *A Pattern for Joint Operations: World War II Close Air Support, North Africa.* A volume in the *Historical Analysis* series. Washington, D.C.: Office of Air Force History and the U.S. Army Center of Military History, 1987.

Munson, Kenneth. *Bombers between the Wars, 1919–1939.* New York: Macmillan, 1970.

———. *Bombers, Patrol and Reconnaissance Aircraft, 1914–1919.* London: Blandford Press, 1968.

———. *Fighters, Attack, and Training Aircraft, 1914–1919.* New York: Macmillan, 1969.

———. *Fighters between the Wars, 1919–1939.* New York: Macmillan, 1970.

Murray, Kenneth Malcolm. *Wings Over Poland: The Story of the 7th (Kosciuszko) Squadron of the Polish Air Service, 1919, 1920, 1921.* New York: D. Appleton & Company, 1932.

Murray, Williamson. *Luftwaffe.* Baltimore: Nautical & Aviation Publishing Co., 1985.

Nalty, Bernard C. *Air Power and the Fight for Khe Sanh.* Washington, D.C.: Office of Air Force History, 1973.

Nami, Paul, trans. "Two-Seaters in Battle: The Recollections of Oscar Bechtle," *Cross & Cockade Journal* 20, no. 2 (Summer 1979).

Nassaes, Jose Luis Alcofar. *C.T.V.: Los legionarios italianos en la Guerra Civil Española, 1936–1939.* Barcelona: DOPESA, 1972.

OACS-USAFE [Office of the Assistant Chief of Staff, U.S. Air Forces in Europe], *Fighter Operations of the German Air Force; Tactical Employment.* Wiesbaden: HQ USAFE, Dec. 10, 1945.

Pacquier, Pierre, et al. *Les Forces Aériennes Françaises de 1939 à 1945.* A volume in the series *L'Aviation Française au Combat.* Paris: Editions Berger-Levrault, 1949.

Pahenyanik, G. "Mass Employment of Aviation on the Approaches to East Prussia." Translation by the Army War College from *Vestnik Vozdushnogo Flota,* no. 22 (November 1944). Document no. R-17077.1-2, catalog Number D 792 R9P75, in the MHI library.

The Palestine News. *A Brief Record of the Advance of the Egyptian Expeditionary Force Under the Command of General Sir Edmund H. H. Allenby, G.C.B., G.C.M.G., July 1917 to October 1918.* Cairo: Government Press and Survey of Egypt, 1919.

Parotkin, Ivan, et al. *The Battle of Kursk.* Moscow: Progress Publishers, 1974. Copy in the MHI library.

Passingham, Malcolm, and Waclaw Klepacki. *Petlyakov Pe-2 and Variants.* Windsor, Eng.: Profile Publications, 1971.

Pegg, Martin. *Luftwaffe Ground Attack Units, 1939–1945.* A volume in the *Aircam/Airwar* series. New York: Sky Books Press, 1977.

Pennell, C. R. *A Country With a Government and a Flag: The Rif War in Morocco, 1921–1926.* Cambridgeshire, Eng.: Middle East & North African Studies Press Ltd., 1986.

Pentland, Geoffrey. *Commonwealth Boomerang Described.* Surrey, Eng.: Kookaburra Technical Publications, 1965.

Peterson, A. H., G. C. Reinhardt, and E. E. Conger. *Symposium on the Role of Airpower in Counterinsurgency and Unconventional Warfare: The Algerian War.* Memorandum RM-3653-PR. Santa Monica, Calif.: The RAND Corporation, July 1963.

Pickert, Wolfgang. "III. AA Corps in the Normandy Battles." Manuscript B-597. Allendorf, Ger.: Historical Division, Headquarters, U.S. Army, Europe, Apr. 20, 1947. Copy in the MHI library.

Plocher, Herman. *The German Air Force versus Russia, 1943.* Number 155 in the *USAF Historical Studies* series. Maxwell AFB, Ala.: Air University, June 1967.

Proctor, Raymond L. *Hitler's Luftwaffe in the Spanish Civil War.* Number 35 of *Contributions in Military History.* Westport, Conn: Greenwood Press, 1983.

Profile Publications Research Staff, *The Commonwealth Whirraway.* Leatherhead, Eng.: Profile Publications, 1967.

Quesada, Elwood. "Oral History Memoir." An interview by Steve Long and Ralph Stephenson. MHI Senior Officers' Debriefing Program, n.d. Transcript in the MHI archives.

———. "Tactical Air Power." *Air University Quarterly Review* 1, no. 4 (Spring 1948).

Raff, Edson D. *We Jumped to Fight.* New York: Eagle Books, 1944.

Ratley, Lonnie O. III. "Airpower at Kursk: The Confrontation of Aircraft and Tank." Monterey, Calif.: Naval Postgraduate School, Sept. 1976. Copy in the MHI library.

Rauch, Georg von. "The Green Hell Air War." *AIR Enthusiast Quarterly Two.* Bromley, Eng.: Pilot Press, n.d.

Raus, Erhard, et al. "The Red Air Force." *In* Historical Division, U.S. Army, *Peculiarities of Russian Warfare,* Manuscript T-22. Washington, D.C.: USA, n.d. Copy in the MHI library.

Reilly, Henry J. "The Aeroplane's Role in Battle in Spain." *The Aeroplane* 56, no. 1457 (April 26, 1939).

Rentz, John N. *Bougainville and the Northern Solomons.* Washington, D.C.: Historical Section, Office of Public Information, Headquarters USMC, 1948.

Richards, Denis. *Royal Air Force: 1939–1945.* Vol. 1, *The Fight At Odds.* London: HMSO, 1953.

Richards, Denis, and Hilary St. George Saunders. *Royal Air Force: 1939–1945.* Vol. 2, *The Fight Avails.* London: HMSO, 1954.

Rosenstone, Robert A. *Crusade of the Left: The Lincoln Battalion in the Spanish Civil War.* New York: Pegasus, 1969.

Rowell, Ross E. "Aircraft in Bush Warfare." *The Marine Corps Gazette* 14, no. 3 (September 1929).

Rudel, Hans-Ulrich. *Stuka Pilot.* Dublin: Euphorion Books, 1953.

Rudenko, Sergei. "The Gaining of Air Supremacy and Air Operations in the Battle of Kursk." *In* Parotkin et al., *The Battle of Kursk,* Moscow: Progress Publishers, 1974. Copy in the MHI library.

Ruge, Friedrich. *Rommel in Normandy: Reminiscences by Friedrich Ruge.* San Rafael, Calif.: Presidio Press, 1979.

Rust, Kenn C., and Walter M. Jefferies, Jr., *The Curtiss Shrike.* Leatherhead, Eng.: Profile Publications, 1966.

Saint, Crosbie E., and Walter H. Yates, Jr. "Attack Helicopter Operations in the AirLand Battle: Close Operations." *Military Review* 68, no. 6 (June 1988).

———. "Attack Helicopter Operations in the AirLand Battle: Deep Operations," *Military Review* 68, no. 7 (July 1988).

Salesse, Lieutenant-colonel [*sic*]. *L'Aviation de Chasse Français en 1939–1940.* A volume in the series *L'Aviation Française au Combat.* Paris: Editions Berger-Levrault, 1948.

Sallagar, F. M. *Operation "Strangle" (Italy, Spring 1944): A Case Study of Tactical Air Interdiction.* RAND Report R-851-PR. Santa Monica, Calif.: The RAND Corporation, Feb. 1972.

Sanders, Liman von. *Funf Jahre Turkei.* Berlin: Verlag von August Scherl, 1919.

Saunders, Hilary St. George. *Per Ardua: The Rise of British Air Power, 1911–1939.* London: Oxford University Press, 1945.

———. *Royal Air Force 1939–1945,* Vol. 3, *The Fight is Won.* London: HMSO, 1954.

Savage, Raymond. *Allenby of Armageddon.* Indianapolis: Bobbs-Merrill, 1926.

Schlight, John. *The War in South Vietnam: The Years of the Offensive, 1965–1968.* A volume in the series *The United States Air Force in Southeast Asia.* Washington, D.C.: Office of Air Force History, 1988.

Schulz, Karl Heinrich. "The Collaboration between the Army and the *Luftwaffe*: Support of the Army by the *Luftwaffe* on the Battlefield." Manuscript B-791a. Neustadt, Germany: Historical Division, Headquarters, U.S. Army, Europe; Foreign Military Studies Branch, Dec. 12, 1947. Copy in the MHI library.

Seaton, Albert. *The Russo-German War, 1941–45.* New York: Praeger Publishers, 1970.

Segal, David. "The Iran-Iraq War: A Military Analysis." *Foreign Affairs* 66, no. 5 (Summer 1988).

Sherrod, Robert. *History of Marine Corps Aviation in World War II.* Washington, D.C.: Combat Forces Press, 1952.

Shores, Christopher F. *Martin Maryland & Baltimore Variants.* Windsor, Eng.: Profile Publications, 1971.

———. *Spanish Civil War Air Forces,* a volume in the *Aircam/Airwar* series. New York: Sky Books Press, 1977.

———. *2 TAF.* Reading, U.K.: Osprey Publications, 1970.

Shuhsi Hsu. *The War Conduct of the Japanese,* Number 2 in the series *Political and Economic Studies.* Hong Kong: Council of International Affairs, Hankow, 1938.

Slessor, J. C. *Air Power and Armies.* London: Oxford University Press, 1936.

Slessor, John. *The Central Blue: The Autobiography of Sir John Slessor, Marshal of the RAF.* New York: Frederick A. Praeger, 1957.

Slyusarev, S. V. "Protecting China's Air Space." In *Soviet Volunteers in China, 1925–1945.* Moscow: Progress Publishers, 1980.

Smith, J. R. *The Henschel Hs 129.* Leatherhead, Eng.: Profile Publications, 1966.

Smith, J. Richard. *The Junkers Ju 87A & B.* Windsor, Eng.: Profile Publications, 1966.

Spenser, Jay. *Focke Wulf Fw 190: The Workhorse of the Luftwaffe.* Vol. 9 of *Famous Aircraft of the National Air and Space Museum.* Washington, D.C.: Smithsonian Institution Press, 1987.

Stanton, Shelby L. "Lessons Learned or Lost: Air Cavalry and Airmobility," *Military Review* 69, no. 1 (January 1989).

Steadman, Kenneth A. *A Comparative Look at Air-Ground Support Doctrine and Practice in World War II.* A Combat Studies Institute Study. Ft. Leavenworth, Kans.: U.S. Army Command and General Staff College, Sept. 1, 1982.

Stroud, William P. "Use and Misuse of Conventional Tactical Air Power," *Airpower Journal* (Summer 1987).

Sullivan, John J. "The Botched Air Support of Operation *Cobra." Parameters: The U.S. Army War College Quarterly* 18, no. 1 (March 1988).

Sunderland, Riley. *Evolution of Command and Control Doctrine for Close Air Support.* Washington, D.C.: Office of Air Force History, March 1973).

Sutherland, L. W. *Aces and Kings.* London: John Hamilton, n.d., probably early 1920s.

Tedder, Sir Arthur. *With Prejudice: The War Memoirs of Marshal of the Royal Air Force Lord Tedder G.C.B.* Boston: Little, Brown and Company, 1967.

Terraine, John. *A Time for Courage: The Royal Air Force in the European War, 1939–1945.* New York: Macmillan Publishing Company, 1985.

Thayer, Thomas C., ed. *The Air War.* Vol. 5 of *A Systems Analysis View of the Vietnam War, 1965–1972.* Washington, D.C.: OASD(SA)RP, The Pentagon, 18 February 1975.

Thomas, Hugh. *The Spanish Civil War.* New York: Harper Colophon Books, 1963 ed.

Thompson, Jonathan W. *Italian Civil and Military Aircraft, 1930–1945.* Fallbrook, Calif.: Aero Publishers, 1963.

Timokhovich, I. V. *The Operational Art of the Soviet Air Force during the Great Patriotic War.* Wright-Patterson AFB: Foreign Technology Division translation FTD-ID(RS)T-2320-77, Dec. 30, 1977.

Tinker, Frank G., Jr. *Some Still Live.* New York: Funk & Wagnalls, 1938.

Tolson, John J. *Airmobility, 1961–1971.* A volume in the *Vietnam Studies* series. Washington, D.C.: U.S. Army, 1973.

U.S. 12th Army Group, *Battle Experiences.* A large bound collection of the accumulated "lessons learned" reports filed by 12th Army Group. Copy in the MHI library.

U.S. War Department. Field Manual FM 100-20, *Command and Employment of Air Power.* Washington, D.C.: GPO, July 21, 1943.

———. Technical Manual TM 30-430, *Handbook on U.S.S.R. Military Forces.* Washington, D.C.: U.S. War Department, Nov. 1945. Copy in the MHI library.

USAAF 42nd Bomb Wing. *Air-Ground Cooperation* (n.d.). Copy in the "L1 Italy 1943–1944" file, archives of the Air Force Museum, Wright-Patterson AFB, Ohio.

USAAF HQ. *Army Air Forces in the War Against Japan, 1941–1942.* Washington, D.C.: USAAF HQ, August 1945.

USAAF XII Tactical Air Command HQ. *XII Tactical Air Command Tactical Operations: A Report on Phase 3 Operations of the XII Tactical Air Command for Army Air Forces Evaluation Board European Theatre of Operations.* XII TAC HQ, 18 April 1945. Copy in the MHI library.

Valle, Pedro A. del. *Roman Eagles Over Ethiopia.* Harrisburg, Pa.: Military Service Publishing Company, 1940.

Wagner, Ray. *American Combat Planes.* Garden City, N.Y.: Doubleday, 1968 ed.

———, ed. *The Soviet Air Force in World War II: The Official History.* Garden City, N.Y.: Doubleday, 1973.

Wagoner, Fred E. *Dragon Rouge: The Rescue of Hostages in the Congo.* Washington, D.C.: National Defense University Research Directorate, 1980.

Waters, Alan Rufus. "The Cost of Air Support in Counter-Insurgency Operations: The Case of the Mau Mau in Kenya." *Military Affairs* 37, no. 3 (October 1973).

Weinraub, Yehuda. "The Israel Air Force and the Air Land Battle," *IDF Journal* 3, no. 3 (Summer 1986).

Werth, Alexander. *Russia at War: 1941–1945.* New York: E. P. Dutton & Co., Inc., 1964.

Wewege-Smith, Thomas. *Gran Chaco Adventure: The Thrilling and Amazing Adventures of a Bolivian Air Caballero.* London: Hutchinson & Co., 1937.

Wheeler, Howard. *Attack Helicopters: A History of Rotary-Wing Combat Aircraft.* Baltimore: Nautical & Aviation Publishing Co., 1987.

Wintringham, Tom. *English Captain.* London: Faber and Faber, Ltd., 1939.

Wintringham, Tom, and J. N. Blashford-Snell. *Weapons and Tactics.* Baltimore: Penguin Books, 1973 ed.

Wolfert, Michael L. "From Acts to Cobra: Evolution of Close Air Support Doctrine in World War Two." USAF Air Command and Staff College Student Report #88-2800. Maxwell AFB, Ala.: ACSC, April 1988.

Zhukov, Georgi. *Marshal Zhukov's Greatest Battles.* New York: Harper & Row, 1969.

———. *The Memoirs of Marshal Zhukov.* New York: Delacorte Press, 1973.

Ziemke, Earl F. *Stalingrad to Berlin: The German Defeat in the East.* A volume in the *Army Historical Series.* Washington, D.C.: Office of the Chief of Military History, 1968.

Zimmerman, John L. *The Guadalcanal Campaign.* Washington, D.C.: Historical Division, Headquarters, USMC, 1949.

Zook, David H., Jr. *The Conduct of the Chaco War.* New Haven: Bookman Associates, 1960.

Index